ENGINEERING ACHIEVEMENTS ACROSS THE GLOBAL VILLAGE

PRODUCT USABILITY COSTING: A STUDY AND COST ESTIMATION MODELS

Dhillon B.S.

Abstract: This paper presents usability costing-related facts and figures, usability engineering activities and costs, cost benefit -analysis of a usability study, cost of ignoring usability and cost effectiveness of usability evaluation methods, models for estimating usability engineering costs.

1. Introduction

Usability engineering is an effective approach to product development and is specifically based on customer feedback and data. Today, billions of dollars are being spent annually to produce new products using modern technologies. The usability of these products has become more important than ever before because of their increasing complexity, sophistication, and non-specialist users. For example, over 30% of all software development projects are cancelled before completion primarily because of inadequate user design inputs; resulting in a loss of over $100 billion annually to the United States economy. Moreover, some studies indicate that 80% of product maintenance is due to unmet or unforeseen user requirements (Dhillon, 2004).
The term "Usability Engineering" was coined in the mid-1980's (Butler, 1996; Rosson, 2002).

Cost is an important factor in product development and other issues and it generally varies with time. In fact, the concept of time value of money is not that new, it began over 2500 years ago in Babylon (Paul-DeGarmo, 1979). In those days interest was paid on borrowed commodities in form of grain or other means. In modern times, the development cost of an item not only varies with time but also due to various other factors including usability. More specifically, effectively usable products are not produced by accident but through a careful consideration given to human factors or usability during their design and development. This costs money.

The cost of each usability engineering activity depends on factors such as follows (Rosson, 2002):
- scope of the product under consideration,
- functionality range,
- number of users to be studied,
- number of scenarios to be studied,
- skill and experience of the usability specialists.

Although, the cost of usability engineering activities may become a significant component of product development cost, but past experiences indicate that each dollar spent by a manufacturer in developing the usability of a product, it receives $10 - $100 in benefits, in addition, to winning customer satisfaction and continued business (Dhillon, 2004).

2. Usability costing-related facts and figures

Some of the directly or indirectly usability costing-related facts and figures are as follows:
- it costs approximately $100 billion annually in lost productivity to American businesses because office worker "futz" with their machines an average of 5.1 hours per week (SBT Accounting Systems, 1997),
- a study reported that the training time for new users of a standard personal computer was around 21 hours as opposed to only 11 hours for users of a more usable computer machine (Nielson, 1993),
- a study revealed that the average cost of end-user computing across 18 major Australian companies for supporting a single workstation was approximately $10,000 (Australian) (Ko, 1995). At least 50% of this cost accounted for "hidden" support (i.e., productivity lost because users stopped their ongoing tasks to help each other with computer-related problems),
- a study conducted by the American Airlines revealed that catching a usability-related problem early in the design can decrease the cost of rectifying it by 60% to 90% (Laplante, 1992),
- a study reported that human factors improvements resulted in $2.5 million savings in training cost (Chapains, 1991),
- a study reported that in ergonomically designed office environment absenteeism dropped from 4% to slightly greater than 1% (Schneider, 1985),
- an Australian insurance company spent around $100,000 (Australian) on a usability project concerned with redesigning its application forms to make customer errors less likely and saved $536,023 (Australian) annually (Fisher, 1990),
- a study revealed that an $800,000 investment in reducing the human factors errors for the Line of Sight Forward Heavy resulted in the saving of $80 million in cost (Booher, Rouse, 1990; Rouse, Boff, 1997),
- a study reported that an $300,000 investment in reducing human factors-related errors for the Pedestal Mounted Stinger resulted in the saving of $61 million in cost (Booher, Rouse, 1990; Rouse, Boff, 1997),
- it is estimated that approximately 63% of all software projects exceed their cost estimates due to factors such as frequent requests for changes from users, users' lack of understanding of their own requirements, overlooked tasks, and inadequate user-analyst communication and understanding (Lederer, 1992),
- it is estimated that roughly 80% of software maintenance cost is due to unforeseen/unmet user needs (Pressman, 1992),
- a study reported that in 1991, design changes to usability work at IBM resulted in an average decrease of 9.6 minutes per task, with projected internal savings of $6.8 million (Karat, 1990).

3. Usability engineering activities and costs

There is a wide range of activities that may be employed in developing a usability-engineering product: end user requirements definition, benchmark studies, user profile definition, surveys and questionnaires, usability objectives specification, style guide development, focus groups, heuristic evaluations, task analysis, prototype redesign, design walkthroughs, usability tests (laboratory or field), studies of end user work context, paper-and-pencil simulation testing, thinking-aloud studies, prototype development (high or low fidelity), and initial design development (Karat, 1994; Mantei, 1988).

The cost of usability engineering for a given product includes costs for one or more of the above activities. Usually, no more than six of these activities are completed for any one project/product (Karat, 1994). Moreover, usability engineering-related work on a project is tailored to factors such as the project's requirements, time frame, and resources. Nonetheless, two important guidelines for calculating the cost of usability engineering activities are as follows (Karat, 1994):
- ensure that in personnel cost, costs for all development team support, other support, or contract services, and all costs associated with participants are included,
- ensure that cost is prorated on the basis of number of usability tests to be performed for a specified period, when a permanent usability laboratory is built.

4. Cost-benefit analysis of a usability study

The performance of a usability study costs money but in turn it generates various benefits. More specifically, the main objective of a usability investment is to accrue maximum returns. Many times a quick estimate of cost and benefit is made before making a decision for performing a usability study. Nonetheless, the following example demonstrates the estimation of cost of a usability study along with the estimation of savings after doing the usability study.

Example 1
Estimate the cost and savings of a usability study concerned with reducing the number of support center calls and improving productivity of employees by producing a better engineered, usability tested product. Assume fictitious values as applicable.

4.1. Usability Study Cost Estimation

In the calculation of this cost consideration is given to in-house usability staff, the product of time spent by staff (i.e., usability individuals and developers) and wage rate (same units), and additional variable costs associated with subcontractors, usability laboratory rentals, travels, etc.

For fictitious values of amount and time the cost of the in-house usability staff is estimated as follows:
- annual loaded headcount amount (this includes salary, benefits, and cost with respect to vacation leave, office space, telephones, and equipment) = $ 151,200,

- annual hours per work year (i.e., 35 hours per week times 48 weeks) = 1680,
- hourly wage rate = $151,200/1680 = $90 per hour,
- amount of time spent on usability test by usability specialist for activities such as planning, analysis, implementation, and recommendations = 140 hours,
- amount of time spent by interface designer on redesign activity = 50 hours,
- total amount of time spent by development engineers on usability-related activities = 20 hours,
- total cost of the in-house usability staff = (140 + 50 + 20) (90) = $18,900.

Similarly, for fictitious values of amount, the cost of a fully equipped usability laboratory is estimated as follows:
- total cost of usability participant recruiting at $200 per participant for 8 participants = $1600,
- total amount of participant compensation at $70 per participant for 8 participants = $560,
- total cost of videotapes at $10 per tape for 8 tapes = $80,
- total amount or percentage of laboratory and equipment cost (more specifically, this is amortized cost of laboratory per hour (i.e., say $250/hour) for 25 hours) = $6,250,
- total cost of the fully equipped usability laboratory = (1600 + 560 + 80 + 6250) = $8,490.

Thus, the total usability study cost is

$$TUSC = \$18,900 + \$8,490$$
$$= \$27,390$$

where: $TUSC$ is the total usability study cost.

4.2. Usability Study Savings Estimation

These savings are the result of performing the usability study. Two major components of these savings are savings in support call cost and in-house increase in productivity.

For fictitious values, total savings in support call cost is estimated as follows:
- average cost of a support call = $100,
- total number of product version A units sold = 150,000,
- total number of support calls = 300,000,
- total cost of support calls due to usability-related problems = (300,000 x $100) = $30 million,
- number of support calls per product version A unit sold = (300,000/150,000) = 2,
- after the application of usability engineering, the total number of product version B units sold = 400,000,
- total number of support calls with version B units = 100,000,
- total cost of support calls due to usability-related problems with version B units (100,000 x $100) = $10 million,
- number of support calls per version B unit sold = 100,000/400,000 = 0.25,
- reduction on support calls (i.e., from version A to version B) = 2 − 0.25 = 1.75,
- total savings in support call cost due to improved usability = (1.75) (400,000) ($100) = $70 million.

Similarly, for fictitious values, the annual improvement (in dollars) in in-house productivity is estimated as follows:
- improvement in time, because of the usability action, in performing Task X = 2 minutes,
- number of times task X is performed per day = 4,
- total number of users perform task X = 300,
- hourly wage of users including overheads = $70,
- total user time saved per day because of the usability action = (2) (4) (300) = 40 hours
- total amount of money saved per day because of the usability action = (70) (40) = $2,800,
- total annual amount of money saved through increased productivity = ($2,800) (240 work days/year) = $672,000 - $0.672 million,
- total savings after performing the usability study = 70 + 0.672 = $70.672 million.

Thus, the total usability study estimated cost and the total estimated savings after performing the usability study are $27,390 and $70.672 million, respectively.

5. Cost of ignoring usability and cost effectiveness of usability evaluation methods

Past experiences indicate that over the years many products from usability or human factors standpoint were poorly designed, but put on the market and have subsequently requited in substantial costs to manufacturers in term of reduced sale, increased customer dissatisfaction, tarnished corporate image, and so on. Two examples of scenario such as this are discussed in Ref. (Chapanis, 1991). Nonetheless, some of the possible principal costs of ignoring product usability are shown in Fig. 1 (Keinonen, 1997). Poor sale cost is associated with dissatisfied users not buying the product in future, even if they are made aware of improvements in the usability of a next version of the product. Moreover, it is estimated that a dissatisfied user roughly influence ten others to avoid purchasing the brand (Keinonen, 1997).

Customer support cost pertains to a customer hotline telephone service provided by product manufacturers for individuals having problems in using the product. Products that are difficult to use generate greater user/customer requests for help, thus require more people to handle the customers/users. In turn, greater the customer support costs. Tarnish corporate image cost is rather difficult to estimate. It occurs when customers/users avoid purchasing not only the current or improved usability version of the product, but also other products manufactured by the same company.

User error cost is associated with the users of professional products making errors. These errors lead to decrease in their productivity. More specifically, the probability of user error increases significantly if products are difficult to use; thus higher user error cost. Poor productivity cost is associated with the additional time spent by users of professional products that are difficult to use. Cost due to poor productivity increases significantly, if these products are used daily by a large number of users.

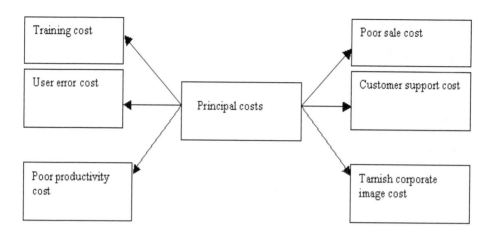

Figure 1. Principal costs of ignoring product usability

Training cost is associated with the training of users when the product or system is first introduced. It increases quite significantly for products that are difficult to use. In the case of user-friendly products, this cost could be very little or negligible.

In usability studies a variety of usability evaluation methods are used. The cost of employing a usability evaluation method is subject to factors such as follows:
- number of involved individuals (i.e., users, developers, usability experts, etc.),
- amount of time required for collecting and analyzing data,
- degree of need for coordination (i.e., whether a given approach needs all the participants to be present together).

Usability evaluation methods may be classified under three distinct cost categories (i.e., low cost, medium cost, and high cost) as presented in Table 1.

Table 1. Classifications of usability evaluation methods with respect to cost

No.	Low cost methods	Medium cost methods	High cost methods
1	Heuristic evaluation	Remote testing	Focus group
2	—	Interviews	Coaching method
3	—	Cognitive walkthroughs	Thinking-aloud protocol
4	—	Questionnaires	Shadowing method
5	—	Logging actual use	Question-asking protocol
6	—	Field observation	Teaching method
7	—	Scenario-based check-lists	Co-discovery learning
8	—	Proactive field study	Performance measurement
9	—	Feature inspection	Pluralistic walkthroughs
10	—	—	Retrospective testing

6. Models for estimating usability engineering costs

This section presents four mathematical models that can be used to estimate various types of usability engineering-related costs.

Model I
This model can be used to approximate usability-engineering cost of a product when usability engineering cost data are available for similar products of different capacities. The usability engineering cost of the desired product can be estimated by using the following relationship [21]:

$$UC_d = UC_0 \left[\frac{K_d}{K_0} \right]^\alpha \qquad (1)$$

where:
- UC_d is the usability engineering cost of the desired product,
- K_d is the capacity of the desired product,
- UC_0 is the known usability engineering cost of a similar product, of capacity K_d,
- α is known as the cost-capacity factor and its value varies for different products. In circumstances where no data are available, it is generally reasonable to assume the value of α to be 0.6.

Example 2
Assume that the usability engineering cost of a 20 GB hard drive personal computer is $200. Estimate the usability engineering cost of a similar 80GB hard drive personal computer, if the value of the cost-capacity factor is 0.7.

By substituting the given data into Equation (1) yields:

$$UC_d = (200) \left[\frac{80}{20} \right]^{0.7} = \$527.8$$

It means the usability engineering cost of the 80GB hard drive personal computer is $527.8.

Model II
This model can be used to estimate the labor cost of correcting usability-related problems associated with a product for a given period. Thus, the labor cost of correcting product usability-related problems is expressed by

$$C_u = (SOH)(LC) \left[\frac{MTTCPUP}{MTTPUP} \right] \qquad (2)$$

where:
- C_u is the labor cost of correcting product usability-related problems for a given period,
- LC is the hourly labor cost of correcting usability problems,
- SOH is the total product scheduled operating hours for a specified period,
- MTTPUP is the mean time to product usability problems expressed in hours,
- MTTCPUP is the mean time to correct product usability problems expressed in hours.

Example 3

A product is scheduled to be used for 3,000 hours annually. Its mean time to usability problems and mean time to correct usability problems are 100 hours and 5 hours, respectively. The labor cost of correcting usability problems is $50 per hour. Calculate the annual labor cost of correcting usability problems. By substituting the given data values into Equation (2) yields:

$$C_u = (3,000)(50)\left[\frac{5}{100}\right]$$
$$= \$7,500$$

Thus, the annual labor cost of correcting usability problems is $7,500.

Model III

This model can used to estimate user error cost over a product's life span. This cost is expressed by

$$UELCC = UEOC + UECC \tag{3}$$

where:
- UELCC is the user error life cycle cost of a product,
- UEOC is the user error occurrence cost associated with the product. This basically is the cost of productivity loss due to user errors over the product life span,
- UECC is the user error correction cost. This is associated with time spent in correcting user errors over the product life span.

Model IV

This model can be used to estimate usability engineering life cycle cost of a product. The usability engineering life cycle cost is expressed by

$$LCC_{ue} = AC_{ue} + OC_{ue} \tag{4}$$

where:
- LCC_{ue} is the usability engineering life cycle cost of a product,
- AC_{ue} is the usability engineering acquisition cost of the product. This cost is an element of a product's procurement cost and is concerned with usability engineering or human factors activities during the product design and development phase,
- OC_{ue} is the usability engineering ownership cost of the product. This cost is a component of a product's ownership (i.e., operational phase) cost and is associated with usability-related problems during the product operational phase.

References

1. Booher H.R., Rouse, W.B.: *MANPRINT as the Competitive Edge*. in MANPRINT: An Approach to Systems Integration, edited by H.R. Booher, Van Nostrand Reinhold Company, New York, 1990, pp. 230-45.
2. Butler K.A.: *Usability Engineering Turns Ten*. Interactions, January 1996, pp. 59-75.

3. Chapanis A.: *The Business Case for Human Factors in information.* In Human Factors for Informatics Usability. Edited by B. Shackel and S.J. Richardson, Cambridge University Press, London, 1991, pp. 39-71.
4. Dhillon B.S.: *Engineering Usability: Fundamentals, Applications.* Human Factors, and Human Eerror, American Scientific Publishers, Los Angeles, 2004.
5. Dieter G.E.: *Engineering Design.* McGraw Hill Book Company, New York, 1983, pp. 324--366.
6. Fisher Pl, Sless D.: *Information Design Methods and Productivity in the Insurance Industry.* Information Design Journal, Vol. 6, No. 2, 1990, pp. 103-129.
7. Karat C.M.: *A Business Case Approach to Usability Cost Justification.* In Cost-Justifying Usability, edited by R.G. Bias and D.J. Mayhew, Morgan Kaufmann Publishers, San Francisco, 1994.
8. Karat C.M.: *Cost-Benefit Analysis of Usability Engineering Techniques.* Proceedings of the Human Factor Society 34th annual Meeting, 1990, pp. 839-843.
9. Keinonen T., Mattelmaki T., Soosalu M., Sade S.: *Usability Design Methods.* Technical Report, Department of Product and Strategic Design, University of Art and Design, Helsinki, Finland, 1997.
10. Ko C., Hurley M.: *Managing End-User Computing.* Information Management and Computer Security, Vol. 3, No. 3, 1995, pp. 3-6.
11. Laplante A.: *Put to the Test.* Computerworld, Vol. 27, July 27, 1992, pp. 75-77.
12. Lederer A.L., Prasad J.: *Nine Management Guidelines For Better Cost Estimating.* Communications of the ACM, Vol. 35, No. 2, 1992, pp. 51-59.
13. Mantei M.M., Teorey T.J.: *Cost-Benefit Analysis for Incorporating Human Factors in the Software Lifecycle.* Communications of the ACM, Vol. 31, No. 4, 1988, pp. 428--439.
14. Nielson J.: *Usability Engineering.* Academic Press, Inc., Boston, 1993.
15. Paul-DeGarmo E., Canada J.R., Sullivan W.G.: *Engineering Economy.* Macmillan Publishing Co., Inc., New York, 1979.
16. Pressman R.S.: *Software Engineering: A Practitioner's Approach.* McGraw-Hill Book Company, New York, 1992.
17. Rosson M.B., Carroll J.M.: *Usability Engineering: Scenario-Based Development of Human-Computer Interaction.* Academic Press, San Francisco, 2002.
18. Rosson M.B., Carroll J.M.: *Usability Engineering: Scenario-Based Development of Human-Computer Interaction.* Morgan Kaufmann Publishers, San Francisco, 2002.
19. Rouse W.B., Boff K.R.: *Assessing Cost/Benefits of Human Factors.* In Handbook of Human Factors and Ergonomics, edited by G. Salvendy, John Wiley and Sons, New York, 1997, pp. 1617-1633.
20. SBT Accounting Systems, 1997, Westlake Consulting Company, Inc., 5444 Westheimer, Unit 1510, Houston, Texas.
21. Schneider M.F.: *Why Ergonomics Can No Longer be Ignored.* Office Administration and Automation, Vol. 46, No. 7, 1985, pp. 26-29.

CONCEPTS OF COLLABORATIVE PRODUCT DESIGN AND DEVELOPMENT: A REVIEW

Janardanan V.K., Radhakrishnan P.

Abstract: This review provides an in depth analysis of the recent research work carried out in the area of collaborative product design and development. This paper is intended for researchers and practitioners in the area of collaborative product design and engineering. The text presented includes the latest developments taken place in collaborative technologies, data modeling, knowledge based design, virtual reality based collaboration, case based collaborative design, feature based design, methods of communication between team members, Software used for collaboration, System architecture design and security requirements for web based collaborative product development system.

1. Introduction

Collaborative product development is an automated environment that enables people (including Designers, Engineers, Managers, Suppliers and Customers) to collaborate and interact, regardless of their geographic location and interaction means. With advent of the World Wide Web, information distributed at different location can be accessed and shared by users anywhere in the world using web tools such as web browsers. The web technologies have also been employed in developing manufacturing systems to associate various product development activities, distributed at different locations into an integrated environment. The integration and collaboration among different partners of the product development team can improve the product quality and reduce the product development lead-time, thus providing better global competitiveness of products in the market place.

In order to shorten the time for development and decrease the number of loops in the development problem, all partners have to work together from the early stages of product development. In recent years open CAD environments have gained more and more importance with the evolution of electronic design tools and wide spread availability of the Internet as a medium for sharing and distributing product models, the constraint that collaboration should be located in geographical proximity has been eliminated. In addition companies are often outsourcing engineering activities performed internally to rapidly design and prototype the product. Almost 50-80% of all components manufactured by Original Equipment Manufactures (OEMs) are now outsourced to external suppliers (Rezayat, 2000b). The most important activity during collaborative design is resolution of design specification conflicts early in the design phase. This reduces the product development lead-time and manufacturing cost to a large extent (Ullman, 1997). In the conventional design and manufacturing process, every activity is transmitted serially. The author conducted a case study of the design and manufacturing of the components of launch vehicle structures and its assembly (Janardanan, 2004). Quiet often some rework is done on these components for smooth assembly of the final product, which increases the cost and lead-time of the product.

Much effort has gone into defining architecture and environments for new open systems. There are many collaboration tools available commercially in the market; however, most of the systems are not suitable for small to medium scale industries due to dedicated functionality that is difficult to customize. This paper presents a comprehensive and critical review of recent research in the area of collaborative product design and manufacturing.

2. Review of recent research

Collaborative design approach to design and manufacturing is evolving rapidly with growing set of technologies and tools for conceptualizing, building, distributing and supporting new processes and products. The designers are very extensively using the web in recent years for a variety of applications. Web reduces the distance between (i) several designers, (ii) between designers and software programs, (iii) between different software programs that need information from each other (Regli, 2000).

2.1. Collaborative product design

A product changes from its birth to its death through a sequence of life cycle phases. First requirements for designing the product are obtained from customer and these requirements are transformed in to design specification. Although the web technology sounds highly relevant and promising it is still not clear how effective it can be used to support collaborative product development and to introduce the kinds of extensions in its present form. The hyperlinks of web pages allow the user to have multiple, simultaneous, and distributed access (Leslie, 2002). By using web browsers the software modules can be down loaded from the server site to the client site by the client. Existing collaborative design tools can be broadly classified into components design tools and assembly design tools. The collaborative design environments for components design allows the designers and engineers to view and modify the component geometry and discuss their ideas through web (Chang, 1999). The team members can access and edit the component models in real time. The different versions of the same model can be represented in a single CSG tree depending upon the applications (Chan, 1999). Collaborative component design tools utilize representation that is also aimed at representing components effectively.

The assembly system tool cPAD (Shyamsundar, 2002) for designers permits to conduct real-time modeling of component geometry and assembly constraints during collaborative design process. In this system, a new assembly representation (AREP) scheme was introduced to improve the assembly modeling efficiency. Many designers at different locations through client browsers using Java 3D can access the AREP model at the server side. The network centric model (Sara McMain, 2000) provides different software tools, manufacturing facilities and analysis services for distributed design and fabrication team members. Manufacturing Advisory Service (MAS) is a web based design tool for fabricating parts via Internet. The MAS generates a dialogue with the designers, engineers about batch size, part size, overall shape, typical tolerance and tools requirements. MAS is a Java applet freely available in the Internet. Personal electronic note book with sharing (PENS) is one in which distributed designers can enter the informal design notes and the project information web automatically grows (Jack Hong, 1996). As it grows, selection can be incrementally shared with collaboration over the World Wide Web. In an era where both network security concerns and distributed collaboration demands are growing together,

PENS has the capability for information sharing that is independent of security firewalls. PENS is client-Server architecture where client software is used for file transmission and Server receives the files (packages) and decides where to deposit. Workflow management is a mechanism to facilitate the teamwork in a collaborative product development environment where remote web based support system TeleDSS is extensively used by team members who are geographically distributed (Huang, 2000).

2.2. Data modeling

The requirement for rapid and effective exchange of accurate product data is being addressed by the International Organization for Standardization (ISO) 10303 series of standards. The series is known as the Standard for Exchange of product model data (STEP). STEP will provide a complete, unambiguous, neutral, computer interpretable representation of the physical and functional characteristics of a product throughout its life cycle. The STEP standard development is intended to provide a means of effectively communicating product data among dissimilar CAD/CAM/CAE/CIM systems and thus greatly facilitates the concept of an integrated data environment (John H Bradham, 1998). In the Artefact Transport System (ATS), different CAD databases are first converted to IGES format and subsequently translated into VRML models for collaboration among designers using Web-browsers (Nidamarthi, 2001).

Extendible Markup language (XML) is used to exchange data between the client and the server in a distributed and collaborative CAD system. This collaborative CAD system employs feature as its collaborative design elements and using XML to define feature operation and communication protocol between the server and the client (Ouyang, 2004). To reduce network load and increase response ability of the system, the feature information is updated incrementally on the client. Virtual Reality Modeling Language (VRML) is a standard language for modeling three-dimensional geometric data. It is being used as de-facto standard modeling language for exchanging geometric data over the web. VRML does not contain any information on solid modeling, complex surfaces and/or shape features that are relevant to manufacturing. Therefore if VRML is to be used significant extensions need to be made. Web-blow is a distributed multidisciplinary design optimization environment, using a number of enabling technologies including software agents, Internet/web, and XML (Wang, 2003). KaViDo is another web-based system for collaborative research and development process (Tamine, 2001). KaViDo is developed to document product development process, manage competencies of distributed experts, exchange user experiences and assist new product development. XML is used to exchange product data and design knowledge between KaViDo and other applications.

2.3. Knowledge based design

Knowledge based product development (KPBD) is the ability to provide the right information to the right person at the right time and in the right format. The concepts of Key Characteristics (KCs) for defining a communication dictionary for humans, or their agents and use XML as an enabling tool for making the dictionary function within the web together will provide a practical ontology and communication mechanism for all levels of the extended enterprise and for the entire product development cycle (Rezayat, 2000a). The knowledge sharing architecture can be classified into three categories: Independent knowledge bases, integrated knowledge bases and interoperable knowledge bases. In the case of independent knowledge bases, the strength of the knowledge is just a sum of these knowledge bases (Tomiyama, 1995). Whereas integrated knowledge bases can be applied to various situations and the strength of knowledge is near maximum. Most web-based collaborative system also use this integrated knowledge base approach, since it requires a platform with uniform language. The entire knowledge base is a federation or a set of loosely coupled components that are usually called intelligent agents. This approach is being widely used in agent-based collaborative system. A multi-agent system is used for knowledge management in research and development project with the objective of providing a system for helping team members to share explicit knowledge as well as experiences, i.e. lessons learned with out much overhead (Cesar, 2003).

2.4. Virtual Reality (VR) based collaboration

The determination of standard geometric shape abstractions of product models is essential for variety of Internet based Virtual Prototyping (VP) models. VRML is necessary for data modeling in particular because different applications modules require geometric information at different levels of abstraction. For example DFM software module requires geometric information at a high level in term of manufacturing features such as ribs, bosses, slots and holes, on the other hand a rendering program requires lower level geometric information such as normal on surfaces or tangents at an edge (Gadh, 1998). Virtual city for engineering is a distributed collaborative geometric modeling module implements sharing information, communication and manipulation of design objects, which are key elements for this system (Quili, 2001). This module is comprised of two parts (i) client side application, which is the geometric modeling application (ii) server side application, which is a multi-user server. The client side application performs the geometric modeling, and the server side application takes care of communication and co-ordination.

An Internet based system called Virtual reality based collaborative environment (VRCE) is implemented using Vnet, Java and VRML to demonstrate the feasibility of collaborative design to small and medium size companies (Kan, 2001). In the CyberEye, Java 3D is used to implement an Internet enabled, platform independent, 3D modeling browser called Cyber eye viewer (Zhuang, 2000). In the Cyber eye viewer, Java servelet connects to the database at the server side through Java database connectivity (JDBC), in which java applets contribute to building the products using Cyber eye viewer. The VRML files are stored in the server and can be directly accessed by the Java applets through Uniform Resource Locator (URL). The geometric data modeled in the SDRC/CAEDS model is directly translated into VRML model in a Collaborative product design environment (Roy, 1999).

2.5. Case based collaborative design

In engineering, it is conventionally estimated that more than 75% of design activities comprises case based design. Reuse of previous design knowledge to address a new design problem, CAD knowledge bases are vital for engineers, who search through vast amount of corporate legacy data and navigate on-line catalogues to retrieve precisely the right components for assembly into new products. Future of the digital libraries for computer-aided design, is to develop structures for capturing the relationship among design attributes and symbolic data representing other critical engineering and manufacturing data like tolerance, process plan etc. (Regli, 2000).

2.6. Feature based design

Traditional Computer-Aided Design (CAD) applications use low-level geometrical entities to represent a product design. It is difficult to extract other design information (e.g. required manufacturing operations) from such product representation. On the other hand, feature-based product representation can integrate information across various design activities. Features not only can represent geometric shapes, but can also describe other product information, such as material properties, technical parameters, and required manufacturing precision (Kun-Hur Chen, 1998). Feature-based design provides the advantage of using high-level entity features, which are closely related to their manufacturing and assembly operations, as element of a part. A feature, which represents a portion of a part, usually has a particular meaning or function for some design application. Multiple applications sharing a part model should be allowed to construct and edit in terms of the features of their own views (Ko, 1993).

2.7. Communication methods

Two communication methods, synchronous and asynchronous communication, have been used for implementing the collaboration functions of Internet based manufacturing systems (Tay, 2000). Synchronous communication allows more than one designer to access the same product database simultaneously. The synchronous communication provides an electronic blackboard for various designers to conduct collaborative design simultaneously. In the synchronous collaborative design, different users can also lock different parts of the model for exclusive editing while the changes are visible to all other users (Chan, 2001). A 3D collaborative mechanical part modeling system, called 3D Syn, is developed to associate part suppliers and buyers by synchronizing the activities of all participants (Huang, 20001). In this system, Java applet server architecture is used to allow a users manipulations to part objects to be transmitted to the others transparently and instantaneously. The asynchronous communication mechanism is used to modify a common design database independently to generate different design candidates. These different versions of design are brought together at the end to identify the optimal design process and object.

2.8. Related software

Many computer and web/Internet technologies, including computer languages such as Java, C++ and C#, Script languages such as Perl and VB Script, Web mark-up languages such as HTML, DHTML and XML, Web-based client Server programming tools such as ASP and JSP, distributed object modeling methods such as Distributed Component Object Model (DCOM), Remote Method Invocation (RMI), Common Object Request Broker Architecture (CORBA) and Vnet, etc., have been employed for developing Web-based manufacturing systems (Yang, 2003). Many commercial software packages have also incorporated some Web/Internet technologies into their systems. CORBA introduced in 1991, is a specification that defines interoperability rules between distributed objects on clients and servers. DCOM, introduced in 1996 by Microsoft, is like CORBA in that it separates interface from functionality by using an IDL(Interface Definition Language). However, currently, the IDL used by CORBA is quiet different from the one used with DCOM, which causes severe interoperability problems. Java applets allow development of web-based application using object oriented Programming technique. Virtual Reality

Modeling Language (VRML) Integrates 3D geometric function in to Web-browsers. Java3D improves 3D geometric capabilities by providing series of java applet classes and allowing users to implement subclasses of these java classes. XML allows the information to be organized in a data structure and the required information to be extracted and displayed using Web-browsers (Rezayat, 2000b).

2.9. System architecture design

Architecture of most Web-based manufacturing systems were designed based upon the Client-server communication method. In these systems, the modules and database are distributed at both the server sites and the client sites. Contrary to the traditional client-server applications, the client program is usually downloaded automatically from the server sites to the local computers through Web-browsers. Depending on the sizes of the programs and databases at the client and server sites, the web-based application are classified in to two categories: fat client and thin server application when the major computation is conducted at the client side, and thin client and fat server applications when the computation is primarily carried out at the server side.

The server side module, implemented by using CGI scripts and other programming languages, is usually responsible for database modeling and database analysis, while a client side module is implemented using Web page, is responsible for obtaining input from designer and displaying result to these designers. In a web-based collaborative product design system, there are number of product models from various designers. A collaboration tool, which allows the designers to generate their concepts in real time and to select the best one among designs through web, is preferred (Roy, 2001). Web-based design tools are implemented on the Internet or Corporate intranets to incorporate expertise and knowledge (Huang, 2001). 'Virtual Consultants' is the name for these web based design tools that are accessible on the Internet. Some typical Virtual Consultants in product development include Quality Function Deployment (QFD), Failure Mode and Effect Analysis (FEMA), Design for Manufacture and Assembly (DFMA) and Design for X (DFX). A Multi-Resolution Collaborative Architecture (MRCA), based on multi-agent co-ordination mechanism, is used as a foundation for the development of Web-centric cooperative applications in global manufacturing (Ulieru, 2000).

2.10. Security

Web-based design and manufacturing system allow the client to download client application program from server side, accessing the client program using local Web-browsers, and execute programs at the remote server side. Security management mechanism are required to be specified at different levels of accessibility permissions for different users for preventing the local machines from being damaged by poor programs and viruses from the remote server sides and preventing the server machines from being visited by the unauthorized clients. In a collaborative environment, security is achieved through password authentication and data encryption. Two types of encryption methods are usually used. Symmetric encryption uses the same secret key to encrypt and decrypt a message, while asymmetric encryption uses one key to encrypt a message and a different key to decrypt the message (Foo, 2000).

In the TeleRP system (Jiang, 2001), the general security strategy is mainly concerned with (1) collection of information from clients and confirmation of clients certificates at different levels (2) clients access security consideration to web servers, directories, and files inside the servers, including individual web page and (3) secure encrypted data transaction methodologies including RSA, and secure HTTP were investigated. In a system of collaborative design environment employs a user login mechanism to control the user activities with enhanced security. This mechanism was implemented using ASP coupled with a SQL server database. The database is used to store the information necessary for user login mechanism and communication methods (Kim, 2001).

3. Summary

This review paper brings out the recent research carried out in the area of collaborative product design elsewhere. The main topics discussed are Collaboration among product development partners, data modeling, knowledge based design, virtual reality based collaboration, case based and feature based design, communication methods, system architecture and security management.

Many new web based product modeling methods have been developed to integrate the CAD systems, such as IGES and STEP, and web based product modeling schemes, such as VRML and XML to improve the efficiency of web based product modeling. The long-term goal of collaborative environment developers has been to create a system that will enable engineers, and other managing groups to conduct and manage product design and development activities as a unified and collaborative process. This review paper brings out the recent research carried out in the area of collaborative product design elsewhere. The main topics discussed are Collaboration among product development partners, data modeling, knowledge based design, virtual reality based collaboration, case based and feature based design, communication methods, system architecture and security management. Many new web based product modeling methods have been developed to integrate the CAD systems, such as IGES and STEP, and web based product modeling schemes, such as VRML and XML to improve the efficiency of web based product modeling. The long-term goal of collaborative environment developers has been to create a system that will enable engineers, and other managing groups to conduct and manage product design and development activities as a unified and collaborative process.

Acknowledgements
The authors are grateful to Director, Vikram Sarabhai Space Centre, (ISRO), Trivandrum, India, for giving permission to publish this review paper.

References

1. Cesar A. Tacla, Jean-Paul Barthes: *A multi-agent system for acquiring and sharing lessons learned*. Computers in Industry, 2003; 52: 5-16.
2. Chang H.C., Lu W.F., Liu X.F.: *WWW based collaboration system for integrated design and manufacturing*. Concurrent Engineering:Research and Applications, 1999; 7: 319-334.
3. Chan S.C.F., Lee P.S.H. Ng V.T.Y., Chan A.T.S.: *Synchronous collaborative development of UML models on the Internet*. Concurrent engineering: Research and applications, 2001; 9: 111-119.
4. Chan S., Wang M., Ng V.: *Collaborative Solid modeling on the WWW*. Proceedings of the ACM Symposium in Applied computing, 1999; 598-602.

5. Foo S., Hui S.C., Leong P.C., Liu S.: *An integrated help desk support for customer service over the Worldwide Web-a case study.* Computers in Industry, 2000; 41: 129-145.
6. Huang G.Q., Huang J., Mak K.L.: *Agent based work-flow management in Collaborative Product Development on the Internet.* Computer-Aided Design, 2000; 32 : 133-144.
7. Huang G.Q., Mak K.L.: *Web-based electronic product cataloging.* International Journal of Computer Application in Technology, 2001; 14: 26-39.
8. Huang G.Q., Huang J., Mak K.L.: *Issues in the development and implementation of Web applications for product design and manufacture.* Int. Journal of Computer Integrated Manufacturing, 2001; 14: 125-135.
9. Hong J., Toye G., Leifer L.J.: *Engineering design notebook for design and sharing.* Computers in Industry, 1996; 29: 27-35.
10. Janardanan V.K., Radhakrishanan P.: *A Collaborative Approach in Manufacturing Deviations/Errors in Launch Vehicle Structures – A Case Study.* Proc. of Int. Conf. on Manuf. & Management, GCMM-2004.
11. Jiang P., Fukuda S.: *TeleRP-an Internet Web-based solution for remote rapid prototyping service and maintenance.* International Journal of Computer Integrated Manufacturing. 2001; 14: 83-94.
12. Bradham J.H.: *STEP-driven manufacturing.* SME Blue Book Series, Computer and automated systems Association of the Society of Manufacturing Engineers, 1998.
13. Kan H.Y., Duffy V.G., Su C.J.: *An internet virtual reality collaborative environment for effective product design.* Computers in Industry, 2001; 45: 197-213.
14. Kim Y., Choi Y., Yoo S.B.: *Brokering and 3D collaborative viewing of mechanical part models on the web.* International Journal of Computer Integrated Manufacturing, 2001; 14: 28-40.
15. Ko H., Park M.W., Kang H., Sohn Y., Kim H.S.: *Integration methodology for feature based modeling and recognition.* ASME, Computer in Engineering, 1993; 0, 23-34.
16. Chen K.H., Chen S.J., Lin L., Changchien S.W.: *An integrated graphical user interface(GUI) for concurrent engineering design of mechanical parts.* Computer Integrated Manufacturing Systems. 1998; 11: 91-112.
17. Monplaiser L., Singh N.: *Collaborative Engineering for Product Design and Development.* American scientific publishers, 2002.
18. Nidamarhti S.S., Allen R. H., Sriram R.D.: *Observations from supplementing the traditional design process via Internet based collaboration tools.* Int. Journal of Comp. Integrated Manufacturing, 2001; 14: 95-107.
19. Ying-Xiu O., Min T., Jun-Chang L., Jin-Xiang D.: *Journal of Zhejiang University Science*, China, 2004; 5(5): 579-586.
20. Sun Q., Gramoll K.: *Internet based distributed collaborative environment for Engineering Education and Design.* Proc. of the American Society for Engineering Education, Annual Conference & Exposition, 2001.
21. Regli W.C., Cicrello V.A.: *Managing digital libraries for computer aided design.* Computer-Aided Design, 2000; 32: 119-132.
22. Gadh R., Sondhi R.: *Geometric shape abstraction for Internet based virtual prototyping.* Computer-Aided Design, 1998; 30: 473-486.
23. Rezayat M.: *Knowledge-based Product Development using XML and KCs.* Computer-Aided Design, 2000a; 32: 299-309.

24. Rezayat M.: *The enterprise-Web portal of life cycle supports.* Computer-Aided Design, 2000b; 32: 85-96.
25. Roy U., Kodkani S.S.: *Product modeling within the framework of the World Wide Web.* IIE Transactions, 1999; 31: 667-677.
26. Roy U., Kodkani S.S.: *Collaborative product conceptualization tool using web technology.* Computers in Industry, 2001; 41: 195-210.
27. McMains S., Squin C., Smith C., Wright P.: *Internet based design and manufacturing.* Final report, 1999-2000; 99-106, MICRO Project, University of California.
28. Shyamsundar N., Gadh R.: *Collaborative virtual prototyping of product assemblies over the Internet.* Computer-Aided Design, 2002; 34: 755-768.
29. Shyamsundar N., Gadh R.: *Internet based collaborative product design with assembly features and virtual design spaces.* Computer-aided design, 2001; 33: 637-651.
30. Tamine O., Dillmann R.: *KaViDo-A web based system for collaborative research and development process.* Computers in industry, 2001; 45: 197-213.
31. Tay F.E.H., Chin M.: *A shared multi media design environment for concurrent engineering over the Internet.* Concurrent engineering: Research and applications, 2001; 9: 55-63.
32. Tomiyama T., Umeda Y., Ishili M., Yoshika M., Kiriyama T.: *Knowledge systemizations for knowledge intensive engineering framework.* Knowledge intensive CAD-1, Chapman & Hall, 1995; pp 55-80.
33. Ullman D.G.: *The mechanical and design process.* New York, MacGraw Hill, 1997.
34. Ulieru M., Norrie D., Kremer R., Shen W.M.: *A Multi-resolution collaborative architecture for Web-centric global manufacturing.* Information Sciences, 2000; 127: 3-21.
35. Wang Y.D., Shen W., Ghenniwa H.: *WebBlow: a Web/Agent-based multidi-sciplinary design optimization environment.* Computers in Industry, 2003; 52: 17-28.
36. Yang H., Xue D.: *Recent research on developing Web-based manufacturing systems: A review.* International Journal of Production Research, 2003; 41: 15, 3601-3629.
37. Zhuang Y., Chen L., Venter R.: CyberEye: *An Internet-enabled environment to support collaborative design.* Concurrent Engineering: Research and Applications, 2000; 8: 213-229.

MANUFACTURING DECORATIVE SHEET METAL COMPONENTS FOR USE IN THE TOURISM INDUSTRY IN TRINIDAD AND TOBAGO

Lewis W.G., Ameerali A.O.

Abstract: This paper highlights the use of shallow drawing in the manufacture of decorative sheet metal components for use in the tourism industry in Trinidad & Tobago. These components were made from aluminium, brass and copper using the Guerin forming process and did not exceed a depth of more than 2cm. To preserve surface finish and to distribute uniform pressure over the entire blank area, a polyurethane pad was used in the design. The die was cast using Aluminium and the design was that of a steel pan, the national instrument of Trinidad & Tobago. Evaluation of the formed components took the form of strain analyses and visual inspection of the drawn components. Graphs of average % thickness strain vs. distance from the centre of the blank were plotted to observe trends in the components and to make comparisons between the different materials used. The graphs showed that the strains were all within the allowable limits for each material tested. The results showed that the Guerin forming technique ideally suited the forming of these components. Thus this process proved valuable since components could be formed quite quickly with minimal time and effort for change over of tooling for the various components.

1. Introduction

The demand for sheet metal components and products will increase in the near future. Trinidad, and Tobago in particular, are fast becoming one of the Caribbean's ideal tourist attractions. As such new production processes have to be implemented so as to optimize the benefits that can be gained from this industry. These benefits may include:
- larger varieties of products for the incoming visitors,
- large volumes of components, which would effectively reduce the average cost of individual components, makes the price for these decorative components more competitive. It will also encourage customers to purchase more of these products. The ideal production system for these decorative components would be one that caters for batch production,
- the setup costs for manufacturing decorative sheet metal products would be greatly reduced since complex tooling and equipment would not be required.

Sheet metal forming is a highly complex process due to the many parameters involved and the non-linear interaction between them. There are also many different forming processes currently available, thus the type of process employed will depend on the product being formed. The specific characteristics of these products are very important in allocating the right forming process. These characteristics include the size of the product being formed, the depth of formation and the intricacy of the part.

Sheet metal forming includes the sheet metal forming operations such as shearing, bending, drawing, spinning, roll forming, stretch forming and bulging to form a variety of products. The distinguishing characteristics of sheet metal products are that they have a large surface area and a low modulus (ratio of volume to surface area) as compared to products produced by casting, bulk deformation or machining. In general, sheet metal forming is a two dimensional deformation, because the thickness change is generally small, but the elastic recovery can be significant. Since the operations are often performed on presses, sheet metal forming is sometimes referred to as press working.

This paper makes use of a flexible polyurethane pad as the punch, instead of the conventional metal component. Flexible tools such as polyurethane have been used in metal forming since the latter part of the 19th century. In the early days natural rubber was used but with the number of synthetic elastomers, such as urethane, with superior properties available today the scope for the use of flexible tools has increased. Urethane pads, rods and tubes are commonly used in dies as pressure pads, stiff springs and cushions.

Rubber pad forming, also known as flexible die forming, employs a rubber pad as one tool half, requiring only one solid tool half to form a component into its final shape. The solid tool half is usually similar to the punch in a conventional die, but it can also be a die cavity. It is assumed that the rubber acts somewhat like a hydraulic fluid in exerting nearly equal pressure on the work piece surfaces. Rubber pad forming was designed to be used on moderately shallow components having simple flanges and simple patterns. The production rates are relatively high, with cycle times averaging less than one minute. Advantages of using this process compared to conventional forming processes include:

1. The elimination of the need for a matched set of male and female dies. Only a male or female die is needed with the elastomer forming the other part thus tooling costs are low.
2. The work piece surface is not marred. The surface of the work piece in contact with the elastomer can be pre-polished, pre-printed or pre-plated eliminating the need for a protective coating.
3. The process can accommodate several thicknesses of the sheet metal being formed.
4. Alignment and mismatch problems are eliminated.
5. Spring back after forming is minimal.
6. Thinning of the work metal, as occurs in conventional deep drawing, is reduced considerably; and
7. The forming radius decreases progressively during the forming stroke, unlike the fixed radius on conventional dies.

On the other hand, the disadvantages of the rubber pad forming process are:
1. Rubber wears out quickly or tears on sharp projections.
2. Shrink flanges may have wrinkles.
3. Lack of sufficient forming pressure results in parts lacking proper definition and sharpness; and
4. The main piece of equipment used in most rubber pad forming processes is a hydraulic press, which is very expensive and relatively slow.

The die rig in this project made use of the principles of the Guerin forming process. The Guerin process is synonymous with the term rubber pad forming. The Guerin process owes its origin to Henry Guerin of the Douglas Aircraft Company who in the late thirties developed the first rubber press tool using rubber for straight flanging (Morris 1955; Johnson 1979; Venkatesh 1986; Thiruvarudchelvan 1993). It is the simplest of all the drawing processes and is used for drawing shallow components. In this operation a female die is mounted on the ram of a hydraulic press. As the press descends, the pressure on the pad builds up and deforms the sheet metal to take the shape of the die. As the ram retracts, the pad returns to its original shape.

Little (1977) stated that under high-pressure rubber might serve as a female die, which is capable of forming a metal blank around a solid punch. It can be assumed that the rubber behaves as a hydraulic fluid; it exerts equal pressure over the entire area. With the Guerin process, rubber pads can be used for blanking and piercing as well as for forming because rubber pads produce better edges on the work piece than band sawing. The hardness of the rubber pad used in the Guerin process was about Durometer Å 72. Begeman (1963) pointed out that a major factor was the considerable reduction in the cost of dies in blanking and forming when the Guerin process is used. The major advantages of this process include simplicity, low cost tooling, minimum wastage of material and uniform pressure on the blank. (Doyle, 1969) noted that the forming pressures are between 1000-2000 psi. Researchers (Thiruvarudchelvan and Gan 1994; Thiruvarudchelvan and Lewis 1988) have carried out improvement works towards deeper draw ratios.

3. Experimental drawing RIG

Strengths of material calculations were carried out in order to size all the components for the drawing rig (Ostergaard, 1967; Querishi 1978; Benham and Crawford 1987; De Garmo 1988). The experimental drawing rig consists of five plates that are mounted and fastened together. These are: 1) shank plate, 2) punch holder, 3) punch plate, 4) stripper plate and 5) bolster plate. Figure 1 gives a diagrammatic representation of how the plates are assembled together. The other major part of the assembly is the punch i.e. the part that carries the pattern that will be pressed on the sheet metal. The punch was made using the casting process.

Descriptions of the five main plates are as follows:

1. The shank plate is the plate that attaches itself to the hydraulic press. The shank plate contains two holes at the centre, which allow the mounting onto the press via the use of two grub screws. The four holes at the extreme ends contain a milled circular section. The bolts pass directly through these holes. These bolts attach the shank plate to the punch holder.

2. The punch holder is a large rectangular plate with numerous holes. There are two holes on the opposite sides of the plate located along a diagonal. The guideposts pass through these holes. The four holes with a diameter 19.05mm (3/4-inch) enable the shoulder screws to pass through. These four holes also contain a circular milled section. The shoulder screws pass directly through these holes and attach themselves to the stripper plate. The underside of the punch holder contains two holes that allow the punch plate to become affixed to it.

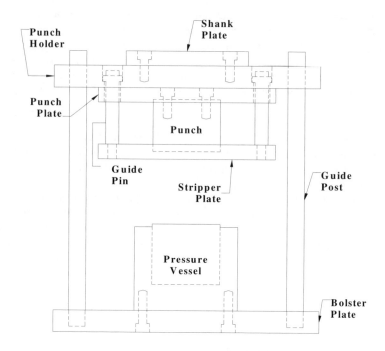

Figure 1. A front view of the drawing rig

3. The punch plate is located directly below the punch holder. The punch plate houses the punch containing the desired form. The punch is attached to the punch plate via the use of two bolts located in the centre region of the plate. The punch plate contains four milled sections of diameter 33.33 mm (1 5/16 inch) at the four corners of the plate. These enable the springs to be held in place and to prevent any movement in the x-y plane.

4. The stripper plate is suspended via the use of the shoulder screws. The shoulder screws are responsible for supporting the entire stripper plate. The stripper plate also contains four milled sections that prevent the movement of the springs in the x-y plane. The centre of the stripper plate contains a hole. The hole is the region where the punch is located. The stripper plate will be responsible for pushing down the stock material after the specimen has been formed, when the die is moving in the upward position. This is accomplished by the use of the springs, which push the stock material back down onto the pressure chamber after the forming process. The pressure vessel is a circular cylinder with a large portion of the inner region removed. The top surface contains two grooves on either side of the chamber. These groves serve as a guide for the sheet metal during the forming process. The under side contains two holes which are threaded. Bolts are engaged into these holes from the bolster plate. This ensures that the pressure chamber does not move from the fixed position.

5. The bolster plate supports the entire weight of the die rig. The two holes at the extreme opposite ends (diameter 24 mm) allow for the attachment of the guideposts. The holes

at the center of the plate enable the bolts to pass from the underside and affix themselves to the bottom of the pressure vessel.

4. Data collection

For the rig effectiveness to be evaluated it was necessary to collect data from specimens formed on the drawing rig using the Guerin process. The purpose of this data collection procedure is to gain information about the degree of strain that each product undergoes during the forming process.

The strain that a material undergoes is the extension per unit length and is given by the following equation:

$$\varepsilon = \frac{\Delta t}{t_o}$$

where:
- ε strain,
- Δt change in thickness of material,
- t_o original thickness of material.

Before performing the strain analysis on the products, the blanks have to be prepared. The method of doing this is as follows:
1. The blanks had to be cut from the stock material and then machined on the lathe to the correct dimensions, 100mm diameter.
2. The specimens were cleaned so as to remove the scale build up on the surface.
3. The specimens were mounted on supporting blocks and three concentric circles were marked off at distances of 40mm, 30mm and 20mm from the centre.
4. A line was drawn from the centre to the edge of the blank.
5. Readings of thickness were taken at the points of intersection of the concentric circle and the radial line using the micrometer (see Figure 2).

The thickness value is the measurement taken on a scribed circle around the point of intersection and an average obtained. The original thickness is the average of three measurements taken anywhere on the blank. By taking measurements at these points, A (40mm), B (30mm), C (20mm), Thickness strain can be calculated and then compared to check for the most heavily stressed points. The circumference of the initial blank and the blank after forming was measured so as to calculate the circumferential strain.

Circumferential strain: $\varepsilon_c = \ln \dfrac{new\_circumference}{orginal\_circumference}$

Thickness strain: $\varepsilon_t = \ln \dfrac{new\_thickness}{orginal\_thickness}$

After the blanks were prepared, the drawing rig was tested and a strain analysis was performed on the drawn components at forming pressures of 1000 & 1500 psi.

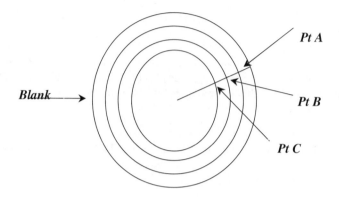

Figure 2. A blank prepared for readings

4.1. Specimen calculation

For this, consider the steel pan punch and a specimen of unannealed aluminum formed at 1000 psi:
- initial thickness = 0.996 mm (Position A),
- initial circumference = 314.16 mm,
- thickness after drawing at 1000psi = 1.03 mm,
- circumference after drawing at 1000psi = 299.14 mm

Now Thickness strain = $\varepsilon_t = \ln\left(\dfrac{1.03}{0.98667}\right)$

$\varepsilon_t = 0.033$ or $\varepsilon_t = 3.3\%$

Circumferential Strain = $\varepsilon_c = \ln\left(\dfrac{299.14}{314.16}\right)$

$\varepsilon_c = -0.049$ or $\varepsilon_c = -4.9\%$

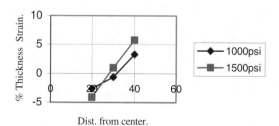

Figure 3. Thickness strain variation

Calculations of the other strains were carried out in a similar fashion. A sample of these results is listed in the strain analysis table for the steel pan punch (see Table 1) and a sample graph of thickness strain variation is shown in Figure 3.

Table 1. Strain analysis of steel pan punch at 1000 psi

Specimen	Initial Conditions		1000 psi	
	Thickness /mm	Circumference /mm	Thickness Strain	Circumferential Strain
Unannealed Aluminium	0.996	314.16	A) 0.033 B) –0.006 C) –0.026	-0.049
Partially Annealed Aluminium	0.983	314.16	A) 0.027 B) 0.017 C) –0.003	-0.026
Copper	0.703	314.16	A) 0.064 B) 0.051 C) 0.051	-0.046
Partially Annealed Copper	0.706	314.16	A) –0.008 B) –0.008 C) –0.023	-0.038
Partially Annealed Galvanized Steel	0.996	314.16	A) –0.220 B) –0.280 C) –0.290	-0.058
Brass	0.860	314.16	A) –0.072 B) –0.072 C) –0.084	-0.030
Partially Annealed Brass	0.830	314.16	A) 0.047 B) –0.049 C) –0.075	-0.029

5. Analysis and discussion

Experimental work was conducted on a 200-ton hydraulic press. The materials used for the specimens were Aluminium of thickness 1 mm, Brass of thickness 0.85 mm, copper of thickness 0.70 and 0.50 mm and Galvanized steel of thickness 1 mm. All the blanks were of diameter 100 mm. The annealed blanks were cleaned and emery papered to remove any scale build up on them.

Raw data was then collected from various specimens and put forward to give a visual perspective of the formed specimens. The quality of the surface finish and the strain on the components could now be ascertained. The thickness strain is of importance in ensuring the quality of the product. The thickness strains were measured directly to see the variation of the wall thickness of the drawn components. The thickness strains during the forming process had the same trend as in conventional drawing. The thickness strain is almost always negative in the bottom of the component i.e. the sheet is thinned in this region. Graphs were also plotted to observe the trends in thickness strain vs. distance from the centre of the blank. (See Figure 3 showing a sample graph).

5.1. Strain analysis on the steel pan components

It was noted from the results of unannealed Aluminium that the strain at 1500 psi was higher than that at 1000psi (note that positive strain indicates thickening of the material and negative strain indicates thinning of the material). One possible explanation for this could be that the blank holding force may be too high and is retarding the flow of material into the punch. This would result in thinning of the base of the component and thickening of the flange. This explanation is supported by the fact that at 40mm from the centre and beyond, the strain values are high and positive, indicating thickening. From the general trend of the line, increasing pressure would increase the strain.

For the case of partially annealed aluminium, the situation is somewhat opposite. The table indicates that at the higher pressure, the component undergoes more thinning at the flange area, which indicates that the material was drawn into the punch better than the unannealed aluminium, i.e. less retardation to movement. The reason for this is that the annealed specimen has better ductility and formability than the unannealed specimen and is able to flow and stretch to take the shape of the pattern more easily.

For the copper specimen, there was a region that had uniform strain. The area maintained constant thickness due to the fact that the rubber pad exerted equal pressure over the entire blank surface. At both pressures there was slight thickening of the flange. This could be because the blank holding force was a bit high for this material. The partially annealed copper also exhibited an area of uniform thickness strain in the flange area and wall. This sample formed quite well since the strain did not exceed 2%. This supports the assumption that the component maintains its thickness during the forming process. This specimen was far from fracture since it is a fact that copper can withstand an elongation of up to 60%. By comparing the two graphs for copper, it is clear that the annealing process gave the copper better ductility and formability since the strains were smaller than the unannealed specimen. The galvanized steel exhibited the largest strains of all the specimens. This is due to the fact that galvanized steel is a tough material and does not possess good ductility/formability as compared to aluminium or copper due to the carbon content present in it. As carbon content increases, the percentage elongation decreases and this accounts for the high strains present in the galvanized steel.

One way to reduce this high strain is by the application of a different lubricant such as molyslip. This may help to reduce the friction between the specimen and rubber pad. The brass showed quite a large strain near to the base of the component. Since brass has good formability and ductility, this thinning could be due to too high a blank holding force. If the blankholder force were too high, the material would be restricted from flowing into the centre of the punch, hence the material at the centre would be subjected to stretching or thinning out, as the case may be, to fill the cavity. The partially annealed brass had smaller strains than the unannealed brass, which was expected because of the effect of the annealing process.

5.2. Production/ change over times

The rubber pad forming technique produced products with a very high quality. The reject rate for this process was extremely low. The production time for rubber pad forming was about 10 seconds to form a single component from a blank. The setup time for the die rig was about 30 minutes as compared to the setup of Spinforming, which is about 15 minutes. The change over time for the punch was approximately 10 minutes because after change over the punch and pressure vessel had to be re-aligned.

6. Conclusion

The literature review brought into light the main types of sheet metal forming techniques with possible applications in Trinidad and Tobago. After drawing various components on the die rig, utilizing the Guerin Process, the strain analysis proved that the process works well. The results showed that fairly good components were formed. For the strain analysis, partially annealed copper had the lowest strain value of all the specimens tested. This indicated that the best material to use for the components is copper. Aluminium could work as well but it tends to lose its lustre after some time. Galvanized steel was used as the control in this project. Brass showed good forming properties and can be considered for use as a decorative component.

References

1. Begeman L.M., Amstead H.B. (1963): *Manufacturing Processes*. John Wiley & Sons, pp.216-226, 230-243.
2. Benham P.P., Crawford, J.R. (1987): *Mechanics of Engineering Materials*. Longman Singapore Publishers, pp. 56-57, 393-401.
3. DeGarmo E.P., Black T.J., Kohsler A.R. (1988): *Materials and Processes in Manufacturing*. Macmillan Publishing, pp. 476-480.
4. Doyle E.L. (1969): *Manufacturing Processes and Materials for Engineers*. Prentice-Hall, pp.287-289.
5. Johnson V.H. (1979): *Manufacturing Processe*. Bennett Publishing, pp. 198-200.
6. Little L.R. (1977): *Metal Working Technology*. McGraw-Hill, pp. 236-240.
7. Morris L.J. (1955): *Modern Manufacturing Processes*. Prentice-Hall, pp.477, 498-503.
8. Ostergaard E. (1967): *Advanced Die making*. McGraw-Hill, pp.22-32, 51.
9. Querishi H.A. (1978): *Mechanics of sheet metal bending with confined compressible dies*. Journal of Mechanical Working Technology, Vol.1, pp.261-275.
10. Thiruvarudchelvan S. (1993): *Elastomers in metal forming*. Journal of Material Processing Technology, Vol.39, pp. 55-82.

11. Thiruvarudchelvan S., Gan, J. (1994): *Drawing of hemispherical cups using Friction actuated blank holding*. Journal of Mechanical Working Technology, Vol.40, pp. 327-341.
12. Thiruvarudchelvan S., Lewis W. (1988): *Fiction actuated blank holding for deep drawing*. Journal of Mechanical Working Technology, Vol.17, pp.108-112.
13. Venkatesh V.C., Goh, T.N. (1986): *Mathematical models of cup drawing by the Guerin and Marform processes*. Journal of Mechanical Working Technology, Vol.13, pp. 273-278.

A STUDY ON THE EFFECT OF SINTERING SPEED AND HATCH SPACING ON THE STRENGTH OF COMPONENTS PRODUCED IN DMLS PROCESS

Radhakrishnan P., Jayaprakash M., Naiju C.D., Gajendran C.

Abstract: There is a general interest today to explore the potential of layer manufacturing as an alternative to conventional subtractive manufacturing. One of the promising processes in this regard is direct metal laser sintering (DMLS) which is generally used to build metal prototypes and tooling for plastic injection moulding, and patterns for rubber moulds and investment casting. The work reported in this paper is part of a study to assess the influence of process parameters of the DMLS process on the part quality and dimensional accuracy. Different tensile specimens are produced by varying only the hatch spacing and keeping all other process parameters constant. The study shows that the less the hatch spacing, the more the tensile strength and the hardness. In order to understand the effect of sintering and hatch spacing on the structure of the parts produced a study was carried out to analyse the microstructure to assess its effect on the tensile strength and hardness. The results are reported and discussed in the paper.

1. Introduction

While comparing various RP technologies, the direct metal laser sintering shows the great promise for direct production of functional prototypes and tools. (Simchi et al, 2003). Among the parts made using powder metal sintering copper base materials rank second only to iron base parts in terms of commercial applicability. As with other parts made using this process, the final properties and related performance of copper-base parts depend on successful sintering techniques (Alain Marcotte, 2000).

Laser beam binds the metallic particles in the DMLS process by liquid phase sintering whose main advantage is very fast initial binding. This binding is based on capillary forces, which might be very high. The reaction speed in this stage is determined by the kinetics of the solid-melt transformation. This transformation is several orders of magnitudes faster than physical diffusion (Kruth et al 1996).

There are two main forms of liquid phase sintering. When a liquid phase is obtained by inducing melting in a mixture of powders and is persistent throughout the high temperature portion of the sintering cycle, the process is termed as persistent liquid phase sintering. A persistent liquid can also be obtained by partially melting a pre alloyed powder above its solidus temperature (termed as super-solidus LPS or SLPS) and is widely used for processing tool steels, stainless steels and super alloys. In some systems with a low inter-solubility even in the presence of a persistent liquid, an activator can be used. This is termed as activated liquid-phase sintering.

Densification during liquid phase sintering commonly occurs in three stages after the liquid forms: rearrangement, solution-reprecipitation, and final-stage sintering (Rajiv Tandon et al). Super solidus LPS involves heating a pre-alloyed powder between the solidus and liquidus temperatures to form a liquid phase. The fundamental difference between classic LPS and SLPS lies in the sequence of events leading to densification. Figure 1.

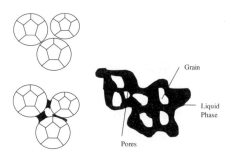

Figure 1. Sequence of stages leading to densification via super solidus sintering

In the first stage, liquid forms as a pre-alloyed powder is heated above the solidus temperature. The commonly observed liquid formation sites are the grain boundaries within a particle, the inter particle neck region, and the grain interior. These sites depend on several factors, such as powder microstructure, alloy chemistry, particle size, and the heating rate. As the liquid volume increases, at a critical temperature above the solidus, a threshold amount of liquid exists along the grain boundaries. Above this, the grains possess enough mobility to rearrange leading to particle fragmentation and capillary-induced rearrangement. Thus, densification is analogous to viscous flow sintering, because the semi-solid particles turn mushy and flow once the liquid spreads along the grain boundaries. Subsequently, continued densification occurs by solution reprecipitation, grain shape accommodation, and pore removal, as in classic LPS.

Khaing, et al, (2001) describes the DMLS process in detail with all the process steps involved in it like part design and data preparation, fabrication and epoxy filtration. Delamination between the base plate and the part was observed by them. It was also found that the humidity of the environment affected the sintering process. The quality of the bronze coating may affect the bonding between the base plate and the part. The hardness of the sintered parts ranged from 26 to 33 on Rockwell B scale whereas that of the infiltrated part increased to 65 to 69 HR B. By applying the low melting point infiltration of material such as silver or lead-tin alloy, the hardness can be improved, but the process becomes more complex.

The review of the literature shows that the process of layer manufacturing should be understood better in terms of the influence of process parameters on the reliability of the parts produced. The objective of the study is to optimize the process parameters like hatch spacing and sintering speed so that better microstructure, more hardness and higher tensile strength can be achieved in parts built using DMLS process.

This work was carried out at the Rapid Product Development Centre (RPDC) of Central Manufacturing Technology Institute, Bangalore, India.

2. Direct metal laser sintering

Direct Metal Laser Sintering is a laser-based rapid tooling and manufacturing process developed and is carried out on a EOSINT-250 M machine. This is done by using a CO_2 laser (200 W, 10.6μm) to sinter special non-shrinking steel or bronze-based metal powders layer by layer. The powder material used in this study is "Direct metal 50" which is a Bronze-Nickel powder with the following composition:

The basic material consists of the structural metal which is a high melting point metal and a binder (low melting point metal). Applying heat from a laser beam to the system causes the binder to melt and to flow into the pores formed by the non-molten particles. During liquid phase sintering the copper and phosphorous (Copper Phosphide) matrix formed in the powder is wetted to nickel. The parts are built up in thin layers of 50 μm successively by locally sintering metal powder. Figure. 2 shows a schematic arrangement of the layer manufacturing process using DMLS.

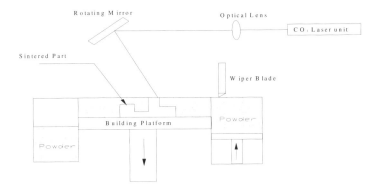

Figure 2. Schematic arrangement of DMLS process

Out of the several variables of the DMLS process, sintering speed and hatch spacing are the key variables influencing the bonding of metal powder particles. This study was therefore taken up to determine the values of these which will give the best tensile strength and hardness on the part produced. This will certainly be an important input to both design and manufacturing engineers.

2.1. Hatch spacing

Hatch spacing or hatching distance is the distance between successive laser scan lines. Figure 3 shows successive hatched lines within a contour.

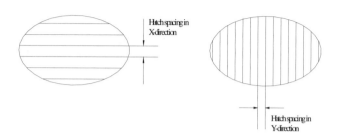

Figure 3. Hatch spacing

2.2. Energy density

Energy density is given by the equation (Neilbling .F, 2002):

$E_d = P_L / (h_S \cdot V_S)$ J/mm^2

where: P_L = Laser Power   in W

h_S = hatch spacing   in mm

V_S = sintering speed in mm/s

Hatch spacing has an inverse relationship with the energy density, which directly influences the heat intensity and hence the mechanical properties of the part built.

3. Methodology adopted

The methodology adopted in this study is discussed below.

3.1. Determination of strength

Tensile strength is an important criterion in determining the end use of a part in a mechanical assembly. Bronze-Nickel powder blend, used for producing metal prototypes, is basically a copper rich alloy and will behave as a ductile material after it is sintered. The major focus of this work, hence, is determining the tensile strength. The cross section of the test piece was selected as per the Indian Standard for tensile testing of copper and copper alloys (IS 2654 – 1977). The test piece was modelled using Magics RP software. The CAD model was sliced in layers of 0.05 mm thickness using Magics and the layer data was loaded onto the process software (PSW) of the RP machine. Test pieces of same cross section and dimension were built with four different levels of hatch spacing; 0.2 mm, 0.25 mm, 0.3 mm, 0.35 mm and keeping all other parameters constant. They were machined to conform to the Indian standards and tested for tensile strength and microhardness. The tensile test was carried out in a Universal Testing Machine of make VEB Werkstoffpruffmaschinen, and microhardness test was done in Mitutoyo Vicker's Microhardness Tester.

3.2. Study of microstructure

Four test pieces of dimensions 6 mm length x 4 mm width x 5 mm height were built with the hatch spacing levels 0.2 mm, 0.25 mm, 0.3 mm and 0.35 mm. These were polished using emery papers of grades 100, 320, 1/0, 2/0, 3/0 and 4/0 and finally polished in diamond polishing machine using diamond paste in kerosene. Microstructure was obtained for two cases: one without etching and another with suitable etchant – Ferric Chloride. Photographs of microstructures were captured using an image analyser. Scanning electron microscope (SEM) was used to find the micrographic structure of the powder for comparison.

3.3. Results

Table.1 shows the values of the tensile strength and microhardness obtained in the study.

Table 1. Tensile strength and hardness for different hatch spacing

Sl No	Hatch spacing (mm)	Tensile Strength (MPa)	Microhardness (HV)
1	0.2	135.44	216
2	0.25	126.99	213
3	0.3	108.17	142
4	0.35	101.87	108

Figure 4 shows a plot of tensile strength against hatch spacing. As hatch spacing is increased there appears to be significant reduction in the tensile strength. Minimum hatch spacing appears to give the best tensile strength.

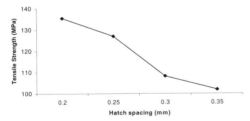

Figure 4. The influence of hatch spacing on tensile strength

The variation in microhardness with increase in hatch spacing is shown in Figure 5 Appreciable reduction in microhardness is observed with increase in hatch spacing.

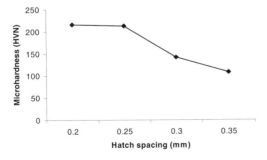

Figure 5. The influence of hatch spacing on microhardness

4. Metallographic studies

Metallographic studies were carried out using a scanning electron microscope. First the powder which is a combination of bronze, nickel and phosphorous was studied. Figure 6 shows the picture of the bronze nickel powder used in the trials prior to sintering.

Figure 6. SEM picture of bronze – nickel powder

Microstructure studies were carried for all hatch spacings. The paper includes the microstructures obtained for hatch spacing of 0.2 mm and 0.3 mm.

4.1. Microstructure for hatch spacing of 0.2mm

The structure of the sintered part shown in Figure 7 shows non-uniformity in dispersion.

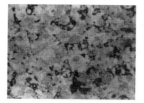 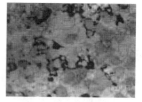

(a) Unetched 100 X   (b) Unetched 200 X

Figure 7. Photomicrographs at a hatch spacing of 0.2 mm

Porosity is not clearly distinguished in this picture even with a higher magnification. This may be because of the 50 % overlap resulting from a hatch spacing of 0.2 mm.

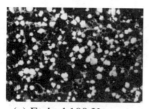 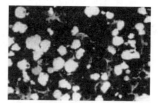

(a) Etched 100 X   (b) Etched 200 X

Figure 8. SEM Photographs of etched specimen

Figure 8 indicates a three-phase structure, a bright irregular phase, brown regular phase and a porous phase. The bright irregular phase is indicative of nickel, brown regular shape, unmelted bronze (circular) and black spaces pores.

4.2. Microstructure for hatch spacing of 0.3 mm (Figure 9 and 10)

Figure 9a. Unetched 100 x

Figure 9b. Unetched 200x

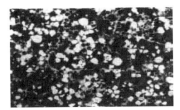

Figure 10a. Etched 100x

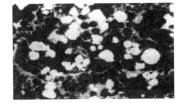

Figure 10b. Etched 200x

The sintering in this case appears to be uniform at the speed employed. It is observed that the material dispersion is uniform for this structure. For magnification of 200X too, the dispersion appears uniform. The nickel and bronze phases are more in this structure. This structure is found to be the most uniform phase.

4.3. Discussion

The influence of hatch spacing on tensile strength and microhardness is appreciable when hatch spacing is less. In this case the overlapping area in the sintering zone is more. This results in parts with better strength. When hatch spacing is increased, there appears to be a reduction in the tensile strength of the part.

Microstructure study shows that the uniform dispersion of structure obtained at a hatch spacing of 0.3 mm. From the point of view of obtaining uniformity in the structure, a hatch spacing of 0.3mm appers to be optimum.

As the strength was found to be more for 0.2 mm hatch spacing and uniform microstructure was obtained for 0.3 mm hatch spacing, it was decided to repeat the test for a smaller value of hatch spacing i.e. 0.1 mm. Hence, one more component with a hatch spacing of 0.1 mm was built keeping all other process parameters constant. It is found that the tensile strength and microhardness remains closer to the value of the hatch spacing level of 0.2 mm. So, further reduction of hatch spacing does not seem to improve strength.

5. Conclusion

The chemical composition of the metal powder is found out to be copper rich powder. The sintering phase that the DMLS process follows is found to be liquid phase sintering.

Hatch spacing has considerable influence on strength. From the tests conducted it is concluded that the optimum hatch spacing for building the components is 0.2 mm from tensile strength consideration and 0.3 mm for better microstructure.

Acknowledgement

The authors wish to acknowledge the support of Mr. G. Viswanathan, Chancellor, Vellore Institute of Technology and the Department of Science and Technology, Govt. of India for providing a research grants to carry out this work.

References

1. Marcotte A.: *Sintering of Copper-base alloys*. Production Sintering Practices – Shaping Consolidation Technologies, (2000), No 487.
2. Khaing M.W., Fuh J.Y.H., Lu L.: *Direct metal laser sintering for Rapid Tooling processing and characterization of EOS parts.* Journal of Material Processing Technology, (2001),Vol. 113, pp. 269-272.
3. Kruth J.P.,. Leu M.C., Nagakawa T.: *Progress in Additive Manufacturing and Rapid Prototyping*. Annals of the CIRP, Vol. 17/2, (1998).
4. Niebling F., Otto A., Geiger M.: *Analysing the DMLS –Process by a macroscopic FE Model*. Proceedings of SFF 2002, The University of Texas, Austin, pp. 384.391.
5. Tandon R., Johnson J.: *Liquid Phase Sintering.* Shaping and consolidation Technologies, pp. 565-573.
6. Simchi AQ., Petzoldt F., Pohi H.: *On the development of direct metal laser sintering for rapid tooling*. Journal of Materials Processing Technology, (2003), Vol. 141, pp. 319-328.

THE DESIGN PROCESS WITHIN ARTEX - RAWLPUG

Balfour C. S., Harrison D.K., Wood B.M., Temple B.K.

Abstract: This article looks at how Artex-Rawlplug has devised a process for designing and developing new products. It gives an insight into how this process works and the control measures needed to manage a project efficiently. It also looks at how technologies have had an influence on the design process and how it has improved products time to market. This report concludes by reviewing a product that was launched this year

1. Introduction

Artex Rawlplug are a medium to large scale manufacturing company specialising in fixings for the construction and D.I.Y market. they also manufacture their own D.I.Y. products including textured finishes, coving and decorative fixtures for domestic use. They are owned by the British Plaster Board (BPB) company who specialise in the manufacture of plasterboard, plaster and accessories and tools relating to their main products.

They were formed in 2001. The reason for this was both companies shared many channels into the market with routes in both trade and retail sectors. Bringing the companies together to form Artex-Rawlplug allowed for cross fertilisation of market knowledge through the sales and marketing teams, best practice to be shared across operational activities and increased opportunity for benefits to both Artex-Rawlplug and the combined customer base.

Product Development procedures were introduced and developed over 10 years ago within the company. The purpose of these procedures was to generate ideas, develop new products and product proposals in order to create a systematic, pro-active approach to New Product Development (NPD) and Product Design Proposals (PDP), which would result in a constant programme of work and a continual flow of successful products to maximize commercial opportunity. This strategy had enabled Artex-Rawlplug to grow and increase its profits, reduce product time to market and gain customer satisfaction through cost competitive, fit for purpose product.

2. NPD process

NPD is a design process and project management tool in which measures have been set up so that there is more control and clear concise instructions on how a product is developed whether it be a totally new design or simply a product change. NPD covers all new product ideas from original new ideas, incremental development to existing products, range extensions, cost reductions, own brand ranges and life cycle extension. (Waizeneker, 2000).

The aim and mission of NPD was to create an owned and measurable process, with company wide commitment for a systematic, market driven approach to idea generation to drive sales growth through the NPD programme. The process and ideas must have synergy with the business objectives and current product ranges respectively. The new products must be sellable and meet market needs through the existing routes to market.

NPD was developed essentially as the process of deriving new products and bringing them to market in the shortest time within planned budgets and planned cost of ownership. It was compiled in a way so that it was systematic and organised, but not cumbersome.

NPD is a project management process, therefore the process must apply effective project management techniques to ensure complex projects are coordinated and controlled in order to increase the chances of success i.e. project planning, project leader, project teams and implementation of the plan.

All projects are slightly different; however, the key defining characteristics are defined in the NPD chart (Figure 1). This chart is a graphical representation of the process, and it is made up of four stages which will be discussed in length in the next section. The NPD chart does not stand alone; there are many methods in which the way the NPD is controlled for example a data base. Concurrent engineering (continuous consultation) methods should be applied to all activities which requires the support of all functions for the project to be successful and completed in the shortest period of time.

3. Key stages of NPD

As mentioned before there are four key stages within the NPD process. They consist of idea generation, concept feasibility, Project development stage and finally the review stage. In the next few sections the stages are broken down and explained.

3.1. Idea generation

An annual target of 50 ideas has to be achieved; therefore, all idea generation methods are encouraged and all ideas generated must be processed through these procedures. Ideas fall into three main categories:
- stated needs - e.g. customer own brand range, range extension, cost reduction, legislation,
- perceived needs - incremental development to an existing product, product and brand enhancements,
- original needs such as a new "new" or "blue sky" products.

An idea can come from any source i.e. internal – shop floor and staff, customers, end users, suppliers, external innovators etc and should be recorded onto the Idea Form ID001. Also if the idea is successful and launched then the originator of the ideas receives a cash benefit. Ideas are weighted quickly in days on market, technical and financial risk and prioritised on potential benefit. Experienced individuals with company wide knowledge carry out this weighting process for both sides of the business.

The design process within ARTEX – Rawlpug 59

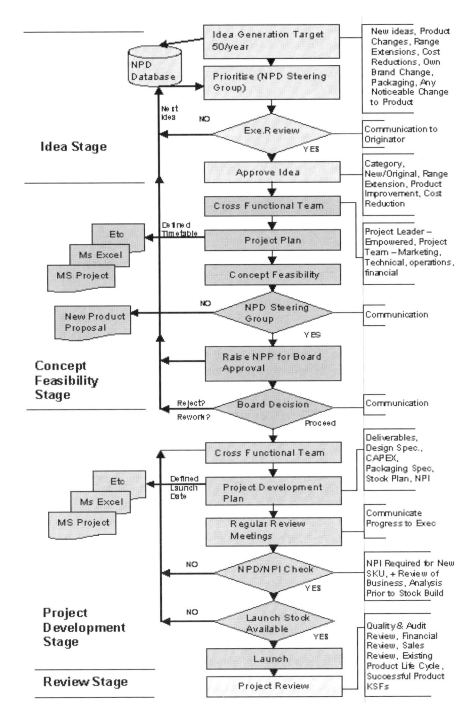

Figure 1. NPD Process Diagram (Artex Rawlplug, 2004)

The prioritised ideas are quickly assessed by an NPD steering team consisting of Technical Director, Sales Director and Operations Director in accordance with the business objectives. Approved ideas will then be progressed to the Concept Feasibility stage for formal evaluation by dedicated resource. Prioritised ideas are split into two groups:
- lower risk/lower benefit, quick win – "me too" products, range extensions, own brand etc.,
- higher risk/higher benefit – incremental development, "new", blue sky ideas.

Ideas get discussed as an agenda item at the monthly campaign planning meetings to ensure they are given commercial exposure and focus. Typical format is:

- generation of new ideas,
- review of other/previous ideas,
- agree priorities,
- feedback on progress.

3.2. Concept feasibility

Approved ideas are formally evaluated (days/weeks) as to their suitability of meeting the business strategy criterion and achieving at least £100K gross margin per annum for the first 3 years.

In order to ensure concepts are developed, evaluated and completed in a full and timely manner, a dedicated project leader and cross-functional team representing all functions is nominated and will be responsible for the completion of the concept plan. A project plan is compiled with clearly defined tasks, allocated resource, milestones and goes-no go decisions Concept Feasibility Plan template. On completion of the Concept Feasibility stage the Technical Director will review the findings and decide:
- to request more work, modify and repeat the feasibility,
- to progress to a product proposal for board approval to commit further expenditure and to ensure the project development is adequately resourced to meet the business objectives,
- to recommend make or buy,
- to scrap the concept,
- to select the next prioritised idea for concept evaluation.

3.3. Project development

In order to ensure projects are developed in accordance with the product proposal a dedicated project leader and cross-functional team representing all functions will be nominated and will be responsible for the full and timely completion of the project plan. A project plan will be compiled with clearly defined tasks, allocated resource, milestones and go-no go decisions Project Development Plan template:
- stage 1 - the first critical stage is to ensure the new product proposal is technically feasible. Acceptance will ensure the project is progressed to full development where prototypes have been available for testing, performance has been estimated, material specifications agreed with commercial in accordance with the product proposal,
- stage 2 - packaging specification,
- stage 3 – logistics,

- stage 4 - financials and costing,
- stage 5 and 6 - test market and Internal Pre-launch,
- stage 7 - commercial planning – launch.

All stages should be resourced as concurrently as possible to reduce time to market and maximise opportunity. Rigorous regular reviews are carried out at relevant milestones. If the original requirements are not met, key decisions must be made to continue, stop or change the process.

3.4. Project Review

It is essential that a quality review and audit of all stages of this procedure takes place twice yearly in order to ensure proper execution and to identify improvements to the process.

A detailed financial review takes place for all product launches on a quarterly basis where actual performance will be compared to the forecast given in the product proposal. A Project Review Form will be completed. If the sales trend deviates from the expected, customer research should be initiated to identify if the deviation is product performance related or due to the market opportunity. Suitable corrective action as will be taken to ensure the product achieves the forecast.
Existing products will also be reviewed to identify how the product life cycle can be extended. Successful product launches and the key critical success factors shall be identified to ensure launches are continuously improved.

3.5. Records/Controls

In order to properly utlise the NPD process records and measured controls must be properly utlised and complied. There is a range of accompanying forms and matrixes in which to document all information.
All the idea forms are held and maintained in the Technical Dept. All ideas and concept feasibilities are maintained in a database, summarising the needs, means and opportunity for continual review.
The Project leader issues all concept feasibility and project development plans to the circulation list at the start of each phase (cross-functional teams) and updated at regular intervals throughout the duration of the plan.
The completed project files are suitably located in each business centre and are in accordance with BS EN ISO 9000. This gives the ability to look back at old projects and see how they proceeded and were developed. This can be helpful if new staff is brought in, as it allows them to look back at how projects were run and if any to avoid past mistakes.

4. Plug case study

Up to now when deciding on a fixing to use the first step had to be to find out what material it was going to be fixed into and then find a plug that was suitable for that type of substrate. For example for board a plaster board plug was required , sheet material such as panel doors required a micro plug and for materials like concrete and block work the standard plastic plug was required (Figure 2). Most trade and D.I.Y users tended to use this standard plug for most substrate materials although it wasn't specifically designed to be a universal plug.

Figure 2. Plastic Plugs 5mm-&mm
(Artex-Rawlplug, 2004)

Figure 3. Substrate Range
(Artex-Rawlplug, 2004)

This standard plug was developed in the 1980s and at this time was considered a breakthrough design. However over the last few years technology has advanced and the standard plug has been assimilated by competitors, which meant Artex-Rawlplug lost some of their share of the market. So for the average D.I.Y it was difficult to find the right fixing for the job unless they had some experience. As a result of a range of issues from an outdated design, improvement on technologies and the lack of intellectual property rights on their old plugs.

Artex-Rawlplug decided that they had to develop a new plug that was universal. I.e. it could be made from different materials, and be used in all types of substrate. They also needed to break into the relatively un-captured European market. Europe tends to use a different substrate from the U.K. they tend to use hollow block and nylon plugs. However there is an increasing number in the British construction industry now using hollow block (Figure 3).

The major issue in the design was not just designing the plug that was universal but making it so that the design was registered and could be patentented in order to gain a hold in this new market.

4.1. Uno plug

The Uno plug has many features and benefits some of which are that it fixes light –medium objects to any wall type. It is an easy choice as only one range of plugs is needed which saves times finding the right fixing and space in the toolbox. The expansion along the whole of the plug allows it to fix to thin and hollow walls. It is not connected at the top so it provides instant expansion, grip and a quick confident grip. The plug does not spin in the whole due to the extra fins. The screw passes through the end of the plug and collapses hen fixed for extra security.

The plug comes in a handy sized strip which is useful for the trade side of the business and is more economical (Figure 4, Figure 5) The plug is also available in a range of different sizes all coming in a different colour range, 5mm is yellow, 6mm is red, 8mm is blue and 10mm is grey (Figure 6).

Figure 4. The 6mm Uno Plug (Artex-Rawlplug, 2004)

Figure 5. The 6mm Strip Uno (Artex-Rawlplug, 2004)

Figure 6. Uno Range (Artex-Rawlplug, 2004)

The plugs were originally designed using a 3d modeling package which allowed the designer greater freedom to experiment with the product and design before it was prototyped and manufactured thus saving thousands of pounds in development.

Due to the nature of the product the plug can only be tested in the real material as it is used for load bearing other products. The material it is made from then has a significant influence. One of the ways in which they can make small tools is to simply send off the 3d modeling drawings and files to a small tool maker and gets them to produce a tool which is capable of producing a low batch quantity in order to be tested.

This is only possible if the tool maker has the right 3d software and the knowledge and skills required to carry out and interpret the drawings and files. To produce these small tools can take only a short time from days to a few weeks to produce. Thus making the design process faster and efficient, which in turn speeds up the development process and the products time to market?

5. Summary and future work

In this report the design process within Artex-Rawlplug has been analysed and broken down into the various stages in order to gain a better understanding of how a manufacturing company develops and manufactures their own products. It has also explained why the NPD process is so important to the companies continuing growth in an ever expanding market.

The use of computers for design and manufacture has extended from 2D drawings for workshop tooling to full 3D capabilities to guide machine tool paths (T.I.C, 2005). Parts can now be designed, assembled and tested before tool manufacture. Simulations can be performed to test machining, clash of part assemblies and more complex stress analysis.

References

1. Artex-Rawlplug, Product Specification and Design Guide, 2004.
2. CAD/CAM/CAE, Accessed 20/02/2005, http://www.tic.ac.uk/technologies/technology.asp?method=tech&techID=69.
3. Waizeneker J. et al.: *Product Design and Development*. Croner CCh Group Ltd, London rd, Kingston upon Thames, Surrey, 2000.

Table 2. Percent parity weights of QMP and objectives of ISO 9004:2000 (example)

No	QMP	Item	Objective	Company A % Parity	Company A Global Parity	Company B % Parity	Company B Global Parity
1	Customer focus	1	Customer satisfaction	13.4	30.2	2.7	13.4
		2	Customer focus	11.7		5.4	
		3	Customer feedback	5.1		5.3	
2	Leadership	1	Management commitment	0.5	7.5	2.4	18.5
		2	Strategic quality planning	0.8		1.3	
		3	TQM vision	0.7		1.2	
		4	Refining scope, objectives, methodologies	0.5		1.2	
		5	Strategy process	0.5		2.3	
		6	Leadership training	0.6		2.1	
		7	Resource management	1.2		2.1	
		8	Plan for implementation	1.0		1.1	
		9	Modifying organisation structure	0.4		0.9	
		10	Financial barriers	0.8		2.7	
		11	Competitive strategy	0.5		1.2	
3	Involvement of People	1	Training and education	0.8	5.5	2.0	21.6
		2	Communication	0.4		2.2	
		3	Human resource utilization	0.7		3.0	
		4	Social responsibility	0.4		1.6	
		5	Behavoir	0.6		2.0	
		6	Cultural barriers	0.7		3.3	
		7	Formation of steering committee	0.7		3.9	
		8	Positive attitude towards quality	0.4		1.2	
		9	Identifying advocates and resistors	0.8		2.4	
4	Process approach	1	Management of process	3.4	13	2.2	8.8
		2	Procedures	5.4		4.4	
		3	Product quality design	4.2		2.2	
5	System approach to management	1	Role of quality department	1.5	7.3	2.4	14.9
		2	Business outcomes	1		2.0	
		3	Quality awareness	1.4		1.6	
		4	Competent Project teams	1.8		3.6	
		5	Business characteristics	0.8		1.7	
		6	Resource management	0.8		3.6	
6	Continual improvement	1	Continuous improvement	1.8	9.5	1.1	7.8
		2	Review status of TQM adoption	1.2		1.1	
		3	Determining improvement projects	1.2		1.1	
		4	Operational results	2.7		2.6	
		5	Structure	2.6		1.9	
7	Factual approach to decision making	1	Quality data and reporting	1.7	13.7	1	7.2
		2	Information and analysis	4.2		1.8	
		3	Internal business performance	2.1		1.0	
		4	Competitive benchmarking	3.8		2.5	
		5	Performance rewards	1.9		0.9	

For Company A the top four objectives were Customer satisfaction (13.4%); Customer focus (11.7%); Procedures (5.4%) and Customer feedback (5.1%). The weakest objectives were Modifying organizational structure (0.4%); Communication (0.4%); Social responsibility (0.4%) and Positive attitude towards quality (0.4%). These weaknesses suggest that Company A needs improvement in the softer QMPs objectives. For Company B, the top four objectives were Customer focus (5.4%); Customer feedback (5.3%); Procedures (4.4%) and Formation of steering committee (3.9%). The weakest objectives were Factual approach to decision making (7.2%); Continual improvement (7.8%); and Mutually beneficial supplier relationship (7.8%). These weaknesses suggest that Company A needs improvement in the harder QMPs objectives.

Table 3. Ranked objectives in descending order of percent parity weight

Company A		Objective in descending percent priority	Company B		Objective in descending percent priority
%	Rank		%	Rank	
13.4	1	Customer satisfaction	5.4	1	Customer focus
11.7	2	Customer focus	5.3	2	Customer feedback
5.4	3	Procedures	4.4	3	Procedures
5.1	4	Customer feedback	3.9	4	Formation of steering committee
4.4	5	Contact with suppliers and professional associates	3.6	5	Competent Project teams
4.2	6	Product quality design	3.6	6	Resource management
4.2	7	Information and analysis	3.3	7	Cultural barriers
3.8	8	Competitive benchmarking	3	8	Human resource utilization
3.8	9	Supplier management	2.7	9	Customer satisfaction
3.4	10	Management of process	2.7	10	Financial barriers
3.2	11	Integrating the voice of the customer and supplier	2.6	11	Contact with suppliers and professional associates
2.7	12	Operational results	2.6	12	Operational results
2.6	13	Structure	2.5	13	Competitive benchmarking
2.1	14	Internal business performance	2.4	14	Management committement
1.9	15	Supplier assessment	2.4	15	Identifying advocates and resistors
1.9	16	Performance awards	2.4	16	Role of quality department
1.8	17	Continuous improvement	2.3	17	Strategy process
1.8	18	Competent Project teams	2.2	18	Product quality design
1.7	19	Quality data and reporting	2.2	19	Management of process
1.5	20	Role of quality department	2.2	20	Communication
1.4	21	Quality awareness	2.2	21	Supplier management
1.2	22	Determining improvement projects	2.1	22	Leadership training
1.2	23	Review status of QMP adoption	2.1	23	Firms characteristics
1.2	24	Resource management	2	24	Training and education
1	25	Business outcomes	2	25	Behavior
1	26	Plan for implementation	2	26	Business outcomes
0.8	27	Firms characteristics	1.9	27	Structure
0.8	28	Business characteristics	1.8	28	Integrating voice of customer and supplier
0.8	29	Identifying advocates and resistors	1.8	29	Information and analysis
0.8	30	Training and education	1.7	30	Business characteristics
0.8	31	Financial barriers	1.6	31	Quality awareness
0.8	32	Strategic quality planning	1.6	32	Social responsibility
0.7	33	TQM vision	1.3	33	Strategic quality planning
0.7	34	Human resource utilization	1.2	34	Supplier assessment
0.7	35	Cultural barriers	1.2	35	Competitive strategy

5. Conclusions

Small and medium enterprises throughout the world are using the ISO 9001:2000 Standard as the preferred vehicle to quality management system implementation. This standard however must be used in conjunction with the ISO 9004:2000 in order to have an effective and efficient QMS that continually improves the performance of the organization. The ISO 9004:2000 uses the fundamental rules and beliefs of eight quality management principles and an efficient and effective QMS depends on the extent to which these QMPs are implemented.

In order to continually improve its operations SMEs must be aware of its strengths and weaknesses with respect to the extent of implementation of the QMPs. This paper has used to the analytic hierarchy process to empirically determine the extent to which the QMP were implemented in two SMEs that were recently certified to the ISO 9001:2000 Standard. This was represented by percent parity weightings of the QMP and their respective objectives.

Both SMEs were approximately 58% compliant to the QMPs and objectives. This suggests that at certification there is still much to achieve with respect to compliance to the requirements of ISO 9004:2000. However, there were contrasting areas of strengths and weaknesses. Involvement of people was Company A's weakest areas while it represented Company B's greatest strength. This suggests that although both had almost identical degree of implementation, the areas needed for continual improvement was starkly different.

The objectives of Customer satisfaction, Customer focus and Procedures were greatest areas of strength for both SMEs which is in line with the general theory that SMEs have good customer relations. The QMP that were ranked first and second for Company B were Involvement of people and Leadership respectively, while for Company A Involvement of people and Leadership were ranked eight and sixth respectively. This suggests that there may be a proportional relationship between involvement of people and leadership

References

1. Antony J., Leung K., Knowles G., Gosh S. (2002): *Critical success factors of TQM implementation in Hong Kong industries*. International Journal of Quality & Reliability Management, Vol.19, No.5, pp. 551-566.
2. Amar K., Zain M.Z. (2002): *Barriers to implementing TQM in Indonesian manufacturing organisation*. The TQM Magazine, Vol.14, No 6, pp. 367-372.
3. Baidoun S. (2003): *An empirical study of critical factors of TQM in Palestinian organisations*. Logistics Information Management, Vol.16, No.2, pp. 156-171.
4. Chin K.S., Pun K.F. (2002): A proposed framework for implementing TQM in Chinese organisation. International Journal of Quality and Reliability Management, Vol.19, No 3, pp. 272-294
5. Crowe T.J., Noble S.J., Machimada S.J. (1998): *Multi-attribute analysis of ISO 9000 registration using AHP*. International Journal of Quality & Reliability Management, Vol.15, No 2, pp. 205-222.

6. Decision Support Software (2000), *Expert Choice User Manual*, Version 10, Decision Support Software, McLean, VA.
7. Ghobadian A., Gallear D. (1997): TQM and organisation size. International Journal of Quality and Reliability Management, Vol.17, No 3, pp. 121-163.
8. Gotzamani K.D., Tsiotras, D.G. (2001): *An empirical study of the ISO 9000 Standards' contribution towards total quality management*. International Journal of Operations & Production Management, Vol.7, No 4, pp. 247-260.
9. Lee Y.C., Zhou X. (2000): *Quality management and manufacturing strategies in China*. International Journal of Quality & Reliability Management, Vol.17, No. 8, pp. 876-899.
10. Prasad S., Motwani J., Tala J. (1999): *TQM practices in Costa Rican Companies*. Work Study, Vol. 48, No 7, pp.250-256.
11. Prochno P.J.L.C., Corrêa H.L. (1995): *The development of manufacturing strategy in a turbulent environment*. International Journal of Operations & Production Management, Vol.15, No 11, pp. 20-36.
12. Quazi H.A., Padibjo S.R. (1998): *A journey toward total quality management through ISO 9000 certification: a study on small and medium-sized enterprises in Singapore*. International Journal of Quality &Reliability Management, Vol.15, No 5, pp. 489-508.
13. Russell S. (2000): *ISO 9000:2000 and the EFQM excellence model: competition or co-operation*. Total Quality Management, Vol.11, No 4/6, pp. 657-665.
14. Saaty T.L. (1996): *Multi-criteria Decision Making: The Analytical Hierarchy Proces*. RWS Publications, Pittsburgh, Pennsylvania.
15. Saaty T.L. (eds.) (2000): *Decision Making for Leaders*. RWS Publications, Pittsburgh, PA.
16. Yang J., Shi P. (2002): *Applying analytic hierarchy process in a firm's overall performance evaluation: a case study in China*. International Journal of Business, Vol.7, No 1, pp. 31-46.
17. Sohal A.S., Terziovski M. (2000): *TQM in Australian manufacturing: factors critical to success*. International Journal of Quality & Reliability Management, Vol. 17, No 2, pp. 158-167.
18. Tannock J., Krasacholand L., Ruangpermpool S. (2002): *The development of total quality management in Thai manufacturing SMEs: a case study approach*. International Journal of Quality & Reliability Management, Vol.19, No 4, pp. 380-395.

COMPUTER SUPPORT FOR INSTALLATIONS IN INTELLIGENT BUILDINGS – INTEGRATED ENERGY MANAGEMENT SYSTEMS

Rak B., Kurcz L.

Abstract: This is an analysis of intelligent systems installed and developed in intelligent buildings. Text presents a short description and advantages of using solutions based on intelligent systems. Article also includes information about applications of intelligent management systems such as HVAC.

1. Introduction

Intelligent building as a construction object should secure achievement of maximum exploitation comfort and minimum construction and exploitation costs at the same time. Many high quality devices are being used to realize this idea. These devices are technologically advanced and require computer support. Devices or installations do not decide about "Intelligence" of a building, it's only a part of building's equipment. But if connected in a one system, they create an environment that monitors and reacts if anything happens inside a room to assure optimal conditions [1-3]. This connection should secure communication between autonomous working devices and a failure of one device shouldn't cause a failure of entire installation. System should also be „open", which means that it can be developed and expanded during exploitation of a building. A necessary element of that kind of system is computer support of energy management, which enables an „intelligent" control of parameters determining comfort conditions.

2. Exploitation comfort

The main goal of intelligent systems installation is to ensure and improve exploitation comfort. This is a complicated issue, which can be divided into a few categories:
1. Thermal comfort.
2. Psychological comfort and safety.
3. Simplicity and easiness of device handling.

2.1. Thermal comfort

Thermal comfort is determined by conditions, which cause that a human being doesn't feel cold or hot. Heat balance of a human body is balanced and heat exchange is performed by radiation, convection, insensible perspiration and vaporization of water from respiratory tracks. The temperature of a body when no activity is performed is about 37^oC, and mean temperature of skin surface is between $32-34^oC$ [1]. Parameters that influence thermal comfort are: temperature, humidity and air movement in a room, and also heat conduction resistance through clothes and activity of a person inside that room. Control of air parameters in a modern building is impossible without computer support.

2.2. Psychological comfort and safety

These two issues can't be described without considering personality and character of human being. However it's possible to describe parameters that improve psychological comfort and safety:
- adequate placement of a building (far from noise and pollution sources),
- friendly architecture (visual sensations often decide about sympathy for a place),
- inside decorations,
- installed security systems,
- fire signalling and protection,
- supervision and control of owed systems,
- fault-free working systems (a fast removal of a possible fault),
- access control for some users,

Computer support makes possible to control some of these parameters.

2.3. Simplicity and easiness of device handling

Using modern technologies in buildings and places designed for common use caused that it's necessary to design them in a friendly way. Easiness of use can be achieved when a device is designed according to rules written below:
- clear menu with organized structure,
- simple handling with well-thought number of buttons,
- interface adapted to human body (steering by voice, remote, control panel, personal computer),
- ability to modificate and adapt steering menu.

3. Economical consideration

In the time of balanced development it's very important to calculate building exploitation costs during designing phase. It's caused by the fact that construction costs are only small part of all costs, which are in whole period of buildings life. Also the amount of money spent on systems installed in a building determinates costs of exploitation. Automation costs is an investition that pays back in a short time period because installed systems and devices help save energy, which is a big part of exploitation costs. Besides building automation can be divided into several phases if the installed system is open, the only condition is to consider it in project phase so it can be developed in many ways.

Changing work modes of systems can create energy and money savings in heating, ventilation and air conditioning. For example periodic action of systems or connection of heating, ventilation and air conditioning. Perfect example of that connection is Commerzbank Tower in Frankfurt [4]. Tests showed that in this building 2/3 time of year air conditioning can be turned off and substituted by natural ventilation, which generates large money savings.

AN AHP ANALYSIS OF THE QMPS OF ISO 9004:2000 IN ISO 9001:2000 CERTIFIED SMES IN TRINIDAD AND TOBAGO

Lewis W. G., Lalla T.R.M.

Abstract: Ever since it was introduced in 1987, small and medium enterprises (SMEs) worldwide have increasingly adopted the ISO 9001 Standard as their preferred quality management system. As such, many researchers have sought to determine the extent to which certified organizations have been able to attain total quality management (TQM). This paper examines empirically the extent to which the quality management principles (QMPs) of the ISO 9004:2000 have been implemented in two SMEs operating Trinidad and Tobago. It employs the Analytic Hierarchy Process approach in determining the percent weightings of the eight QMPs. The findings suggest that SMEs have different strengths and weaknesses and that special attention should be paid to the areas of weaknesses in order to attain greater TQM status. It was also found that there was a proportional relationship between the QMPs of involvement of people and leadership.

1. Introduction

The International Organization for Standardization (ISO) is a worldwide federation of national standards bodies. The work of preparing International Standards is normally carried out through ISO technical committees. ISO 9001:2000 and ISO 9004:2000 were prepared by the technical committee ISO/TC 176. They were developed together in order to promote ease of transition and efficiency within an organization. ISO 9001 is the quality management standard; ISO 9004 is the quality management system (QMS) standard.

ISO 9004 is recommended as guide for organizations whose top management wishes to move beyond the requirements of ISO 9001, in pursuit of continual improvement of performance (ISO 2000). It is based on eight quality management principles (QMPs). In order to move beyond a state where ISO 9001 certified companies are just compliant with the Standard's requirement to one in which the QMPs are fully implemented and functional, it is essential for them to know the extent to which the QMPs are operational. Although the ISO 9004 Standard provides self- assessment questions to evaluate the maturity of the QMS for each major clause in the Standard, there is no such tool to measure the strengths and weaknesses of each of the QMPs on which the clauses are based. The paper addresses this issue by using the Analytic Hierarchy Process (AHP)and the Expert Choice (EC)software to determine the percent weightings of each QMP and their respective objectives in two small and medium enterprises (SMEs) which had been certified to the ISO 9001:2000 Standard.

2.1. Criteria and objectives of ISO 9004:2000

In practice, the words objectives and criteria are used interchangeably. An objective is something that is sought after or aimed at, whereas a criterion is a principle or a standard that an idea or objective is judged by (Decision Support Software, 2000).

2.1.1. Criteria

The ISO 9004:2000 Standard's criteria consist of eight QMPs that reflect best business practices (Russell, 2000). QMPs can be defined as comprehensive and fundamental rules or beliefs for leading and operating an organization, aimed at continually improving performance over the long term by focusing on customers while addressing the needs of other stakeholders. These rule or beliefs are as follows:

1. *Customer Focused Organisation:* Small and medium enterprises (SMEs) must understand that they depend on their customers and therefore determine current and future customer needs, meet customer requirements and strive to exceed customer expectations.
2. *Leadership:* SMEs top management must establish unity of purpose and direction. They must create and maintain the internal environment in which people can become fully involved in achieving the organization's purpose.
3. *Involvement of People:* Employees at all levels must be recognized as the essence of the organization and strategies put in place to ensure their full involvement.
4. *Process Approach to management:* A desired result is achieved more efficiently when related resources and activities are managed as a process.
5. *System Approach to Management:* Identifying, understanding and managing a system of interrelated processes for a given objective improves SMEs effectiveness and efficiency.
6. *Continual Improvement*: The "Plan- Do-Check-Act" (PDCA) is applied to processes. The Plan establishes the objectives and processes necessary to deliver results in accordance with customer requirements and the organizations policies; the Do implements the processes: the Check monitors and measures processes and product against policies, objectives and requirements for the product and report the results; and the Act take actions to continually improve process and system performance.
7. *Factual Approach to Decision Making (FADM)*: Effective decisions are based on the analysis of data and information.
8. *Mutually Beneficial Supplier Relationship (MBSR):* SMEs and its suppliers are interdependent, and a mutually beneficial relationship enhances the ability of both in order to create value.

2.1.2. Objectives

An attempt was made to consolidate a list of objectives that reflect the rule or beliefs of the eight QMPs. This was derived from a review of the existing literature and empirical evidence based on practitioners' reflections (see Table I). The relative importance of these attributes/factors is mapped to the findings of 10 respective recent studies reported in the literature. These studies have focused on SMEs and were conducted in different countries and regions, such as Australia (Sohal and Terziovski, 2000); Brazil (Prochno and Correa, 1995); China, (Lee and Zhou, 2000; Chin and Pun, 2002); Costa Rica (Prasad *et al.*, 1999); Greece (Gotzamani and Tsiotras, 2001); Hong Kong (Antony *et al.*, 2002); Indonesia (Amar and Zain, 2002), Palestine (Baidoun, 2003); Singapore (Quazi and Padibjo, 1998); and Thailand (Tannock *et al.*, 2002). Although exploratory in nature, the findings provide some indication of the importance of the attributes/factors that would affect the success and benefits of the QMPs in SMEs.

3. The AHP-based study of QMP implementation

The AHP approach involves the decomposition of a complex problem into a multi-level hierarchical structure of characteristics and criteria with the last hierarchical level constituting the decision alternatives. These alternatives are compared to determine the objectives of the problem (Crowe *et al.*, 1998; Saaty, 2000). The objective and subjective judgments of top management from two SMEs were involved in order to make a trade-off and to determine priorities.

After the AHP goal had been established, relevant and important performance objectives which adequately represent the QMPs of ISO 9004 were identified. These objectives were then structured into a hierarchy descending form the overall goal to the QMP and objectives in successive levels. Organizing in a hierarchy serves two purposes: 1) it provides an overall view of the complex relationship inherent in the situation; and 2) it helps decision makers assess whether the issues in each level are of the same order of magnitude, so homogeneity in comparisons is preserved (Yang and Shi, 2002). Saaty (2000) suggests the guidelines for selection of the different levels of criteria and construction of the hierarchy. These are: 1) representing the problem as thoroughly as possible; 2) considering the environment surrounding the problem; 3) identifying the issues or attributes that contribute to the solution; and 4) clarifying the necessary participants associated with the problem.

An AHP framework was developed for facilitating the study, as depicted in Figure 1. The framework consists of three levels, namely the goal (i.e. level 1), the QMPs of ISO 9004:2000 (i.e. level 2) and their respective objectives (Level 3). The goal was to assess the effectiveness of QMPs' implementation in ISO 9001 certified SMEs. Each stage represents QMPs' criterion that comprises several objectives and benefits towards the attainment of ISO 9000 certification.

Empirical information and data was obtained through the combined judgments of individual evaluators from two specially chosen SMEs in Trinidad and Tobago. Invited evaluators in each SME under study were asked to carefully evaluate the criteria of each hierarchy level by assigning relative scales in a pair-wise fashion with respect to the goal of the model. With a set of semi-structured questions, the interviewees were asked to assess a pair-wise comparison among eight QMP and 46 objectives and benefits. A nine-point scale was used to assign the relative scales and priority of weights of criteria (Saaty, 2000). Experience has confirmed that the scaling mechanism reflects the degree to which one could distinguish the intensity of relationships among the levels of decision criteria and elements (Saaty, 1996, 2000). Pair-wise comparison is a key step in an AHP model to determine priority weights of factors and provide a rating for alternatives based on qualitative factors (Saaty, 2000; Yang and Shi, 2002). After the acquisition of evaluators' views, Step 5 followed with the computation of normalized weight priorities of the different hierarchies of criteria of the AHP model. This was done using computer software, Expert Choice (Decision Support Software, 2000).

Level:								
1	Effectiveness of ISO 9004:2000 Implementation through ISO 9001:2000 in SMEs							
2	Customer focus	Leadership	Involvement of people	Process approach	System approach to management	Continual improvement	Factual approach to management	Mutually beneficial supplier relationship
3	Customer satisfaction	Management Commitment	Training and education	Management of process	Role of quality department	Continuous improvement	Quality data and reporting	Supplier assessment
	Customer focus	Strategic quality planning	Communication	Procedures	Business outcomes	Review status of TQM adoption	Information and analysis	Supplier management
	Customer feedback	TQM vision	Human resource utilization	Product quality design	Quality awareness	Determining improvement projects	Internal business performance	Integrating the voice of the customer and the supplier
		Refining scope, objectives, methodologi	Social responsibility		Competent Project teams	Operational results	Competitive benchmarking	Contact with suppliers and professional associates
		Strategy process	Behavior		Business characteristics		Structure	Performance rewards
		Leadership training	Cultural barriers		Firms characteristics			
		Resource management	Formation of steering committee					
		Plan for implementation	Positive attitude towards quality					
		Modifying organisation structure	Identifying advocates and resistors					
		Financial barriers						
		Competitive strategy/						

Figure 1. An analytical framework for AHP analysis

The relative importance of each factor was rated on a measurement to provide numerical judgments corresponding to verbal judgments. Priority means the relative importance or strength of influence of a criterion in relation to other criteria that is place above it in the hierarchy. The normalized eignenvalues method is recommended when the data is not entirely consistent (Crowe *et al*, 1998; Saaty, 1996).

Since different levels or hierarchies were interrelated, a single composite vector of normalized weights for the entire hierarchy was determined, using the vector of weights of the successive hierarchy. The geometric mean of evaluators' scores then combined the pair-wise comparison judgment matrices. Both local priorities (i.e. relative to the parent elements) and global priorities (i.e. relative to the goal) were generated. These were represented by total and sub-total of priority scores (i.e. Step 6). Each set of comparative judgments would be entered into a separate matrix to derive the 'local priority' (i.e. the preferences with respect to the specific criterion). The weights of the criteria and its sub-criteria would be derived in a similar fashion. The process would continue until all comparison judgment matrices were obtained. All acquired data and information were then computed and analyzed using the computer software, Expert Choice (Decision Support Software, 2000).

4. Data analysis and findings

Two SMEs hiring less than 100 people and operating in Trinidad and Tobago (T& T) were selected for this study. Table 1 depicts the characteristics of the two SMEs (i.e. Companies A and B) that were analyzed against a list of 25 selection criteria based largely on Ghobadian and Gallear (1997). These criteria were chosen to reflect the unique characteristics of SMEs, in terms of their structure, procedures, behavior, culture, processes, people and contacts endemic to T&T. A "Y" means that the company possesses the characteristic, while an "N" means that it does not. Company A was less than 5 years old and operated as a job shop, and Company B was greater than 15 years and was involved in batch production. None of them had previous knowledge of quality management systems. Both had a fluid organizational structure with key personnel performing many functions. In both cases, top management was in close contact with the main customers and had direct input and contact with the key suppliers.

In total 8 interviews were conducted with senior personnel including Chief Executive Officers, general managers, production managers and customer service managers of the two SMEs. These personnel are responsible for and/or involved in quality management practices and performance measures in their organizations. Their views provide a wide spectrum of experience and expertise within their organizations and across various industry sectors in T&T.

The overall percent parity of ISO 9004 effectiveness for Company A was 58% and 58.4% for Company B. This reflects the level of achievements of the SMEs towards the level -1 goal of effective QMPs implementation through ISO 9001:2000 (see Figure 2). The inconsistency indices of the AHP analysis for Companies A and B were 0.03 and 0.04, respectively. These fall within the acceptable level of 0.10 as recommended by Saaty (1996), which indicated that the evaluators assigned their weights consistently on examining the priorities of decision criteria and assessing the effectiveness of QMPs implementation.

The local and global priorities of the different levels of criteria and objectives are depicted in Table 2. Each is implemented to varying degrees at certification. At this point, an independent certified auditor from an internationally recognized body determined if the compliance requirement of ISO 9001:2000 have been effectively implemented. This would be reflected by the QMPs and objectives with the highest percent parity. Similarly, by the QMPs and objectives with low percent parities would represent would not have been properly implemented. Those with the lowest can be considered as areas of weaknesses therefore represent areas for improvement. For Company A the QMPs of greatest strength were Customer focus (30.2%); Mutually beneficial supplier relationship (13.3%); and Process approach (13.0%); The QMP which should be targeted for improvement were

Table 1. Selection of SMEs for the study

No.	List of Selection Items	Company A	Company B
1	Willingness and resources available to implement ISO 9001	Y	Y
2	Had little prior knowledge of quality management systems	y	Y
3	Require upgrade from ISO 9001:1994 to ISO 9001:2000	N	N
4	Number of employees less than 100	Y	Y
5	Company less than 5 years old	N	Y
6	Company between 5 and 15 years old	Y	N
7	Company greater than 15 years old	N	N
8	Job shop type	N	Y
9	Batch production type	Y	N
10	Continuous flow	N	N
11	Service	N	N
12	Project type	N	Y
13	Family owned	Y	N
14	Government owned	N	N
15	Sole proprietary	N	Y
16	Public company	N	N
17	Division	N	N
18	Skilled based	N	Y
19	Established by entrepreneur	Y	Y
20	Based in an industrialised area	N	Y
21	Based in the East West Corridor	Y	N
22	Based in Central	N	N
23	Based in South	N	N
24	Certified at present	Y	Y
25	HSEQ requirements	N	Y
	Remarks		
	Y means yes ; N means No		

Involvement of people (5.5%); System approach to management (7.3%) and Leadership (7.5%). The QMP of greatest strength for Company B were Involvement of people (21.6%); Leadership (18.5%); and System approach to management (14.9%); while those which needed improvement were Factual approach to management (7.2%); Continual improvement (7.8%); and Mutually beneficial supplier relationship (7.8%).

A closer examination of the percent parity of the objectives helped identify specific areas of strength and weaknesses. Table 4 ranks the objectives in descending order of percent parity weightings.

Intelligent system is more expensive than standard automation system [9, 10] and can be adapted to investor requirements, but when it comes to combining heating, ventilation, air conditioning, power management, lights etc. then an open intelligent system that manages these sources is very comfortable and relatively cheap solution. Savings are generated by: reduced number of cables, which is necessary to create a system, low costs of co-ordination, activation, exploitation and development of that kind of system.

4. Examples of intelligent systems

Building's energy management system can be categorized into three groups (Figure 1):

Figure 1. Building management system

- management level – analysis and presentation of results,
- automation level – groups of devices that secure data transmission between object devices, also information processing and regulation of automation processes,
- object level – devices, which control other devices (controllers) or devices, which gather information from surroundings.

Nowadays to manage intelligent system in intelligent buildings generally is used BMS [5, 6] (Building Management System) based on *peer to peer* network architecture. That solution allows using many different devices produced by separate companies and add them to the system depending on user requirements. It allows also monitoring and managing of building resources from any place in the system. Connection can be performed in many different ways: cable (strand, concentric, light pipe), power line, radio waves, and other. Currently there are many communication standards and each proposes many smart solutions. The goal of using open systems is the same but there are differences between them in ways the devices are connected and how they communicate. From among existing systems few of them can be featured for interesting solutions and shares in global market:
- EIB - KONNEX (European Installation Bus),
- LONWORKS (Local operating Network),
- BACNET(Building Automation Control Network).

4.1. EIB – KONNEX

System EIB [2] was developed in the eighties by leading European producers. It found its practical application in building industry and is used instead of classic wiring system to control and monitor work of electrical devices in a building. In this system traditional switches that handle circuits are substituted by digital devices, which share information through an additional connection. The connection is called bus and it connects different systems or units, and also it allows using various combinations of devices depending on user idea [7].

Another advantage of this system is standardization, which guarantees easiness of substitution and development of devices in the system, and also accessibility of service. In 1990 EIBA (European Installation Bus Association) was created. This association gives EIB mark and develops rules of certification and trainings, and also prepares tests of devices and technical conditions, which have to be met. EIBA developed a system of building wiring and technical equipment digital control, by using sensing and acting devices showed on Figure. 2. Parameters that influence steering are for example time, weather, activity inside building and other. There are few variations of this system; difference between them is the way of communication [8]: EIB INSTABUS – bus is a cable with safe voltage 24 V. This kind of solution gives safety and is cheap, important matter is that wiring should be installed in construction phase of a building:

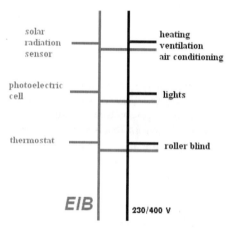

Figure 2. EIB open system

- POWERNET – bus is combined with power line 230 V, this creates a possibility of existing buildings automation without additional wiring. It is also used in places where additional wiring cannot be applied. Devices also have inputs for traditional switches so local steering can be also enabled,
- FUNK BUS EIB – devices in this type of EIB system communicate through radio waves using installed emitters and receivers powered by batteries (energy consumption is very low so devices can work for a long time). They work on frequency 433,42 MHz which fits in ISM (Industry Science Medical) frequency band. This sort of system connection can be applied in existing buildings or to expand existing systems.

4.2. LONWORKS

LONWORKS [7, 9, 10, 11] was designed by few leading electronic companies and entered market in 1991. Local operating network is introduced in a whole new way and eliminates

solutions of bus wiring developed so far. It's a new technology brought up to build distributed intelligent control system. It provides computer application for measurement and control and also data transmission. It also contents every other advantage of open distributed intelligent control systems:
- ability of applying the same communication project in multiple applications,
- reduced quantity of wiring,
- increased flexibility of system,
- increased reliability.

System has a flat topology (showed on figure 3). The most important element of this is multi-processor integrated circuit "Neuron Chip" 3120 or 3150 with built-in multipurpose operating system. Processor has a built-in LONTALK function which makes information exchanged between devices liable to stable rules, and that gives the possibility of communication between devices from different producers. Every "Neuron Chip" is a network node and performs a specific function. Another important element is Transceiver/Receiver a device that enables "Neuron Chip" to communicate with LONWORKS network based on various ways of communication. This solution makes this system unique and gives the possibility of connecting devices placed in distant locations – tunnelling. Only physical connection of devices will not cause the system to work. It is necessary to connect devices logically – program a device which signals it is suppose to receive and which are reserved for other devices. A special tool is required to make this connection – LONWORKS system starting program ran from a computer connected to the network. In LON system two kinds of configuration types can be created:
- a system with very strong device dispersion where every element is equipped with Neuron Chip and performs simple measuring or acting tasks,
- a system where Neuron Chip is responsible for actions and reactions of many devices placed in a room connected with them through developed functional input and output solution cooperating with other elements of the network.

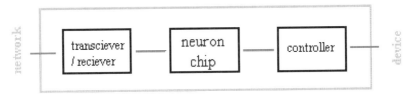

Figure 3. Network node

LONWORKS is a BMS (Building Management System), used to manage systems installed in a building such as: heating, ventilation, and air conditioning.

4.3. BACNET

BACNET protocol [10, 11] is a set of communication rules related to software and devices in the system making possible a connection of different systems. Information's can be transmitted through a local computer network. The protocol was published in 1995 and in 2001 an updated version was released. In 2003 BACNET protocol was given an ISO certificate and became European Union standard. BACNET is not an automation system it is a communication system. The core of this system are communication protocols Ethernet and ARCNET (computer network protocols). BACNET offers a layer topology of network. Buildings automation system uses BACNET protocol to transmit data and its connection is created with routers. To create a connection where systems using different protocols can communicate it is necessary to use translators – gateways (showed on Figure 4). Existence of many different communication systems in one building is often economically and technologically reasonable. A perfect solution would be a data transmission in one standard but nowadays only BACNET offers a solution where there is a possibility of connecting systems with different communication protocols.

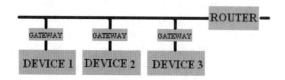

Figure 4. BACNET connection scheme

Another advantage of this system is that it is independent from existing technology, which allows substituting devices in any time of system exploitation. According to this BACNET gets more and more practical applications and is perfect for integrating resources of big buildings.

5. Heating, ventilation and air conditioning

Because the point of building automation is energy saving it is important that heat supply and removal should be precisely planned in an economically reasonable way. This was the reason of creating HVAC – Heating Ventilation Air Conditioning [4,6] a system, which is responsible for keeping specified values of air parameters: temperature, humidity, and quantity of fresh air. This kind of solution based on an open system is followed by many benefits such as:
- full integration of devices in form different,
- flexible programming system and access to all data and all controlled parameters in building ensuring energy savings,
- automatic optimalization of device work parameters,
- visualisation of system work and signalling of problems,
- ability to divide and separate control of air conditioning and to turn off any parts of the system when it is not necessary for them to work,
- ability to save history of device usage,
- ability to change settings of HVAC system by determination of individual air parameters form every node of the network,
- alarm phase service and information about condition of devices.

This solution perfectly fits for big buildings where system load is constantly changing in time. Similar solutions are applied in small buildings; differences are only in size of the system and quantity of devices. Very important matter in system control is hierarchy of steering signals. Integration of heating, ventilation and air conditioning prevents contradicted work of systems. HVAC based on an open system gives user freedom in choosing solutions for heat and energy supply. An example of HVAC conception for one room is showed on Figure 5.

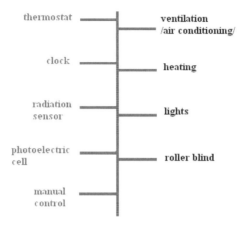

Figure 5. HVAC scheme

Every measuring or controlling device (left side of the scheme) gives data for acting devices controlling various systems (right side of the scheme). Often changes of load in monitored environment cause that it is reasonable to use flexible systems and devices. A good result gives a system where few sources of heat and energy work together:
- gas furnace,
- heat pump,
- electric heaters,
- solar collector, solar cell,
- radiant heater,
- ventilation /air conditioning/ VAV divided into zones.

It is also recommended to use devices that reach their highest performance in short time and to secure safety of using this sort of devices and systems. Also it is essential to create a possibility for manual control of these devices so the system or some of devices can be shut down if needed.

8. Conclusions

Analysis of data and information that concern intelligent buildings and integrated energy management allows to say that:
- monitoring and integrated management of buildings resources seriously affects on energetic performance of a building and secures comfort of exploitation and safety of users,
- rationalization of energy usage through integration of heating, ventilation, air conditioning and renewable energy sources with computer support is not only a luxury but also a necessity,

- it is important for management systems to secure compatibility of various devices and systems,
- system should be reliable and simple for a regular user,
- nowadays intelligent solutions for buildings automation are very expensive but in time their cost will reduce. Considering environment protection, energy saving and short time of investition cost return, the prevalence of intelligent systems is unavoidable.

References

1. Fanger P.O.: *Komfort cieplny*. Warszawa, 1974.
2. Petykiewicz P.: EIB *Nowoczesna instalacja elektryczna w inteligentnym budynku*. Warszawa, 2001.
3. Szepietkowski M., Ruszniak P.: *4.3 Inteligentne budynki*. Inteligentne budynki, Weka IB, 2003.
4. Szepietkowski M., Ruszniak P., Stoch A.: *4.3.1 Główne systemy inteligentnych budynków*. Inteligentne budynki, Weka IB.
5. Szepietkowski M., Ruszniak P.: *4.3.2 Systemy kontroli i zarządzania w dużych budynkach*. Inteligentne budynki, Weka IB.
6. Mull T.E.: *Systemy dystrybucji powietrza i systemy HVAC*. Utrzymanie ruchu, kwiecień 2005.
7. Nieckuła P.: *Commerzbank Tower – nowatorskie rozwiązanie wentylacji*. Rynek instalacyjny, nr 3, 2004.
8. http://www.eib.pl działy: Inteligentny dom, Opis systemów EIB, Projektowanie
9. http://www.echelon.com dział Solutions
10. http://www.bacnet.org dział Toutorials
11. http://www.strataresource.com dział Documents plik Investigating Open Systems

The International Journal of **INGENIUM** 2005 (1)

ENGINEERING ACHIEVEMENTS ACROSS THE GLOBAL VILLAGE

edited by

Janusz SZPYTKO

Computer Aided Engineering

Cracow - Glasgow - Radom, 2005

TABLE OF CONTENTS

page

2.	**Computer Aided Engineering**	**91**
2.1.	Optimization of directional control valve utilizing Pro\Engineer software and its fast prototyping, *Gromala P., Lisowski E.*	93
2.2.	Modeling of mechanical systems, when FEA techniques is used, *Suárez H.E.J.*	101
2.3.	A mathematical model for controller synthesis of the electro chemical discharge machining process, *Mediliyegedara T.K.K.R.., De Silva A.K.M., Harrison D.K., McGeough J.A.*	109
2.4.	Modeling of the electro chemical discharge machining process with a neuro fuzzy pulse classification system, *Mediliyegedara T.K.K.R.., De Silva A.K.M., Harrison D.K., McGeough J.A.*	117
2.5.	SISREDIS: a 3D digital scan, *Salas A., Márquez M.*	123
2.6.	A fuzzy-neural system for cut-make-production estimation, *Li Q., Walker M., Ranil S.*	131
2.7.	Design and fabrication of surgical titanium implants using CAD and SLA, *Harun W.A.R.W., Devadass V.*	137

All rights reserved. No part of this book may be reproduced, stored in a retrieval system, or transmitted, in any form or by any means, without prior written permission from the Publisher.

The International Journal of INGENIUM
Chief Editor: Professor David K. Harrison, Glasgow Caledonian University, UK

© GCU Glasgow

ISSN 1363-514x

A CIP catalogue record for this publication is available from British Library

Publishing cooperation: Instytut Technologii Eksploatacji – PIB w Radomiu

OPTIMIZATION OF DIRECTIONAL CONTROL VALVE UTILIZING PRO\ENGINEER SOFTWARE AND ITS FAST PROTOTYPING

Gromala P., Lisowski E.

Abstract: One of the features of modern engineering CAx software is a complex structure. Nowadays such programs make possible to create geometric models, preparing technical documentation, conducting different types of analysis such as kinematical, FEM in one single package. In this paper method of optimization of directional control valve as well as preparing of prototype using Rapid Prototyping utilizing the modest engineering software is presented. Parametric model of directional control valve's body has been built in Pro/Engineer. Then optimization of the flow channel has been conducted using Pro/Mechanica. Finaly two prototypes were built with Rapid Prototyping method. Now these prototype models can be used for laboratory experiments of flow through considered valve.

1. Introduction

At present moment the development of computer technology and CAD 3D systems is so high that requirements of designers are satisfied, especially it applies to designing in CAD 2D. In the scope of 3D designer have such interesting software as Pro/E, Unigraphics, Catia, etc. Using this software requires deep knowledge about it as well as the engineering knowledge. Practically new version appears annually, theirs usage is much simpler; help is more technically developed and have better design capabilities. Everything this intensify creative possibilities of design engineers. Now design process with the aid of CAD systems can start from creating geometry model with omitting technical documentation. It creates possibilities to presenting products that does not exist and make possible to conduct different analyses which concerns strength of materials, flow simulations and even some economical analysis. Also building a prototype now is much simpler as well as cheaper than it was few years ago. Now, before engineer/scientist decides to build prototype can conduct wide variety of virtual tests before choosing the most promising solution. Prior, building a prototype was based rather on engineering's experience and sometimes on a guess. Nowadays because of utilizing engineering tools such choice is much reliable.

Large international corporations are interested in using of new generation CAx systems as well as medium and small companies which have to be prepared for often change of market demand and adaptation of the product to current needs. In large corporations usually a complex systems are used, in small companies only individual workstation are met. In spite of different industry scale in both cases CAx system are the base of the development and improvement of quality of project and product. It requires possessing appropriate engineering resources and knowledge of computer techniques. Using these techniques is possible in the more extensive range when the more extensive knowledge and skills of engineer is.

Designing in the solid form is one of the most important activities in case of modeling parts using 3D techniques. Prepared model is then use in different virtual studies (Fig. 1) for instance optimization, FEM or CFD analysis, kinematical analysis collision study, presentation of the product, etc. before making the prototype. During such studies it is possible to check results and change the geometry and other parameters of the model. In such way the number of errors decrease in the design process and the quality of product is raising. Finally at the end it is possible to build a plastic prototype and conduct laboratory study. In this paper author presents creating the prototype of directional control valve. Fig. 1 presents the steps that were used before building final prototype. Authors first prepared geometric models in Pro\E, than conduct optimization of flow channels in Pro\M. Subsequently analysis of flow has been conducted using CFD code. Finally when the results of optimization and CFD analysis confirmed that pressure drop decrease the prototype was built using Rapid Prototyping method.

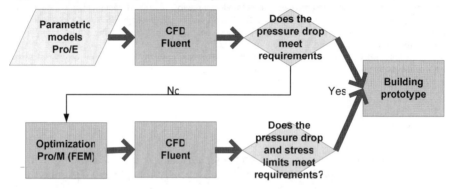

Figure 1. Phase of prototyping of the directional control valve

2. Solid modeling of directional control valve using pro/engineer software

Hydraulic valves are construction with complex geometry. It especially applies to body of the valve. Theirs complex geometry follows from necessity of creating flow channels and different kinds of spool chambers as well as other steering mechanism. Designing and modeling of hydraulic elements such as directional control valve is connected with possibilities of conducting additional FEM analysis with respect to high working pressure and required tightness. Additionally in case of hydraulic elements simulation of flow is required. Usage of lumped parameters models in modeling flow through control valve, when the subject of analysis is shape of channels gives very poor results.

The object of the analysis is six-way control valve with nominal diameter 6. The valve consists of control sections, covers, cartridge valves and steering mechanism. Sections are installed in layer system with covers outside. Receiver ports "A" and "B" are in control section, the supply port "P" is in supply cover, whereas the tank port "T" in outlet cover. Control sections can be also equipped with relief valve so as to control pressure in receiver ports independently. During analysis was assumed, that control valve is equipped in typical hydraulic components such as: relief valve, check valve, etc.

The geometrical model of directional control valve was prepared in Pro\E, at the beginning, particular components such as body of the valve, spool, and closing covers were performed, then other equipment such as valves, steering mechanism and finally units. Performed models are presented in Fig. 2; body of the control valve is showed in Fig. 2a. Cross-section of control valve is presented in Fig. 2b. Hydraulic connections between parts of control valve are realized by high pressure channel "P", channels connected to spool chamber and outlet channel "T". Sections are mounted by screws located in holes "M". Fluid flows to receiver ports "A" or "B" from port "P" in supply cover through channels in control section.

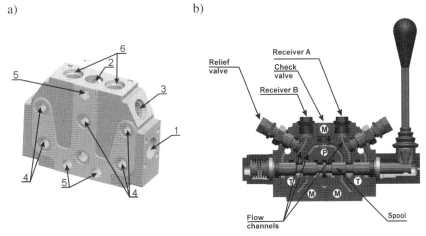

Figure 2: a) Model of the body of the directional control valve: 1 – spool chamber, 2 – check valve chamber, 3 – relief valve chamber, 4 – channels between sections, 5 – assembly holes, 6- receivers; b) Cross section of the directional control valve body

3. Optimization of the construction

In the construction of directional control valve it is meaningful from strength point of view minimization of the deformation of mating parts and edges. Even very small deformation can cause arising of leakage or spool seizure. Other problem is shaping of the flow channels in such a way to minimize pressure drop. Searching of optimal solution can be realized by single simulation or in automatic way through definition of optimization task. Optimization is the process of searching of optimal solution at assumed limits. In the paper optimization of the control valve body had been undertaken using Pro/Mechanica software. Minimization of pressure drop can improve efficiency of the valve. Unfortunately it is not possible to conduct optimization study with Pro/M in such a way that assumed goal function will be minimum of pressure drop. However it is well known that the larger flow channels are the pressure drop decrease. So in considered paper as a goal function minimization of valve's mass has been assumed. Analysis was focused on areas of flow way where large pressure drop occurs. Such area is check valve chamber. Dimensions that decided about channel shape and check valve were set as a parameters that were change. In Fig. 3 are showed parameters that were used in the optimization. Also new auxiliary coordination system was introduced (Fig. 3).

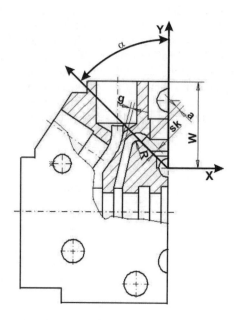

Figure 3. Parameters used during optimization process

Optimization starts (Mechanica/Structure/DesignStudy – Optimization) with definition of goal function, limits and parameters. One of the limits was thickness of the wall "g". It should be thicker than 3.0 mm. Additional parameter g that was connected with other parameters using relations was created. Next it was used in the section *Limits: g > 3 mm*. Rest of the assumed limits in the optimization looks as below:

$R >= r_1$
$a >= a_1$
$sk >= sk_1$
$w >= w_1$
$w <= w_2$
$g >= g_{min}$ (minimal thickness of the wall)

Parameter	Value
r_1	1 mm
a_1	0.5 mm
sk_1	3.5 mm
w_1	35.25 mm
w_2	38.25 mm
g_{min}	3.0 mm

Simulations were conducted with following assumptions:

Working pressure	25 [MPa]
Modulus of elasticity E	126174 [MPa]
Poison's ratio υ	0.25

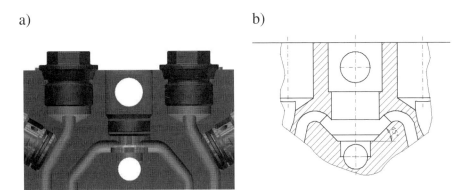

Figure 4. Solution of check valve chamber: a) original solution, b) modified solution

Fig. 5 depicts map of radial displacements distribution for original version of the directional control valve's body. Peak values occur in the lower area of spool chamber but theirs values are lower than required clearance and reached to 2.5 μm for compression and 5.0 μm for extension. Presented results showed area where deformations of spool chamber are the biggest.

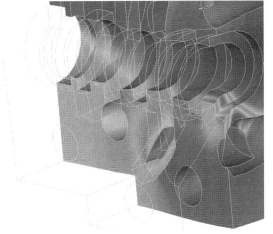

Figure 5. Map of radial displacements after optimization

Presented results of numerical calculation showed that stress levels as well as displacements are in the acceptable range. Change of the channel shapes and check valve chamber made insignificant decrease of radial displacements in the area where the largest displacements occurs. As a result of optimization new values were received:
$R = 2.198$ mm, $sk = 3.81$ mm, $g = 3.021$ mm

Assumed values:
$R = 2.15$ mm, $sk = 3.8$ mm, $a = 0.5$ mm, $w = 38$ mm
Now the thickness of the wall $g = 3.097$ mm, and the mass of the valve body was the smallest.

4. CFD analysis of flow

When the optimization study has been finished models of optimized channels were transferred to the CFD FLUENT package. Mesh was prepared in GAMBIT software using tetrahedral element. In the regions were large pressure drop were suspected mesh was refined. Next in FLUENT appropriate boundary condition were applied and finally numerical simulations were conducted. Analyses were carried out for different values of volumetric flow rate in the range up to 100 dm^3/min. k-ε turbulence model have been used.

Results of analyses were prepared in the graphical form of pressure and velocity distribution. Two different designs were compared with each other: original one with the optimized one. In both cases the highest pressure drop occurs in the areas of check valve and spool chamber. Simulation results confirmed decreasing of pressure drop for optimized channel shape. Fig. 6 depicts pressure distribution in a flow channel of directional control valve.

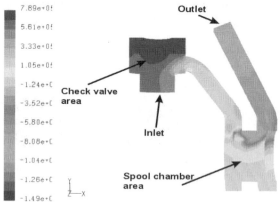

Figure 6. Pressure distribution in a flow channel

5. Building a prototype

Conducted optimization studies as well as CFD analysis of pressure drop allows for preparing solid models of directional control valves with new shape of flow channels. Subsequently these solid models were used for preparing prototypes using Rapid Prototyping method. This method is used for building a physical models of parts based on the CAD data. This technique is perfectly suited to building prototypes of parts with complex geometry such as flow channels of directional control valve. Considered prototype models were built at Bergen University Collage, Norway (www.hib.no). It was possible thanks to cooperation of Cracow University of Technology with Bergen University. Used method was an additive one which means that prototype was built by adding next layers of material every 0.2 mm.

Fig. 7 presents process of putting of a liquid epoxy layer that later solidified. In last stage prepared prototype required some additional surface cleaning.

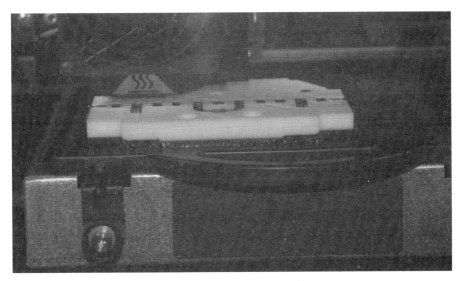

Figure 7. Process of creation of prototype

Fig. 8 presents created prototypes of directional control valve. These elements will be subsequently use in laboratory study for determination of pressure drop. Control sections were prepared for both solutions: original and optimized one.

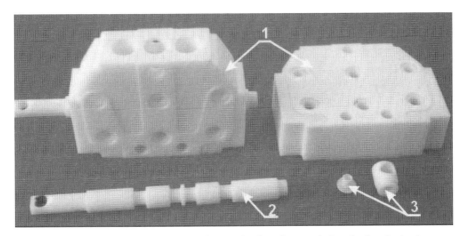

Figure 8. Created prototypes: 1 – body of control valve; 2 – spool; 3 – check valve's elements

6. Summary

In the paper the example of utilizing of computer engineering tools for creating a prototype of directional control valve has been presented. At the beginning parametric solids were created using latest version of Pro/E Wildfire. On the base of designed models optimization study of the valve body has been conducted. Next some additional analysis of flow in the CFD code has been conducted using Fluent code. Finally for the most promising model prototype has been built. Process of prototyping was multistage and required transferring data from one program to another in different formats. During data exchanging process it was necessary to control it in order to not to lose important information. Our experience shows that even in case of such complex geometry as body of control valve the quality of standard format is satisfactory. Conducted analyses showed that getting of reliable solution is very time consuming and requires use of computers with a large computational power. It is essential especially in case of multiple analyses and optimization. Results of detailed studies on directional control valve confirmed that it is possible to decrease the disadvantageous phenomena that occurs in case of fluid's flow through the valve, especially to decrease pressure drop. Built prototype makes possible to verify our numerical results with laboratory studies. Results of this experiment will be presented as soon as possible.

Application of proposed techniques is a very useful tool which allows forming of hydraulic channels and chambers. Much better is direct exchanging of the data between the modules of Pro/E package than between Pro/E and other CAE systems.

References

1. Del Vasco G., Lippolis A.: *Three-Dimensional Analysis of Flow Forces in Directional Control Valves*. International Journal of Fluid Power, Volume 4 Number 2 July 2003, 15-24.
2. Wu D., Burton R., Schenau G., Bitner D.: *Small Openings*. International Journal of Fluid Power 4 Number 1 April 2003, 31-40.
3. Lisowski E.: *Modelowanie Geometrii Elementów Maszyn i Urządzeń w Systemach CAD 3D*. Wydawnictwo PK, Krakow 2003.
4. Dahlen L., Carlsson P.: *Numerical Optimization of a Distributor Valve*. International Journal of Fluid Power, Vol. 4, No 3, November 2003, 17-25.
5. Gromala P., Rohatyński R., Lisowski E.: *Application of Fluent Packages In Modelling of Flow Through Directional Control Valve*. The 12th International Conference on Fluid Flow Technologies, Budapeszt 2003.
6. Gromala P., Lisowski E.: *Graphical of Phenomena Occurred During Fluid Flow Through Control Valve Based on CFD Analysis*. The 18th International Conference on Hydraulics and Pneumatics, Prague 2003, 490-497.

MODELING OF MECHANICAL SYSTEMS WHEN FEA TECHNIQUES IS USED

Suárez H.E.J.

Abstract: One of the fundamental problems mechanical engineers encounter when using CAD/CAM/CAE techniques in their work, lies on a poor application of the modeling of the mechanical systems upon which analysis is to be carried out through these techniques. Modeling of mechanical systems for the three (3) techniques differs substantially amongst them, due to necessities required for each application. Bearing in mid that the Finite Element Analysis (FEA) is a kind of CAE technique, some general recommendations are presented in this work for modeling of mechanical systems when this is used in such a manner that they serve as guides to those who for the first time have to use this technique and which permits them to minimize application errors during the process.

1. Introduction

The use of computer tools in the work of engineering (Jaramillo, 2003) is currently a fact that can not be denied by engineers. These tools are practically used as of the first stages of the design process. However, its use and application should be carried out by someone with a degree of expertise and domain on the issue.

When FEA technique is used, one of the fundamental problems that could be encountered is that of carrying out inappropriate modeling of the systems for analysis and/or construction. A bad model implies that inadequate results or results that are poorly adjusted to reality could be obtained and as a consequence of such, high computer-process times dedicated to the generation of the routines.

Now, generally, the engineer who for the first time encounters these types of techniques incurs in the error of considering that modeling for each technique is equal. In some way this is fomented by the inter-operational characteristics of the software used, since the great majority of these allow models to go from one technique to another without much difficulty. It is common that the neophyte engineer in the use of this FEA technique tends to consider: 1. that every system or mechanical piece is possible and analyzable through these techniques, 2. that the pieces, components, and systems should be modeled just are they are seen in reality, 3. that a model made through CAD can be transferred, without any type of modification, to a FEA analysis.

The object of this work is to present in a general manner, some recommendations to bear in mind when working with this technique. Hence, in the first part, we broaden on CAD modeling techniques, followed by an analysis of the care to exercise when carrying out a modeling for CAE – especially when working with the technique of finite elements (FEA), which is currently one of the most propagated in analysis software.

2. Modeling for CAD

With CAD, we have gone from simple programs that allow us to make the plans for a single piece to parametric modelers, in which three-dimensional models can be carried out of complex pieces, as well as assemblies of mechanical systems. These in turn carry out associations among the dimensions of the piece for analysis, so any modification done to one of them will automatically be transferred to the other associate dimensions. This allows for the making of pieces that are generally standardized or defined by certain dimensions or input data (screws, gears, connect, etc.) to be easy and quick. In the range of modelers we can find solutions like *Solidworks* (Solidworks, 2002), *Soliedge* (Solidedge, 2003), *Inventor* (Autodesk, 2003), *Mechanical Desktop* (Autodesk, 2003), among others. All these options permit us to find solutions that fit any budget need.

The functionality of these software programs allows us today to model systems in three dimensions, with the virtual model almost equal to the real model – these can even be visualized with textures and presentation environments. However, the person modeling the design at this stage should be aware of the direction the model being made will take; will it continue in a process of analysis or is it simply required for the generation of drafts (blueprints) in such a way that it allows adjusting the model to the needs required in the following process.

Figure 1. Three dimensional model of a valve

3. Modeling for finite element analysis

The analysis for finite elements basically implies three stages of activity: pre-processing, processing, and post-processing (Chandrupatla, 1999). The pre-processing stage implies the generation of models, boundary conditions, and information about loads and materials. In the processing stage we get the solution to the system of simultaneous equations that represent the system's behavior. The post-processing stage introduces results of analysis carried out. Generally, in this last stage we calculate and show the configuration of displacement, model shapes, temperature, and load distribution, among others.

The success of analysis through this technique depends to great extent on the simplifications carried out on the model, as well as on the definition of boundary conditions, form of load application, and type of mesh perform to the model. In finite elements analysis can only yield correct answers if the model is both valid and precise

(Fagan, 1992). Validity depends on that it also represents the physical problem in the computer, while precision depends on how the model closes and its solution converges.

To clear up what we have been discussing in this section, we will discuss a simple example to display the different ways in which a physical system can be modeled, and that the best model is not always that which is most similar to reality whenever working with the theory of finite elements. For the previously stated, suppose we wish to determine the values of maximum stress and deflection of a cantilever girder, as shown in figure 1.

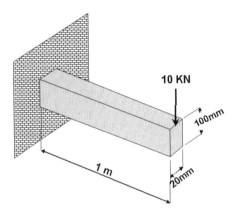

Figure 2. Cantilever girder

Maximum theoretical stress values are presented in the girder and are calculated according to equation (1) (Gere, 1984). The value of the maximum displacement is presented at the free extreme of the girder and such value is calculated according to equation (3) (Gere, 1984), supposing the girder is ASTM A-36 (E=200GPa) structural steel:

$$\sigma_{max} = \frac{Mc}{I} \tag{1}$$

$$\sigma_{max} = 300 MPa \tag{2}$$

where: M, is the flexural moment; c, represent the distance from de base to the centroid of the cross section and I, is the moment of inertia.

$$Y_{max} = \frac{PL^3}{3EI} \tag{3}$$

$$Y_{max} = 0.01m = 10mm \tag{4}$$

Where Y_{max}, is the maximum deflection in the girder; L is the girder length and E, the modulus of elasticity. The girder shown on the figure can be modeled in different ways and using different types of elements see table 1.

Table 1. Detail of types of possible models to be used

Model	Element type
Model 1	Beam
Model 2	2-D
Model 3	Plate
Model 4	Brick

If the desire is to broaden on the characteristics of the elements: Beam type, 2-D, Plate, and Brick, we recommend consultation of Spyrakos (Spyrakos, 1996), given that such are not included in this document so as to not distract with unrelated aspects herein.

Results of stress values for each model are shown in figures 3, 4, 5, and 6. The table 2 shows a listing of maximum stress and maximum deflection values for each case, as well as the number of nodes, number of elements, and time utilized for analysis of each model. Analysis by finite elements for this case was carried out with Algor software (Algor, 2003) (Algor User's guide, 2003).

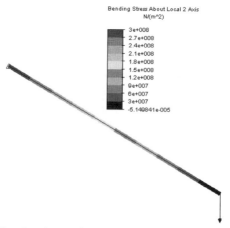

Figure 3. Results of stress for the model using Beam type elements

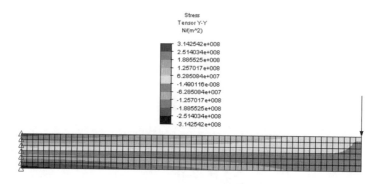

Figure 4. Results of stress for the model using 2-D type elements

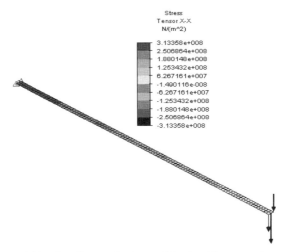

Figure 5. Results of stress for the model using Plate type elements

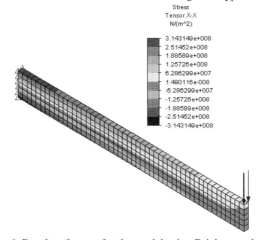

Figure 6. Results of stress for the model using Brick type elements

Table 2. Results of analysis carried out with different types of elements

Mod. No	No Nodos	No Elemt.	t [min]	σ_{max} MPa	Y_{max} mm
1	27	20	0.018	300.00	10.00
2	427	360	0.028	314.25	10.06
3	303	200	0.023	313.35	9.97
4	612	250	0.034	314.31	10.03

Table 3. Determination of error percentile with respect to theoretical values

Model	% of error	
	Stress	Displacement
1	0%	0%
2	4.75%	0.6%
3	4.45%	0.3%
4	4.77%	0.3%

From the previous analyses and from analysis of results consigned in table 3, we were able to determine that the model that comes closest to the theoretical values is the one using Beam type elements, since it is the simplest in its modeling. However, we were also able to determine that the model that comes closest to the real shape of the physical system is model 4, which obtained the highest percentage of errors in the values of normal stress and the longest process times and of utilization of nodes and elements.

It is worth noting that for the specific case of the previous example, requiring the maximum stress and displacement values, the model that best fits these requirements was the one using Beam type elements. But if we wanted, for example, local effects like buckling of the web-girder or deformation of top flanges of an "I" girder, with equal load conditions, this model would no longer be appropriate and we would have to think of a model that best reflects the geometry of the piece. In this case, we could use a model that uses Plate or Brick elements.

As a special recommendation: when modeling any type of mechanical system, it is convenient to carry out a bibliographic revision in specialized journals to inquire whether or not the type of physical event to be modeled has already been addressed by another author. In such case, some works can be cited, shedding broader vision of problems engineers encounter when modeling systems through the theory of finite elements. R. H. MacNeal (MacNeal, 1998) has undergone a revision of how development along time has been for Plate and Shell type elements. Such development has permitted these types of elements to yield better results; thereby, somehow allowing the simplifying of some physical models that were modeled three dimensionally with the use of these types of elements.

A work published by F. Teixeira (Teixeira, 2003) carries out the analysis of a connecting rod, which has been modeled in seven (7) different forms. All these differentiated by the level of detail of the CAD model, but keeping the original basic dimensions. As a result of this work, it was found that for highly detailed CAD models – when the computer model was very similar to the real model – process times were much higher than for models where some details were omitted or simplified; nevertheless, results of stress do not represent a considerable variation. The work concludes by proposing three aspects to follow in modeling: 1) divide the model according to the existence of an eventual symmetry, 2) remove characteristics from the model that do not define the essential properties of such, 3) control the generation of surfaces on the model to allow organizing the surfaces for the application of loads. It is expected that by following these recommendations, analysis times can be reduced by 80%; with 70% of this time associated to CPU processing, and with an error margin of the results below 2%.

In the same manner, Jaramillo and Areiza (Jaramillo and Areiza, 2000), through a published work, analyze different types of models for a compound section, which corresponds to the transversal section of a metallic girder and concrete slab bridge. This work compares five different models for one same type of example. Element types such as: Brick - Plate, Brick - Brick, Plate - Plate, and Plate - Beam were used for the concrete slab and the metallic girder, respectively. Since the objective of this work was to determine the effects of shear force, flexural moment and deflections, the model that best fit these requirements was the one using Plate type elements for the concrete slab and Beam type elements for the girder. However, if other effects are sought – different to what we have discussed – we recommend that another type of modeling be used.

6. Conclusions

Through this work, we have made a revision of some modeling recommendations when work is done in FEA with specialized software. However, these recommendations are but a few, from a great number, that could be mentioned were space limitations not a hindrance. These recommendations solely seek to offer guidance to those starting in this type of work. Only constant work in the application of these types of techniques and in the solution of engineering problems can bring about the necessary expertise to become what is considered a good modeler.

Now, a good CAE modeler is not one who carries out an exact copy of reality, but rather one who can represent reality in the simplest way possible, obtaining adequate results by making adequate and efficient use of process times and computer capacity. Thus, for each physical event to be modeled there is an optimal model that fits the desired requirements or results. Upon a complex system, it is generally recommended to use simple models that somehow represent the real model to be able to determine the viability of the analysis – before carrying out the final model – since valuable time and computer resources could be wasted on an analysis that with the available software possibilities could not be developed.

It is indispensable that those with expertise in modeling mechanical systems, and who are somehow involved in the formation of future engineers in the use of these tools, do not create false expectations to students or on the applications of these types of software, which – as everything else – also has its limitations. Limitations inherent to it or to the resources it uses. In other words, it is necessary to first know the reach and limitations of the software so as to not assume that it is possible to model every mechanical system or physical event through the technique of finite elements, and that oftentimes it is necessary to simplify a great part of the real model for its analysis to be feasible.

References

1. Algor software, Win-27-Jun-2003- Static Stress With Linear Material Models. Algor, User's Guide.
2. Chandrupatla T.R., Belegundu A.D.: *Introducción al estudio del ELEMENTO FINITO EN INGENIERÍA*. Prentice Hall editores, segunda edición, México 1999, Pags. 412.

3. Fagan M.J.: *Finite element analysis: Theory and practice*. Longman Group UK Limited, New York 1992, Pag. 221.
4. Gere J.M.,. Timoshenko S. P.: *Mecánica de materiales*. Grupo Editorial Iberoamérica S. A., segunda edición, México 1984, Pags. 229, 788.
5. Jaramillo H.E.: *Los nuevos paradigmas del diseño en ingeniería Mecánica*. III Congreso Bolivariano de Ingeniería Mecánica, Lima – Perú, Julio 2003.
6. Jaramillo H.E., Areiza G.: *Algunas generalidades sobre el modelado de secciones compuestas usando elementos finitos*. Revista Scientia Et Technica, Universidad Tecnológica de Pereira, Edición No 13, p. 43-50. (2000).
7. MacNeal R.H.: *Perspective on finite elements for shell analysis*. Finite Elements in Analysis and Design, 30 (1998) 175-186.
8. Spyrakos C.C.: *Finite Element Modeling: In Engineering Practice*. Algor Publishing Division, Pittsburgh USA 1996, Pags. 33-74.
9. Teixeira F., Borges C.: *On the determination of the optimal modeling conditions for higher performance finite element analyses*. Finite Elements in Analysis and Design 39 (2003) 207-216.
10. www.solidworks.com
11. www.ugsolutions.com
12. www.autodesk.com

A MATHEMATICAL MODEL FOR CONTROLLER SYNTHESIS OF THE ELECTRO CHEMICAL DISCHARGE MACHINING PROCESS

Mediliyegedara T.K.K.R., De Silva A.K.M., Harrison D.K., McGeough J.A.

Abstract: In this study a mathematical model for the metal removal of ECDM process has been developed. Five distinct types of pulses were identified for various gap conditions. Material removal mechanisms are analysed. Voltage and current waveforms were acquired. A Neuro Fuzzy Pulse Classification System (NFPCS) was utilised to classify pulses. Then, a model for the amount of material removal is developed based on the type of the pulse on the machining gap.

1. Introduction

Electro Chemical Discharge Machining (ECDM) is a hybrid non-conventional manufacturing process which combines the features of Electro Chemical Machining (ECM) and Electro Discharge Machining (EDM) (Mediliyegedara et al., 2004[a]). The ECDM process consists of a cathodic tool and an anodic workpiece, which are separated by a gap filled with electrolyte, and pulsed DC power applied between them. This leads to electrical discharges between the electrodes, thus achieving both electrochemical dissolution and electro-discharge erosion of the workpiece (De Silva et al., 1995). One of the major advantages of ECDM, over ECM or EDM, is that the combined metal removal mechanisms in ECDM, yields a much higher machining rate (De Silva, 1988).

The performance of ECDM, in terms of surface finish and rate of machining, is affected by many factors. Relationships between these factors and machining performance are highly non linear and complex in nature (Mediliyegedara et al., 2004[b]). Due to the complex and stochastic nature of the electrical discharges, it is very difficult to develop a mathematical model to describe electrode gap characteristics. Therefore, the formulation of better control strategies demands a stochastic model which could be used to simulate the various pulses types according to the gap condition. Altpeter et al. (1998) model the Material Removal Rate (MRR) of EDM. They used Arc Voltage, Arching Current and arching period of individual pulses to calculate the amount of metal removal per pulse. Further, Altpeter and Tricarico (2001) discussed the influence of dielectric contamination by machining debris.

It is possible to identify five distinct types of pulses in the ECDM process such as Electro Chemical Pulse (ECP), Electro Chemical Discharge Pulse (ECDP), Spark Pulse (SP), Arc Pulse (AP) and Short Circuit Pulse (SCP). SP and AP also known as Normal Discharge Pulse and Abnormal Discharge Pulse respectively. Open Circuits Pulse (OCP) is not present in ECDM as some electrochemical current flows even with a larger gap. De Silva (1988) has presented a detailed analysis of various pulses in ECDM. Mediliyegedara at el. (2004[b]) developed an artificial neural network pulse classification system to classify pulses in the ECDM process. Further, they used a Fuzzy Logic Approach to classify pulses in the ECDM process (Mediliyegedara at el., 2004[c]). The objective of this study is to develop a mathematical model to estimate the metal removal rate in the ECDM process.

2. Experimental procedure

Figure 1 and Figure 2 shows a block diagram of the experimental setup and the experimental setup respectively. A motion control board, which is capable of producing an analogue output and capable of obtaining an encoder input, is fixed to a Personal Computer (PC). A linear optical encoder, the resolution of which is 1 µm, was used as the linear position feedback device. A motion control algorithm, which is used to count the number of encoder pulses and to produce reference signal r(t) to the tool position control system, was programmed with the LabVIEW 7.0 software package.

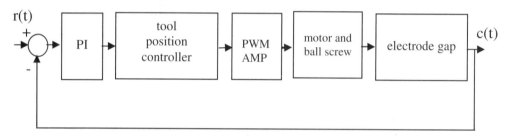

Figure 1. A block diagram of the experimental set-up

Figure 2. Experimental setup

$NaNO_3$ was used as electrolyte. A mild steel work piece and a copper electrode were used. The duty ratio and pulse duration were set to be 50% and 200µs respectively. Gap voltage and working current wave forms were acquired at a sampling frequency of 1 MHz.

3. Metal removal model

The MATLAB 6 software package was used to develop the material removal model a block diagram of which is shown in Figure 3.

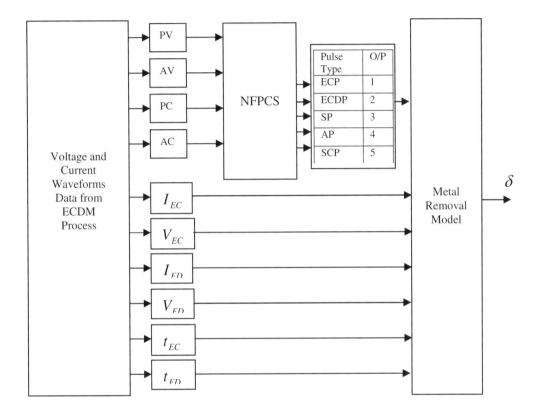

Figure 3. A block diagram of the metal removal rate simulation model

A Neuro Fuzzy Pulse Classification System (NFPCS) was employed to classify pulses. An Adaptive Neuro Fuzzy Inference System (ANFIS) was employed to develop a Neuro Fuzzy Pulse Classification System (NFPCS). A block diagram of NFPCS, which is composed of five ANFISs namely ANFIS1, ANFIS2, ANFIS3, ANFIS4 and ANFIS5, is shown in Figure 4. There are four input variables such as normalised values of Peak Voltage (PV), Average Voltage (AV), Peak Current (PC) and Average Current (AC). There is one output per each ANFIS. Outputs O_1, O_2, O_3, O_4 and O_5 are corresponding to ECP, ECDP, SP, AP and SP. The NFPCS has been train in such a way that the output of O_1 is '1' when the input features are corresponding to an ECP. Similarly O_2, O_3, O_4 and O_5 will be equal to '1' when input features are corresponding to ECDP, SP, AP and SP.

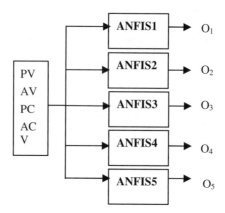

Figure 4. Architecture of the Neuro Fuzzy Pulse Classification System (NFPCS)

4. Amount of metal removal per pulse

There are two metal removal mechanisms in the ECDM process due to hybrid nature of the process. Firstly, metal is removed due to the Electro Chemical (EC) action. Secondly, metal is removed due to the Electro Discharge (ED) action. Author proposed a generalised voltage waveform and current waveform for the ECDM process under following assumptions:
1. Pulsed energy source is static pulse generator in which pulse energy can be changed independently of the pulse repetition frequency (Lauther et al., 2003).
2. Rise time of the voltage pulse is zero.
3. Fall time of the voltage pulse is zero.
4. Conductivity of the electrolyte remains same during machining.

The generalised voltage waveform and generalised current waveform for the ECDM process are shown in Figure 5 and Figure 6 respectively.

4.1. Amount of metal removal due to EC action (δ_{EC})

When a potential difference is applied across the tool and workpiece, machining will commence with electrochemical dissolution provided the working gap is large. In this stage current waveform is approximately proportional to the voltage waveform. The material removal mechanism at this stage can be explained with Faraday's law of electrolysis. The mass dissolved from the workpiece due to EC action is given as:

$$m_{EC} = \frac{AI_{EC}t_{EC}}{ZF} \qquad (1)$$

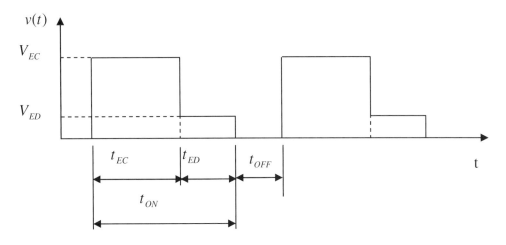

Figure 5. Generalised voltage waveform for the ECDM process

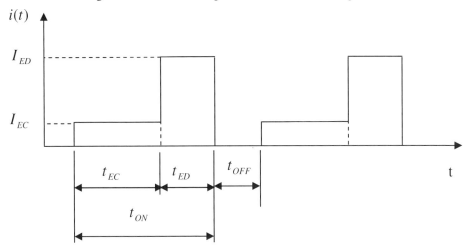

Figure 6. Generalised current waveform for the ECDM process

Where I_{EC} is the current passed for time t_{EC}. The reacting ions have atomic weight A and valency Z, the quantity A/Z being the chemical equivalent. F is a universal constant known as the 'Faraday's constant. It is the amount of electric charge necessary to liberate one gram- equivalent of (A/Z) of an ion in electrolysis. Its commonly accepted value is 96500C. Consider a cathode electrode, lateral surface of which is insulated, having a cross sectional area 'a' and an anode tool having density 'ρ'. The amount of metal removal due to electrochemical dissolution 'δ_{EC}' can be written as:

$$\delta_{EC} = \frac{\text{Vol}_{EC}}{a} \qquad (2)$$

The volume of metal removal Vol_{EC} can be written as,

$$Vol_{EC} = \frac{m_{EC}}{\rho} = \frac{AI_{EC}t_{EC}}{ZF\rho} \tag{3}$$

Therefore, $\delta_{EC} = \dfrac{AI_{EC}t_{EC}}{ZF\rho a}$ (4)

4.2. Amount of metal removal due to ED action (δ_{ED})

The volume of metal removal due to ED action (Vol_{ED}) is proportional to the energy dissipated per pulse (Altpeter and Tricarico, 2001).

Therefore, $Vol_{ED} = \alpha V_{ED} I_{ED} t_{ED}$ (5)

Where, V_{ED}, I_{ED} and t_{ED} are the discharge voltage, discharge current and discharging period respectively. α is a proportional constant. Consider a cathode tool having a cross sectional area 'a' and an anode workpiece with density 'ρ'. The amount of metal removal per pulse due to ED action 'δ_{ED}' can be written as:

$$\delta_{ED} = \frac{Vol_{ED}}{a} \tag{6}$$

Therefore, $\delta_{ED} = \dfrac{\alpha V_{ED} I_{ED} t_{ED}}{a}$ (7)

So, the total metal removal per pulse (δ) can be written as:

$$\delta = \delta_{EC} + \delta_{ED} \tag{8}$$

$$\delta = \frac{AI_{EC}t_{EC}}{ZF\rho a} + \frac{\alpha V_{ED} I_{ED} t_{ED}}{a} \tag{9}$$

4.3. Amount of metal removal per electro chemical pulse (δ_{ECP})

In this stage current waveform is approximately proportional to the voltage waveform. For ECP, t_{ED} is equal to zero. Therefore δ_{ECP} can e written as:

$$\delta_{ECP} = \frac{AI_{EC}t_{EC}}{ZF\rho a} \tag{10}$$

4.4. Amount of Metal Removal per Electro Chemical Discharge Pulse (δ_{ECDP})

During an ECDP, metal removal is taking place due to both EC action and ED action. Therefore, δ_{ECDP} can be written as:

$$\delta_{ECDP} = \frac{AI_{EC}t_{EC}}{ZF\rho a} + \frac{\alpha V_{ED} I_{ED} t_{ED}}{a} \tag{11}$$

4.5. Amount of Metal Removal per Spark Pulse (δ_{SP})

In this situation, t_{EC} is equal to zero since metal removal is taking place only due to ED action. Therefore, δ_{SP} can be written as:

$$\delta_{SP} = \frac{\alpha V_{ED} I_{ED} t_{ED}}{a} \quad (12)$$

4.6. Amount of Metal Removal per Arc Pulse (δ_{AP})

In this situation, t_{EC} is equal to zero since metal removal is taking place only due to ED action. Therefore, δ_{SP} can be written as,

$$\delta_{AP} = \frac{\alpha V_{ED} I_{ED} t_{ED}}{a} \quad (13)$$

4.7. Amount of Metal Removal per Short Circuit Pulse (δ_{SCP})

When the machining gap bridges with the debris or the tool touches the surface of the workpiece, a short circuit pulse can be observed. At this stage, amplitude of the voltage can be assumed to be as zero. Therefore ideally there is no metal removal during a SCP.

$$\delta_{SCP} = 0 \quad (14)$$

5. Evaluation of control strategies of controller

As sown in Figure 1, a PI controller was employed to control the gap width. Average gap voltage is used as the control parameter. Percentages of each type of pulse were obtained under various control strategies. MRR was calculated based on the relationships explained in Section 4. Table 1 shows control strategy Vs theoretical and experimental MRR. It is clear from Table 1 that the MRR depends on the control strategy used in the controller. By maintaining the gap width at an optimum level metal removal rate can be increased. However, it is equally important to consider the surface roughness of the machined part. Therefore there is a need for a model which could be used to estimate the surface roughness of machined parts with the ECDM process.

Table 1. Control Strategy Vs Theoretical and Experimental Metal Removal Rate (MRR)

Control Strategy No.	Percentage of Pulses (%)					MRR (mm/min)	
	ECP	ECDP	SP	AP	SCP	Theoretical	Experimental
1	18.74	64.98	7.77	4.69	3.82	0.45	0.42
2	10.36	55.12	20.10	7.27	7.15	0.77	0.63
3	10.18	38.94	29.35	10.76	10.76	1.36	1.12
4	4.56	28.73	36.13	18.37	12.21	1.61	1.69
5	4.50	21.14	41.88	9.37	13.11	1.82	1.92
6	5.09	28.19	64.19	1.38	1.15	3.39	3.22
7	7.05	29.75	60.47	1.57	1.17	3.44	3.48
8	4.89	24.85	67.71	1.38	1.16	3.46	3.66
9	3.72	20.55	73.39	1.16	1.18	3.49	3.21
10	5.09	24.66	67.71	0.98	1.57	3.51	3.66

6. Conclusions

A mathematical model has been developed to estimate the metal removal rate of the ECDM process. Electrode gap voltage and current waveforms were obtained under various control strategies. The pulse parameters were obtained. Then the pulses are classified using a NFPCS. The MRR was estimated using the proposed model. Experimental results show that the proposed model can be successfully used to estimate the MRR of the ECDM process. Then, the proposed model was employed to asses the performance of various control strategies. Further research is required to develop a model which could be used to estimate the surface roughness of machined parts with ECDM process.

References

1. Altpeter A., Cors J., Kocher M., Longchamp R.: *EDM Modeling for Control*. 12th International Symposium for Electromachining (ISEM-XII), Aachen, Germany, 1998, pp. 149-155.
2. Altpeter F., Tricarico C.: *Modeling for EDM Gap Control in Die Sinking*. 13th International Symposium Elecromachining, 2001, pp. 75-83.
3. De Silva A.K.: *Process Developments in Electrochemical Arc Machining*. PhD Thesis, University of Edinburgh, 1988.
4. De Silva A.K.M., Khayry A.B., McGeough J.A.: *Process Monitoring and Control of Electroerosion-dissolution Machining*. IMechE Conference Transactions, 11th Int. Conf. on Comp.-Aided Production Eng., 1995, pp. 73-78.
5. Lauter M., Shulze H.P., Wollenberg G.: *Modern Energy Source for Eelctrical Discharge Machining (EDM)*. 18th International Conference on Computer-Aided Production Engineering, 2003, pp. 399-348.
6. Mediliyegedara T.K.K.R., De Silva A.K.M., Harrison D.K., McGeough J.A.: *An Intelligent Pulse Classification System for Electro Chemical Discharge Machining (ECDM) – A Preliminary Study*. Journal of Material Processing Technology, 2004[a], 149, pp. 499-503, ISBN 0924-0136.
7. Mediliyegedara T.K.K.R., De Silva A.K.M., Harrison D.K., McGeough J.A.: *An Intelligent Gap Width Controller for Electro Chemical Discharge Machining (ECDM) Machine*. 4th International Seminar on Intelligent Computation in Manufacturing Engineering (CIRP ICME 04), Sorrento, Italy, 2004[b], pp. 473-478.
8. Mediliyegedara T.K.K.R., De Silva A.K.M., Harrison D.K., McGeough J. A.: *Fuzzy Logic Approach for Pulse Classification of Electro Chemical Discharge Machining (ECDM)*. 34th International MATADOR Conference, Manchester, 2004[c], pp. 161-166.

where: p is the set of positions of the 3D points, q are their corresponding positions in the image, s is a scale factor since the coordinates of the points are homogenous, [R, t] represent the extrinsic parameters of the camera with r_i (r_1, r_2, r_3) their corresponding columns, R are the matrix of rotation, t the transferring vector and A is the matrix of intrinsic parameters of the camera.

Being, α and β the scale factors in axis U and V, (Uo, Vo) the coordinates of the main point of the image and γ are the parameter that represents the missed orthogonal axis in the image. The matrix A contains all the parameters necessary to make the calculations of the three-dimensional coordinates of the points. These calculations are made up applying space triangulation.

2.2. Space triangulation

The space triangulation (Merino, 2000) is applied when at least two observers are used, which allows univocal recovering of the points position in the 3D space. This express principle: "Given the base of a triangle, the distance between two of its vertices, and both corresponding angles, the absolute position of the third vertex can be determined univocally". According to this principle and as it is possible to observe in Figure 3, the absolute position of the P point can be known if it is known the distance D and the angles α and β.

Its practical application consists of a projector of structured light (laser beam) as an observant and of a passive observer of capturing images (camera), which will collect the scene as shown in Figure 4 (Gómez, 1997). In this way the points A and B in Figure 3 are defined.

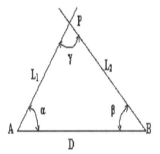

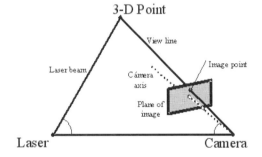

Figure 3. Principle of space triangulation   Figure 4. Principle of the laser triangulation

2.3. Generation of superficial meshes

The simplest method to visualize a curve is based on the approach of the segment by a polygonal line. For its implementation, it is enough evaluating the curve in a sequence of parameters values, creating the polygonal line that joint the obtained points (Figure 5). The previously described method to create a curve will allow approximating a parametric surface by a polygonal mesh (Towers, 2003), where the surface in a subgroup of distributed points of the parametric space in an evaluated grid. By each four neighboring-points two triangles are formed (Figure 6).

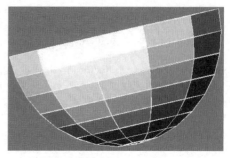

Figure 5. Approach of a curves by a polygonal line

Figure 6. Polygonal Mesh

Each piece of surface that comes near by means of two triangles are denominated *patch*. The quality of the surface reconstruction depends on the number of captured points, which determine the number of patches to be created (Figure 7).

Figure 7. Polygonal Mesh with greater number of patches
Source: AutoCad 2002 Help, © 1982 – 2002 Inc. Autodesk

Most methods of curves and surfaces generation are based on the use of control points from which averages of these points are defined by:

$$P(u) = \sum_{i=1}^{n} P_i \cdot B_i(u) \tag{3}$$

where: P_i is the control points and $B_i(u)$ are the parametric form functions.

A surface is mathematically described using two parameters that establish a system of coordinates on the surface, which allows crossing it. The surface can be defined directly from a mesh of control points, like a curve, with the control points forming a bi-dimensional distribution and the parametric form functions depend on two parameters, B_{ij} (u, v). Alternatively, it is possible to define a surface from one or several curves. Furthermore, meshes of triangles are used like an approach to the surface. The representation and design of surfaces using a rectangular mesh of control points, $P_{ij} = (x_{ij}, y_{ij}, z_{ij})$, will be used in this research work.

The surface will be calculated in intervals, *patches*, in the form:

$$S(u,v) = (x(u,v), y(u,v), z(u,v)) = \sum_{j=0}^{m} \sum_{i=0}^{n} P_{ij} \cdot B_{ij}(u,v) \tag{4}$$

The parametric form functions $B_{ij}(u, v)$ can be obtained by product of the parametric form functions in u and v, used for curves:

$$B_{ij}(u,v) = F_i(u) \cdot G_j(v) \qquad (5)$$

where: F and G are polynomial parametric form functions.

These parametric form functions can be Spline, Bezier, or B-Spline, among others.

3. Experimental work

In this section the application of the previously exposed theories, in project SISREDIS, are described.

3.1. System hardware

The system hardware used is made up of a laser projector with wavelength of 630 –680 nm, power of 1 mW, a digital camera Benq DC1500 with sensor CMOS of maximum resolution 640x480 pixels, a platform with rotational movement and a personal computer (PC) with processor Intel Pentium4 of 1.9Ghz, 256Mb of RAM, 32Mb of Video and Windows2000.

3.2. Geometric calibration of the camera

The first step is the system's calibration, process in which the required parameters are defined to calculate the 3D coordinates of the points of the scanned object (Masmoudi, 1995). In this process, the reference systems used are shown in Figure 8, where $(X_L, Y_L,$ and $Z_L)$ is the laser reference system, $(X_C, Y_C,$ and $Z_C)$ is the camera reference system, $(X_P, Y_P,$ and $Z_P)$ is the object reference system, (R, C) is the monitor reference system, and (U, V) the flat image reference system.

In order to determine the 3D coordinates of a point, is necessary to know the camera's intrinsic parameters, represented in matrix A (Equation 2). The squares flat group images come from different points of view (minimum 3) and the coordinates of the corners of each box in pixels are obtained, as shown in Figures 9. After that, the correspondence is made to its real 3D coordinates for each square.

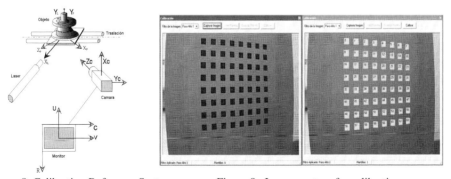

Figure 8. Calibration Reference Systems   Figure 9. Image capture for calibration

3.3.3. Data capture and analysis

After calibrated the system it is come to capture the coordinates of the points that form the object to digitize. For it the video signal is taken through port USB (Universal Serial Bus), the capture surroundings is a dark atmosphere, each frame of video becomes at gray levels. All the levels that are smaller to the value of the chosen threshold will be black and the others will be white. This allows filtering of the image the beam of laser light. The threshold is a value of intensity in the rank [0...255].

The system of capture and analysis is developed in Borland C++ Builder 5. The capture of video and filtrate of the image is made by means of a component of image processing called ImageEn 2.0.7 (HiComponents, 2003).

The data acquisition is made of sequential form, applying a well-known speed rotational movement to the platform where the object is placed, illuminating in each stage with the active element (laser light) a zone of the scene (Figure 10). As the direction of the beam is controlled we know then the triangulation angles.

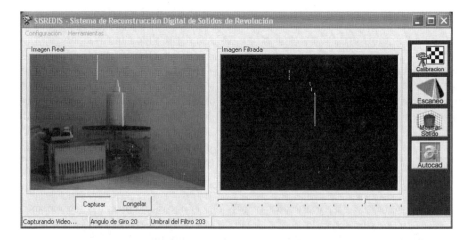

Figure 10. Real and filtered captured images

The value of the corresponding receiver angle is obtained from the projection in the image of the illuminated line. Also, due to the previous calibration it is known the distance D, in Figure 2, between the receiver and the projector. Each projected line is converted into a series of points, which will be used to create the cloud of points representing the scanned surface. To achieve this, a rectangular mesh of control points is applied.

There are 2 methods that can be used in order to obtain pixels (u,v) of the segmented image, which are being analysed to determine which one is the best option for this application. These methods are:
1. *Evaluation by vicinity of pixels:* consists of evaluating each white pixel and using for triangulation only those that have by neighbours others white pixels.

2. *Evaluation by medium of pixels*: It consists of evaluating each row of pixels of the image whatever these are white or not, but using only those pixels from the medium of pixels counted in the row.

The reconstruction process is recursive, which mean that for each captured image a set of control points is selected, and then an instruction is given to the platform to rotate a predetermined value for the next captured image. The value in degrees of the rotation can be user-selected depending on the details of the object.

In the surface reconstruction cycle, only the control points that were taken from the model's cloud of points are used to represent the surface of the piece. Therefore, in the reconstruction process, the control points will be selected interactively until the required surface is obtained. The control points for the reconstruction of the model are ordered and stored in script archives to be executed in AutoCAD for its visualization.

4. Applications

A system of three-dimensional reconstruction of real scenes presents/displays a solution to the engineering problem that rose from the traditional systems of contact surface reconstruction. The developed system finds numerous applications within the industrial sector. Some examples are:

- capture of the dimensional information of objects,
- identification of pieces,
- generation of surfaces,
- manufacture of duplicates,
- reverse engineer,
- quality control of pieces.

Also it is applied in diverse fields like robotics, medicine and art, where obtaining information from certain scene is very important. Some other applications are:

- inspection and identification in industrial processes,
- construction of virtual scenes,
- development of systems of recognition and visual identification of people,
- plastic or dental surgery,
- animation and computerized films,
- data bases of artistic sculptures.

5. Conclusions

The methodology implemented in a system of three-dimensional artificial vision has been described, which allows to take an object from the 3D space into the 2D virtual world of the computer. This work demonstrates that using peripherals of low resolution and/or cost, conventional domestic hardware, it is possible to develop useful artificial vision applications with industrial potentials.

This project is a didactic example due to the low resolution of the used equipment. A methodology for the process has been implemented for the 3D scan, which consists of the following steps: Preparation and assembly of the hardware for data acquisition, geometric calibration of the system, calculation of the 3D coordinates of the points by means of laser triangulation, generation of the mesh of cloud points and control points by means of rectangular surfaces; and preparation of script files for visualization purposes.

In order to improve the results of the system it is recommended to improve the capacities of the physical parts of the project, specially the laser projector and the camera. Independent visualization software should be developed so to increase independence of the system. A new version of SISREDIS should include capture of color and texture of the object surface.

References

1. Autodesk: *AutoCad2002*. Available in: http://www.autodesk.com/
2. Gomez G., Diaz P., Lopez C.: *Joint Obtaining of the Information Three-dimensional and Chromatic. First Approach to the Optical Characterization of Surfaces*. Automatic computer science and, University of Valladolid, 1997.
3. HiComponents: *For ImageEn Delphi and C++Builder*. 2003, Available in: http://www.hicomponents.com/zzdownloads.asp
4. Masmoudi L., Lopez C., Gomez G.: *Precise Calibration of Cameras with Model of Distortion and Robust Reconstruction of Three-dimensional Coordinates*. Automatic computer science and, University of Valladolid, Valladolid, 1995.
5. Merino M., Gomez J.: *Actualization of the architecture of a system of vision for the industrial measurement of three-dimensional pieces and redesign of the reconstruction module and visualization*. University of Valladolid, Valladolid, 2000.
6. Simarro C., Ricolfe Sanchez R.: *Técnicas de Calibrado de Cámaras*. Department of Automation Engineering and Systems, Polytechnic University of Valencia, Valencia, 2003. Available in: http://ja2003.unileon.es/SUBMISSIONS/viar/135.pdf
7. Torres J.C.: *Design attended by computer*. Computer science department of Languages and Systems, Superior Technical School of Computer Science Engineering, University of Granada, Granada, 2003: http://lsi.ugr.es/~jctorres/cad/teoria/Tema3.pdf
8. Zhang Z.: *A Flexible New Technique for Calibration*. Microsoft Research, the USA, 2002. Available in: http://research.microsoft.com/~zhang/

A FUZZY-NEURAL SYSTEM FOR CUT-MAKE-PRODUCTION ESTIMATION

Li Q., Walker M., Ranil S.

Abstract: Systems in the real world are made up of three different "parts", hardware, software, and the human. Humans have always been the "part" of systems that are difficult to engineer. In this paper, an integrated control strategy is discussed, in which the human judgment is used for performance evaluation. This integrated control strategy is achieving a better trade-off between complexity and flexibility, and has been illustrated in a system, implemented in a garment factory, using cut-make-production(CMP) cost-estimation. The real time application showed that the system is suitable for decision making in an uncertain environment.

1. Introduction

Systems in the real world are made up of three different "parts", hardware, software, and the human. Humans have always been the "part" of system that is difficult to engineer. They are an integral part of a system and the engineering process *must* be able to understand how best to fit the human, from the perspective of achieving the overall system goals. Many researches have shown a system were the human is not an afterthought but, rather, an integrated component working in concert with the other hardware, software, and human components [4].

For more than 30 years, researchers and Information Systems specialists have built and studied a wide variety of systems, for supporting information decision-makers, referred to as Decision Support Systems or Management Decision Systems. Dealing with uncertain and imprecise information has been one of the major issues in most Decision Support Systems. There are a variety of approaches to choose, in order to cope with uncertain, imprecise, vague, and even inconsistent information, such as Bayesian and Probabilistic Methods, Mathematical Theory of Evidence, Belief Networks, Fuzzy Methods, Possibility Theory etc [2].

In solving ill-structured, unstructured or semi-structured problems, under some uncertainties, integrating human component is one approach that can achieve a better trade off between complexity and flexibility. Achievement of the goal for human-system integration requires a variety of activities including function allocation, control/display integration, and consideration of human/systems characteristics and limitations [3]. In this paper, a case study of a garment factory Cut, Make and Production (CMP) estimation is used, and the concept of human integration is elaborated. This has shown that human integration can achieve a better trade off between the system complexity and flexibility, and is especially important when there are some resource restrictions.

2. CMP estimation problems

The CMP business is unlike most other businesses. Huge investments are required for fabrics, and the manufacturing process. Timing is critical and dependant on seasonal change television/movie advertising and other promotional campaigns. The CMP process involves a considerable amount of estimation. Buyers specify a garment and the fabric and the CMP Manager has to cost it. Accuracy is essential in a tight price climate as an inaccurate cost can mean that a profit margin is not made, or that an apparent profit is not achieved [5].

In the real situation, unlike fabric and trim costs, the CMP cost varies greatly due to varying style and current orders on hand, causing cost uncertainty. Currently, the investigated factory (South African based garment factory) had no choice but to ignore the CMP's inaccuracy in the *Cost estimation*. Reports showed that the CMP can varied between R0.89 and R6.00 in this factory. CMP inaccuracy had a greater impact on profits generated from mass production (monthly output approximate 1 million pieces).

To estimate CMP costs requires the process, of management information, and the cost also associated with the complexity of garment styles. As there are many uncertainties in the market place, an intelligent system is called to achieve a better estimation for CMP cost.

3. Human integrated approach

As outlined above, it is difficult to find an easy solution due to complexity. Especially for those Small and Medium Enterprises (SME), where resources required for the system is limited making implementation of advanced technology for SME difficult. Taking advantage of human flexibility can overcome some of the problems, as the human as a system component is both powerful and the choice for the system as a whole.

3.1. Implementation method/strategy

The best implementation methods for a company are those that effectively meet the needs of its organization and customers. Methods differ widely among companies and are influenced by the complexity of the organization, sophistication of products, and product mix.

The biggest difference between SME and large enterprise, is limited resources (human power, capital, information, and material), which results in the difference for technology/system implementation. During implementation, the SME had to make sacrifices, due to its limited resources. The flexibility of the human, being the most challenging resource of any company, can make up for the limitations when integrating with technology, though time consuming.

3.2. A data collection system

A CMP data collection system was designed, which is built on an Adaptive statistic query language (SQL). The system gathers data on a daily basis, used to generate the actual CMP cost for each style. Firstly, it is used to provide some basic information, such as daily output, and order summary for the factory manager. The implementation had involved the first line manager. Secondly, it recorded some management rules for production, such as arrangement of production lines and daily targets for each line. It tries to capture some uncertain factors and turn it into a fuzzy figure. Thirdly, it collects some basic information for CMP estimation. A serious problem for SME is the data integrity and information accuracy.

To start CMP estimation, other factors of influence had to be taken into account. Information regarding sales and planning, referred to as delivery sequence had to be considered as it influences CMP estimation (order on hand and the required production outputs is provided). The initial CMP estimation process flow is shown below as figure 1.

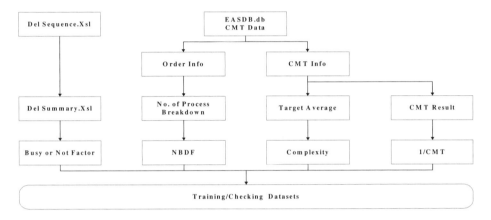

Figure 1. CMP estimation process flow

3.2.1. Analysis of variables

This CMP data collection system is based on management experience and some investigations. Factors working on the CMP costs, is a complex process. From a practice point of view, these datasets can be used as a starting point to conduct CMP estimation.

Taking advantage of human knowledge, the data collected is further processed into three fuzzy values, namely, factory Busy or Not (BON), Number of Process Breakdown Factor (NBDF), and Garment Style Complexity. From the data collected (table 1), a correlation analysis of the three input variables, namely, BON, NBDF and Complexity, is shown in table 1.

Table 1. Correlation analysis result

	BON	NBDF	Complexity
BON	1		
NBDF	-0.1172699	1	
Complexity	-0.00765793	-0.8208222	1

From table 1, a strong correlation for NBDF and Complexity is identified. The more NBDF, the more difficult it is to make the garment. From the analysis, two factors can be taken into consideration (BON gets the lower absolute correlation value with Complexity). This has assisted in showing a procedure to identify the factors of influence for system training. This can be applied when there is a need for manufacturing data analysis.

3.3. Fuzzy-neural system

To conduct the CMP estimation, a Fuzzy-Neural (FM) system based on Takagi Sugeno fuzzy model was set up [1, 6].

A total of 42 datasets were collected from Factory 3, which were ready for use in system training. Two inputs, five hidden node FN architecture was adopted for system training. Training of Factory 3 CMP data, surface view of the decision rules is shown as figure 2. (Optimum Method: backpropagation, Epochs: 200, Average testing Error: 0.0557):

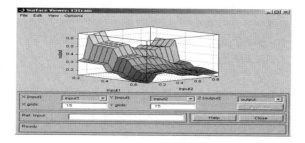

Figure 2. Surface plot of training rules

The above figure shows input 1 as the BON factor, and input 2 as the garment complexity factor. The training result was satisfying, and it can be interpreted by the general knowledge. It shows as the complexity is higher, so is the CMP cost. As the factory BON factor increases, the CMP cost should also increase (NB output is 1/CMP). Some strange points were identified (at input2/BON close to 0 and 1), caused by the inefficient training data. More training data is required, to overcome this problem. Using similar procedure for Factory 1, CMP training with an average testing error of 0.04 was achieved with 150 epochs. Figure 3 surface plot revealed some interesting findings.

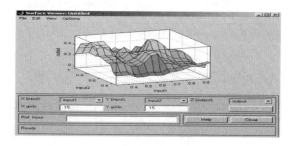

Figure 3. Factory 1 CMP data training rules surface view

The fuzzy surface plot indicates that the CMP should be high when the factory is not busy and in a busy situation. The general rule for complexity is, the higher complexity factor, the higher the CMP cost. (The results shows it costs more when the factory is not so busy. It also means that the factory requires better management when the factory production capacity is not full). The surface rule is not as smooth as Factory 3's, due lack of training dataset. Comparing the two results, factory 1's data is not good for CMP estimation, as its costs are influenced by its management style. Management was informed to take cognisance of the not busy period, to reduce expenditure.

Further modification of the training dataset showed improvement, but did not change significantly. For example, some of the large variance datasets are identified for the first training. If those datasets had been treated as odd/abnormal datasets and taken off the training dataset, the training would achieve a more efficient result, as training with 200 epochs, the average testing error dropped down from 0.056 to 0.04,

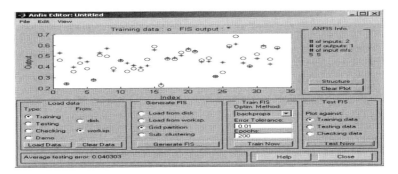

Figure 4. Plot against the training data

This result has provided us some guidance for system improvement, which can be based on a well-prepared training dataset for Fismat (Matlab code) adjustment (combined with the management judgments, and general knowledge). Comparing this approach with the normal theoretical analysis, this approach is easy to implement and understood by management. This is a better choice when there is a trade off between resource limitation and factory needs.

4. System testing

Considering the fact that repeating styles will influence the data, we therefore used 50% historical data and 50% new data (random) - Table 2. The detailed system output variance analysis is shown as in Table 3,

The system-generated results is located in the range from [-0.057, 0.1433] with 95% confidence level. Considering the fact that the training datasets is not adequate, though the result has inspired further investigation [7].

5. Conclusion

In summary, the method chosen for uncertainty estimation, is flexible and can meet estimation requirements. The challenge is to further develop a way to filter the odd datasets thus tuning the system for a robust situation.

Table 2. Checking datasets and generated results

EXP	Complexity	BON	1/CMP	A CMP	Output	E CMP	Variance
218	0.34308	0.34695	0.38697	2.5842	0.5106	1.9585	0.2421
224	0.30191	0.22827	0.50315	1.9875	0.5106	1.9585	0.0146
243	0.329	0.57581	0.49611	2.0157	0.4591	2.1782	-0.0806
253	0.262	0.69175	0.39791	2.5131	0.4495	2.2247	0.1148
296	0.26071	0.74628	0.46633	2.1444	0.4636	2.1570	-0.0059
187	0.29	0.47374	0.38354	2.6073	0.4531	2.2070	0.1535
227	0.22058	0.29422	0.49808	2.0077	0.5335	1.8744	0.0664
248	0.27778	0.57286	0.46691	2.1417	0.4609	2.1697	-0.0130
249	0.30806	0.27958	0.52854	1.8920	0.5192	1.9260	-0.0180
294	0.23197	0.4	0.57584	1.7366	0.5509	1.8152	-0.0453

Table 3. Variance analysis

Variance Analysis	
Mean	0.042859827
Standard Error	0.031744479
Standard Deviation	0.100384858
Sample Variance	0.01007712
Kurtosis	0.12302369
Skewness	0.898887204
Sum	0.428598267
Confidence Level (95.0%)	0.071811056

For most Fuzzy Neutral training, enough good training datasets is vitally important to generate results. More data needs to be collected, as the Neural network architecture requires efficient training and the training dataset needs to cover all the network nodes.

Theoretically the system with complexity is difficult to implement. Taking advantage of the flexible human, the desired result was achieved and the system successfully implemented. The best implementation methods for a company are those that effectively meet the needs of its organization and customers.

References

1. Bontempi G., Bersini H., Birattari M.: *The local paradigm for modelling and control: From neuro-fuzzy to lazy learning*. Fuzzy Sets and Systems, 1999.
2. http://www.infj.ulst.ac.uk/~cbhd23/uncer.html, February, 2000, Available (on-line).
3. Human-Machine Integration, March 2005, http://iac.dtic.mil/hsiac/Human_Machine.htm
4. Human-Systems Integration, March, 2005, Available (on-line), http://www.maad.com/index.pl/human-systems_integration
5. Management 2000 suite CMP, March, 2005, Available (on-line), http://www.aksoft.co.nz/management_2000_suite_CMP.htm
6. Li Q., Walker M.: *An Integrated Quotation Approach for Web-Based Quotation Purposes*. COMA 2004, pp. 355-360.
7. Winchell W.: *Realistic Cost Estimating for Manufacturing*. Published by Society of Manufacturing Engineers, Dearborn, Michigan, 1989.

DESIGN AND FABRICATION OF SURGICAL TITANIUM IMPLANTS USING CAD AND SLA

Harun W.A.R.W., Devadass V.

Abstract: Computer-Aided Design (CAD) and Rapid Prototyping (RP) have been successfully used as planning tools for surgery that have benefited both surgeons and patients. 2D image data of patients is converted to 3D CAD data for further manipulation and anatomic models produced by RP technique provide tactile information that can be used by surgeons to diagnose and plan the treatment and surgical operation. A case study of the application of CAD and RP for design and fabrication of implant used in craniofacial surgery is described. The result of this process is the production of a mould that is used to press form titanium mesh into a ready to use implant, which fits nicely to the patient.

1. Introduction

Computer-Aided Design (CAD) and Rapid Prototyping (RP) have proven to be indispensable and strategic tools in many sectors of industry and have profoundly changed the way designers and engineers do their jobs in the competitive environment. The engineering community has used these tools in design and production processes to speed up the development time, improve quality and reduce cost of products. However, the application of CAD and RP for diagnosis, planning of surgical intervention and treatment planning, which can be considered as an extension of their application in engineering field is a relatively new phenomenon especially in developing countries like Malaysia.

Current CAD and RP systems have paved the way for many new and innovative procedures that benefit surgeons and patients. They complement the traditional use of 2D images from x-ray, CT scans or MRI especially in the understanding and visualisation of complex human anatomy. These tools are able to generate accurate 3D images of bones (hard tissues) or skin surfaces (soft tissues). The 3D virtual and physical models can then be used for diagnosis, studying deformity, designing implants and even carrying out trial surgeries. More advanced application of CAD enables surgeons to visualise the surgical site by providing interactive and intuitive access to surgical or anatomical images during the course of surgical procedure using visualisation and virtual reality systems. This enables the physicians/surgeons and the team to clearly and objectively assess and plan in conjunction with other imaging modalities currently used in hospitals. During the past decade, this combination of medical image processing, 3D CAD and RP has proven to be a very important development, but still to be applied on a big scale in medical and surgery.

This paper reviews the application of CAD and RP in the design and production of custom surgical implants used in craniofacial reconstruction. 3D models are used to reconstruct the skull and the implants that help surgeons realise the benefits of integrating advanced digital technologies with surgical practice.

2. Application of CAD and RP in surgical planning

Surgical planning requires both quantitative and qualitative knowledge of the anatomy prior to the actual surgical procedures. Thus to avoid failure in surgery, preoperative planning is undertaken where the surgeon plans the surgery, an activity similar to the process planning carried out by manufacturing engineers prior to production processes. In conventional practice preoperative planning is done manually using 2D images from various sources such as x-ray films or data from CT scanner or MRI machine. Surgeons have to rely on their past experience and use case studies to envision the actual pathologic processes represented by these images. For example in craniofacial surgeries, surgeons have to study 2D images of skulls in frontal, sagittal or other views obtained from x-ray and CT scan processes and manually plan the steps necessary to undertake surgery and correct the deformities. If the surgery involves the insertion of implants, surgeons have to use their experience to manually design and mould the implants so that they can fit nicely to the patients. As 2D images only provide limited view, surgeons cannot get a definite idea of the surgery well before he/she actually performs it. Due to the trial and error approach to "fix" the patients, it is difficult for surgeons to predict the outcome of surgical operation and consequently, they spend longer time than necessary during the operation, with uncertain results that may jeopardise the patient.

Medical models are nowadays used by clinicians and surgeons for a variety of applications such as 3D visualisation of a specific anatomy, diagnosis, surgical planning, customised implant design, prosthesis production and treatment planning. The application of 3D digital model in visualisation and planning of anatomical parts is further enhanced with the production of RP models. RP techniques allow the building of complex physical models of a patient's anatomy directly from CT or MRI scanners. Several RP processes are available for the production of anatomical models, depending on the intended application. Stereolithography (SLA) is the best known and used RP technique that caters for a variety of applications. SLA builds models layer-by-layer by laser scanning of a light sensitive resin in the area defined by the object's cross-section such as a patient's bone structure. Special needs in surgery has also spurred new and innovative applications in RP such as colour SLA (Wouters, 2001) that enables surgeons to view blood veins and the exact location of tumours in the body parts through physical models.

3D models generated by CAD and RP enable new procedures for difficult-to-treat deformities to be planned more accurately. Examples of work on the use of CAD and RP in surgery have been reported by (Society of Manufacturing Engineers, 1997, James et. al., 2000, Cleary and Nguyen, 2001, Muller et.al. 2003).

3. A case study on using CAD and SLA for design and fabrication of surgical implants

Craniofacial surgery involves surgery of the facial and cranial skeleton and soft tissues to correct congenital deformities or for the treatment of cranial defects or deformities caused by trauma, cranial bone tumours, infected craniotomy boneflaps and external decompression. Other indications for reconstruction of these defects are cosmetic reasons and/or protection of intra-cranial structures to mechanical impact. The aim of craniofacial surgery is to rebuild the head and facial structures of patients with craniofacial problems. It is a complicated and risky endeavour involving intricate procedures that demand the skills and experience of craniofacial surgeons, neurosurgeons, oral surgeons, ophthalmologists, and ear, nose, and throat specialists, among others. The process almost always involves multiple major surgeries, some of which are still experimental and with unpredictable results.

The case study presented in this paper involves a reconstruction of the fractured skull of an accident patient. The patient had a large frontal cranial defect that extended to the left eye orbital, as shown in Figure 1. The objective of the project was to design an implant that fits nicely into the patient's large cranial cavity and produce the implant using a titanium mesh.

In traditional approach, patients are shaved and an alginate impression of the defect is taken. A positive cast is poured in stone in order to fabricate the plate. In most cases due to the complexity of the defect, traditional process is time- consuming as the implant fittings are difficult to produce. Although results have been acceptable, precise fitting of the cranioplasty could not be achieved due to lack of accuracy of the manufacturing process of the implant. This also affects the operation time because of compensation of the bad fitting of the marginal contours.

Using CAD and RP in this case study, the process started with a raw data of the patient obtained from CT scan, with a 1.5mm section interval, 1mm section thickness in spiral mode and a 512 x 512 matrix. The data was converted to a 3D CAD image through DICOM translator and modelled in MIMICS software where anatomical structures of hard tissues of the skull were identified by means of gray level thresholding and were then elaborated for 3D visualisation (Figure 1). Further manipulation of the 3D data was done using MAGICS software in which the "normal" half of the skull was used as a reference and was mirrored to obtain the symmetry of the face (Figure 2).

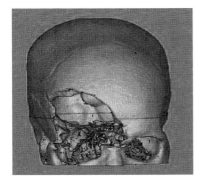
Figure 1. 3D CAD model of fractured skull

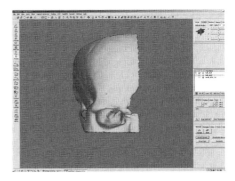
Figure 2. Mirror image of normal part of the face

The resultant mirrored image was used to design the implant that fits nicely to the severed part of the skull (Figure 3). This data was then converted to an STL format to be sent to the SLA machine. The latter produced a 3D physical model of the patient's skull and the implant. Further curing in an UV oven ensures the models were completely solidified (Figure 4).

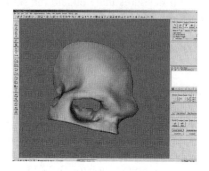

Figure 4. CAD Model of Implant

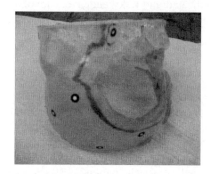

Figure 5. RP model of fractured skull

The implant and the 3D skull models produced by SLA were fitted together and checked for both form and fit to ensure that there was no gap or improper fitting due to cavities on the skull, as shown in Figure 5. If there is any gap between the implant and the skull, the process can be repeated until a good form and fit is obtained. In severe cases as illustrated in this case study, a fit implant could not be obtained directly and thus a manual intervention was introduced whereby the cavities were covered using a hand crafting technique. As the original CAD data of the implant had been modified, the new implant prototype was scanned using a laser scanner and converted into a 3D CAD data. The new data was used to reproduce another implant prototype using the SLA process and the resultant implant was again tested for form and fit with the skull model. This time it showed a near perfect fit without any cavity between the implant and the skull. The implant prototype was used as a pattern to produce an epoxy-based mould (Figure 6). A titanium implant was formed using a hydraulic press that shaped the titanium over the epoxy mould, as shown in Figure 7. The finished titanium mesh implant was tested on the 3D skull physical model, and a good fit was obtained (Figure 8).

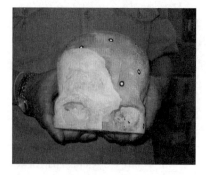

Figure 5. Fitting of implant to skull model

Figure 6. Mould for implant

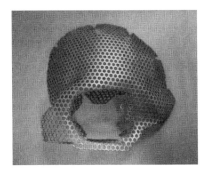 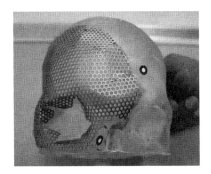

Figure 7. Titanium mesh implant   Figure 8. Fitting titanium implant to skull model

4. Results and discussion

The titanium mesh was successfully fitted to the patient with a good fit that corresponded to the normal facial condition. Due to the accuracy of CAD and RP models the surgeons had a good understanding of the cranial defect and precise fitting implant could be fabricated in order to re-establish skull contours. The excellent fitting also provided the surgeons with the opportunity to attempt new fixation technique of the implant that contributed to significant reduction in operating time. The case study shows that creating 3D CAD and RP models from 2D CT scan image of complex skull structure data provides a valuable tool to surgeons to successfully plan, design and fabricate the surgical implant. Although in this instance the fit could not be obtained the first time, this circumstance could be improved with further practice on the CAD model. Nevertheless, the surgeon and his team were satisfied with the outcome. The surgical rehearsal done by surgeons on the prototype models resulted in acceleration of the surgical planning procedures, reduction in operative time and costs and improved clinical procedures as well as providing higher level of confidence to the surgical team. Through these models, the patient benefited by being able to visualise and understand the surgical process and the expected outcome of the reconstruction process. This improvement in communication provided confidence to the patient prior to the surgery. The quality of the design of titanium implant was very much improved than those produced manually.

5. Conclusion

CAD and RP systems, traditionally tools used in manufacturing activities have enormous potential to be further developed to augment the ability of surgeons to plan and carry out diagnosis, surgical rehearsals and procedures and plan for treatment. Surgeons are able to develop optimised intervention plans, register preoperative data to the actual patients in the operating room and execute accurately the planned interventions. Besides providing an effective tool for preoperative planning, the system have the potential to allow surgeons to plan other aspects of surgery such as the cutting path for manufacturing implants, determine the size of the implants and planning alternative surgical approaches. As shown in this case study, CAD and RP tools allow the design and production of implants that enable surgeons to rapidly produce 3D models to obtain better case information that consequently lead to reduction in operating time, enhancement of patient and physician communication and improvement in patient outcomes. In general, the approach offers an accurate, reliable and practical alternative to the traditional method.

CAD and RP systems offer very promising future and provide good testbeds for future collaborative interdisciplinary work in the application of advanced high technology in clinical and surgical fields. The eventual result is the increase in productivity of the surgery process.

Acknowledgements

The authors would like to acknowledge the work done by the members of RP team and IRPA-PR team at National CAD/CAM Programme, SIRIM Berhad and surgeons from University of Malaya who contributed the patient data. This work is part of the research project under the IRPA-PR fund title "Development of Surgical Planning System for Craniofacial Reconstruction" (project no 06-01-01-0000 PR0042/03) funded by the Ministry of Science, Technology and Innovation Malaysia.

References

1. Cleary K., Nguyen C.: *State of the art in surgical robotics: Clinical applications and technology challenges.* Comput. Aided Surgery; Vol. 6, p 312-328, 2001.
2. Muller A., Krishnan K. G., Uhl E., Mast G.: *The application of rapid prototyping techniques in cranial reconstruction and preoperative planning in nuerosurgery.* Journal of Craniofacial Surgery, Vol. 14, Issues 6, p 899 - 914, 2003.
3. James X., Horace H. S., Samman N., Wang D., Christy S.B.: *Computer-assisted three dimensional surgical planning and simulation: 3D virtual osteotomy.* Int Journal of Oral and Maxillofacial Surgery, Vol. 29 (1), 2000.
4. Society of Manufacturing Engineers, Rapid prototyping Technology: A Unique Approach to the Diagnosis and Planning of Medical Procedures, Rapid Prototyping Association, SME, USA, 1997.
5. Wouters K.: *Colour Rapid Prototyping, an Extra Dimension for Visualising Human Anatomy.* Phidias newsletter, No 6, 2001.

The International Journal of **INGENIUM** 2005 (2)

ENGINEERING ACHIEVEMENTS ACROSS THE GLOBAL VILLAGE

edited by

Janusz SZPYTKO

Computer Aided Manufacturing

Cracow - Glasgow - Radom, 2005

TABLE OF CONTENTS

		page
3.	**Computer Aided Manufacturing**	**143**
3.1.	Determination of machinable volume for finish cuts in CAPP, *Arivazhagan A, Mehta N.K., Jain P.K.*	145
3.2.	Machining lines with multi-spindle stations: a new optimization problem, *Dolgui A., Ignatenko I., Belmokhtar S.*	153
3.3.	Contributions to the design for the computer-aided manufacturing of car components, *Duse D.M.*	163
3.4.	A tool for characterizing manufacturing strategies, *Harrison D.K., Macdonald M.*	171
3.5.	Critical chain scheduling: application to production job order management, *Morandi V., Aggogeri F., Baroni S.*	183
3.6.	Application of simulation and artificial neural network to the problem of scheduling of flexible manufacturing system, *Noor S., Khan M. K., Hussain I., Ullah I.*	191
3.7.	Exploring Web-based collaborative paradigms for manufacturing competitiveness, *Pun K.F., Ellis R., Lewis W.G.*	199
3.8.	Parallel kinematics machines (PKM) and multitasking in aircraft components manufacture, *Selvaraj P., Radhakrishnan P.*	207
3.9.	Holonic theory and modeling applied to MES for metalworking SMEs, *Baqueiro L.G., Cadena M.R., Molina A.*	217
3.10.	Shop floor capacity evaluation, *Haddad R.B.B., Carvalho M.C.H., Bera H.*	227
3.11.	Improving surface quality using MQL machining, *Jung J.Y., Lee C.M., Cho H.C., Cui H., Hwang Y.K.*	235
3.12.	Fuzzy modeling and approximation of surface roughness in drilling *Sivarao P.S.*	243
3.13.	Plant-input-mapping digital redesign with computational delays, *Shimamura H., Hori N., Takahashi R.*	251
3.14.	The integration of higher and lower management control techniques in manufacturing companies, *Perry D.J., Petty D.J., Smith S.A., Harrison D.K.*	259
3.15	Optimizing the vendor selection process of an automobile industry using critical value analysis – a case study, *Parthiban P., Arun J., Ganesh K., Narayanan S.*	271
3.16	Decision support system for the evaluation of enterprise resource planning, *Parthiban P., Narayanan S., Dhanalakshmi R.*	279

All rights reserved. No part of this book may be reproduced, stored in a retrieval system, or transmitted, in any form or by any means, without prior written permission from the Publisher.

The International Journal of INGENIUM
Chief Editor: Professor David K. Harrison, Glasgow Caledonian University, UK

© GCU Glasgow

ISSN 1363-514x

A CIP catalogue record for this publication is available from British Library

Publishing cooperation: Instytut Technologii Eksploatacji – PIB w Radomiu

DETERMINATION OF MACHINABLE VOLUME FOR FINISH CUTS IN CAPP

Arivazhagan A., Mehta N.K., Jain P.K.

Abstract: Identification of machinable volume for finish cut is a complex task as it involves the details not only of the final product but also the intermediate part obtained from rough machining of the blank. A feature recognition technique that adopts a rule-based methodology is required for calculating this small, complex shaped finish cut volume. This paper presents the feature recognition module in a CAPP system that calculates the intermediate finish cut volume by adopting a rule based syntactic pattern recognition approach. In this module, the interfacer uses STEP AP203/214, a CAD neutral format, to trace the coordinate point information and to calculate the machinable volume. Two illustrative examples are given to explain the proposed syntactic pattern approach for prismatic parts.

1. Introduction

Computer Aided Process Planning (CAPP) has come to acquire an extremely important role in implementation of Computer Integrated Manufacturing (CIM) since it serves as the bridge between Computer Aided Design (CAD) and Computer Aided Manufacturing (CAM). Mostly two approaches variant and generative are employed to accomplish this task. Variant approach uses the similarity among the components by means of classification and coding scheme as used in Group Technology (GT) to generate process plans. In the generative approach, knowledge of manufacturing is captured, encoded and supported by manufacturing database to create process plans. Human intervention is needed in the variant approach but the generative system creates process plans automatically. While developing a generative CAPP system for prismatic parts, recognition of machining features from a CAD model is the first and foremost task to plan further activities. To accomplish this task researchers have adopted various feature recognition techniques and CAD neutral formats. Derli and Filiz, 2002, recognized features and constructed a separate rough-cut machinable volume in a CAPP system. Dong and Vijayan, 1997, in their integrated geometric modeling system, have extracted machinable volumes that are required for 3-axis milling operations for prismatic parts. Feature relationship graph approach has been used to extract the application specific features known as high-level features from a CAD model (Fu et.al, 2003). Han and Requicha, 1997, adopted hint-based reasoning approach to recognize slot, step-groove etc in their feature recognizer named IF^2.

Optimal volume segmentation approach is used to extract features and for removing the elementary volume (Huang and Hoi, 2002). Syntactic pattern recognition approach has been implemented to identify the machining features by using DXF format to calculate the size, location and shape of the prismatic part (Jain and Sharavan Kumar, 1998). Features are identified by adopting IGES format to translate into an object oriented data structure for creating process plans (Lee1 & Kim, 1998). Liu et.al, 1996, proposed an approach to identify the removable volume known as delta volume by comparing the final part with the original stock. Sandiford and Hinduja, 2001, adopted STEP AP203 protocol for identifying the features from prismatic parts by attributed adjacency relationship graphs using GT coding scheme. A feature recognition processor has also been developed to recognize manufacturing features from STEP AP224 format (Woo and Sakurai, 2002).

2. Proposed feature recognition module for finish cuts

The feature recognition module for finish cut contains different phases as shown in figure 1. In the system, the prismatic part is designed by using a CAD package called Solid Works. Then, it is transformed into STEP AP203 or AP214 format by using the available STEP translator in Solid Works. This neutral format of STEP is used as an input to the feature recognition module in the CAPP system.

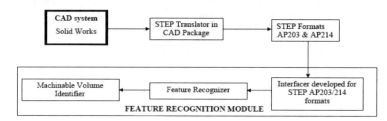

Figure 1. Overview of the Proposed System

The interfacer developed inside the feature recognition module, produces an output file of the prismatic part. The output file contains the coordinate information related to the model. Then, the feature recognizer identifies the features in the part model by syntactic pattern recognition technique and identifies the machinable volumes to be removed for the features obtained after rough cuts. The machinable volume for finish cuts is obtained by deducting the rough-cut feature from the final part input to the system. The identified machinable volume contains the details of all the parameters related to final machining of the part. A schematic representation of the volumes to be removed during the finish cuts is shown in the figure 2, which shows the machinable volumes, including the fillets and chamfers in the final part.

3. Interpreting the neutral format

The first part in the feature recognition module is the interpretation of STEP AP203/214 format. The AP203/214 formats contain two sections (i) Header section (ii) Data Section. The header section starts with the specification of the type of format AP203/214 by a keyword FILE_DESCRIPTION, followed by another keyword FILE_NAME that gives the full details about the file name, person, and organization. The protocols "Configuration Controlled Design" or "Automotive Design" are identified in a key word FILE_SCHEMA inside the header section. The end of this section is denoted by a key word "END SEC". The next section i.e., data section is the core part of the STEP AP203/214 format. It contains all the informations, starting from the time when the part is created to the CAD model's coordinate point information. The data section is started by a key word "DATA".

The representation of the part is given by the entity SHAPE_REPRESENTAION. The information about the CAD model starts from CLOSED_SHELL entity. The closed shell entity contains information about the consecutive lines indicated by their respective IDs such as '#123'etc. Each ID in the closed shell goes directly to another entity ADVANCED_FACE, which is a standard hierarchy in all STEPAP203/214 formats. The ID's in the advanced face are split into many entities depending on the CAD model. The details of entities and the hierarchy of the STEP AP203/214 formats to identify the correct coordinate point information's are different for different shapes.

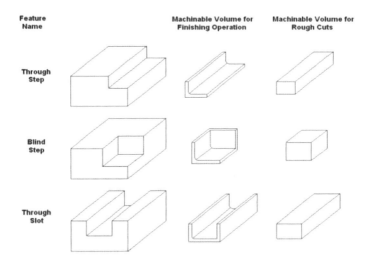

Figure 2. Representation of machinable volumes

4. Steps followed by the interfacer

The interfacer developed for the CAPP system uses the following steps to detect the feature and its coordinate informations. Figure 3 gives the tracing logic applied particularly for prismatic parts with filleted edges as referenced in step 10.

Step1: Read the lines up to semicolon.
Step2: Check for HEADER section
Step3: If HEADER section is found, check for FILE_SCHEMA for finding out the protocol of STEP used.
Step4: If it is Configuration Controlled Design – AP203 or Core data for Automotive Mechanical Design Processes - AP214 then proceed to step 6.
Step5: If it is not AP203/214 then mention the user to input AP203/214 file.
Step6: Check for DATA section
Step7: If DATA section is found, then check for CLOSED_SHELL entity.
Step8: If it is found then read the ID values inside it.
Step9: Now check for the ID's inside the DATA section up to End of File.
Step10: Implement the trace logic developed that suits for the Prismatic part taken as an input for the interfacer and
Step11: Write the appropriate "Cartesian Points" for the whole part and store the information of cylindrical surface or a filleted object separately.

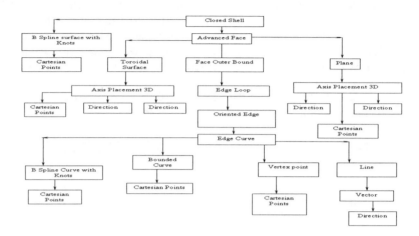

Figure 3. Hierarchy tree for a prismatic part with filleted edges

5. Calculating the intermediate machining volume for finish cuts

In this phase, the arranged points from the interfacer containing the details of all the surfaces present in the rough-cut prismatic part and the final part are used as input to the feature recognition module. Automatically the details of all the features with the edge loop information are taken out in the form of a separate file. In addition, the details regarding the tolerance information and surface finish information of the final part are also taken. After getting this information, the machinable volume identifier calculates appropriate finish cut volume for the features identified by the feature recognition system. Two example parts are discussed below to illustrate the approach to calculate the machinable volumes.

5.1. Illustrative Example: 1: When the Feature Is a Step

The edge loops (e1r, e2r, e3r, .etc.) are shown in figure 4 and the Coordinate Point information obtained from the feature recognizer are used to calculate the machinable volume. The different phases followed by the feature recognizer are given below.

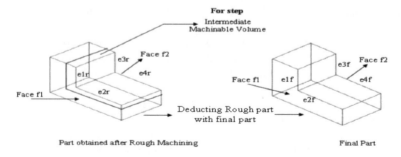

Figure 4. Details of the edge loops for a step type feature

Phase 1: Check whether the coordinate points follow the same string stored in the database. Figure 5 shows the string for a step type feature.

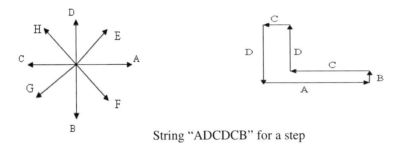

String "ADCDCB" for a step

Figure 5. Syntactic Pattern Recognition approach developed for step

Phase 2: If it is the same then intersect the edge loop of rough machined feature with the final feature.
Phase 3: Calculate the difference in the values for e1r, e2r, e3r, e4r and e1f, e2f, e3f, e4f.
Phase 4: Interpretation of results.

The edge loops on face f1 and on face f2 can be used to determine the thickness (g), length (l), width (w) and height (h) as follows:

(i) edge loops (e1r-e1f) (e2r-e2f) on face f1 and (e3r-e3f) (e4r-e4f) on face f2

$$e1r\text{-}e1f \begin{vmatrix} 10,20,0 & 8,20,0 \\ 10,15,0 & 8,13,0 \end{vmatrix} \quad e2r\text{-}e2f \begin{vmatrix} 10,15,0 & 8,13,0 \\ 20,15,0 & 18,13,0 \end{vmatrix}$$
and

On face f1
$$e1r\text{-}e1f \begin{vmatrix} \text{Dif.of.2mm in X axis} \\ \text{Dif.of.2mm in Y axis} \end{vmatrix} \quad e2r\text{-}e2f \begin{vmatrix} \text{Dif.of.2mm in X axis} \\ \text{Dif.of.2mm in Y axis} \end{vmatrix}$$
and

On face f2 for calculating the thickness (g)
$$e3r\text{-}e3f \begin{vmatrix} 10,20,0 & 8,20,0 \\ 10,15,0 & 8,13,0 \end{vmatrix} \quad e4r\text{-}e4f \begin{vmatrix} 10,15,0 & 8,13,0 \\ 20,15,0 & 18,13,0 \end{vmatrix}$$
and

On face f2
$$e3r\text{-}e3f \begin{vmatrix} \text{Dif.of.2mm in X axis} \\ \text{Dif.of.2mm in Y axis} \end{vmatrix} \text{and} \quad e4r\text{-}e4f \begin{vmatrix} \text{Dif.of.2mm in X axis} \\ \text{Dif.of.2mm in Y axis} \end{vmatrix}$$

Hence the material to be removed is having a uniform thickness of (g) =2 mm in both the XY plane on both the faces f1 & f2.

(ii) The edge loops (e1r-e3r) & (e2r-e4r) and (e1f-e3f) & (e2f-e4f) on XZ plane provide dimensional value width (w) of the finish cut machinable volume

$$\text{e1r-e3r} \begin{vmatrix} 10,20,0 \\ 10,15,0 \end{vmatrix} - \begin{vmatrix} 10,20,50 \\ 10,15,50 \end{vmatrix} \quad \text{and} \quad \text{e2r-e4r} \begin{vmatrix} 10,15,0 \\ 20,15,0 \end{vmatrix} \begin{vmatrix} 10,15,50 \\ 20,15,50 \end{vmatrix}$$

Calculating the thickness (w) in final part

$$\text{e1f-e3f} \begin{vmatrix} 8,20,0 \\ 8,13,0 \end{vmatrix} - \begin{vmatrix} 8,20,50 \\ 8,13,50 \end{vmatrix} \quad \text{and} \quad \text{e2f-e4f} \begin{vmatrix} 8,13,0 \\ 18,13,0 \end{vmatrix} \begin{vmatrix} 8,13,50 \\ 18,13,50 \end{vmatrix}$$

Here a comparison is made for both the parts so that a common difference is obtained in Z-axis from both the rough and final part. Otherwise, the amount of material has to be calculated separately for face f1 and face f2. Here the width (w) is 50 mm.

(iii) The edge loops e2f & e4f on ZY plane provide dimensional value length (l) of the finish cut machinable volume

e2f → 8,13,0   the difference is 10 mm
       18,13,0

e4f → 8,13,0   the difference is 10 mm
       18,13,0

The calculated length (l) is 10 mm including the thickness 2mm.

(iv) The edge loops e1f & e3f on XZ plane provide dimensional value length (l) of the finish cut machinable volume

e1f → 8,20,0   the difference is 7 mm
       8,13,0

e3f → 8,20,50   the difference is 7 mm
       8,13,50

The calculated height (h) is 7 mm including the thickness 2mm.
The schematic representation of the machinable volume is shown in figure 6.

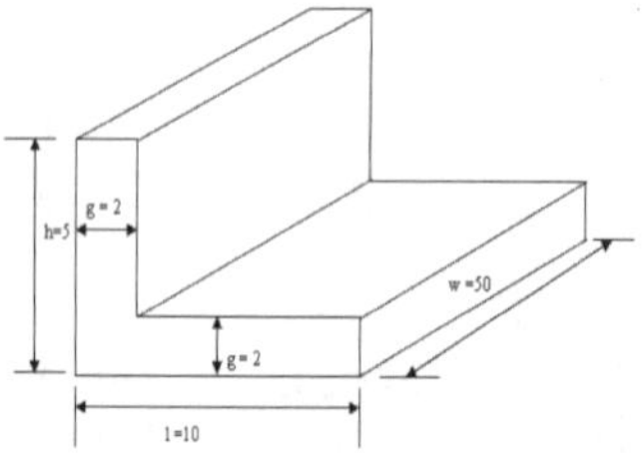

Figure 6. Machinable volume for finish cut with dimensional values

5.2. Illustrative Example: 2: When the Feature Is a Slot

The details of the Edge loops (e1r, e2r, e3r,.etc..) are shown in figure 7 and the Coordinate Point information obtained from the feature recognizer is given below.

Determination of machinable volume... 151

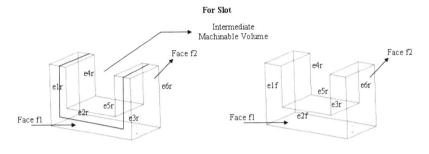

Figure 7. Details of the edge loops for a slot type feature

Rough machined Part: Edge loop on face f1

e1r → 10,60,0   e2r → 10,40,0   e3r → 20,40,0
        10,40,0   20,40,0   20, 60,0

Edge loop on face f2

e4r → 10,60,50   e5r → 10,40,50   e6r → 20,40,50
        10, 40, 50   20, 40, 50   20, 40, 50

Final Part: Edge loop on face f1

e1f → 8,60,0   e2f → 8,38,0   e3f → 18,38,0
        8,38,0   18,38,0   18,60,0

Edge loop on face f2

e4f → 8, 60, 50   e5f → 8, 40, 50   e6f → 18, 40, 50
        8, 40, 50   18, 40, 50   18, 60, 50

Phase1: Check whether the coordinate points follow the same string stored in the database. Figure 8 shows the string for a step feature.

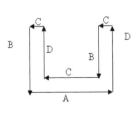
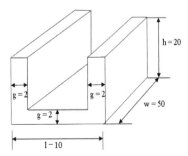

String "ADCBCDCB" for a slot

Figure 8. Syntactic Pattern Recognition approach developed for slot and Machinable volume for finish cut with dimensional values

Phase 2: If it is the same then intersect the edge loop of rough machined feature with the final feature.
Phase 3: Calculate the difference in the values for e1r, e2r, e3r, e4r and e1f, e2f, e3f, e4f.
Phase 4: Interpretation of results.

The procedure followed for the "step" type feature is applied for "slot" and the following dimensional values length (l) = 10 mm, width (w) = 50 mm, height (h) =20 mm, thickness (g) = 2 mm are calculated. The machinable volume is as shown in figure 8. The same methodology is followed also for features such as blind slot, blind step, hole, and pocket.

6. Conclusions

In this paper, a methodology is presented by adopting syntactic pattern recognition approach to calculate the finish cut machinable volume by deducting the rough-cut part from the final part. At present, the feature recognizer is capable of identifying the machinable volume for finish cut from STEP AP203/214 format for both primitive and interacting features such as slot, step, blind slot, blind step, pocket, hole, slot-slot-step, slot-pocket-step etc to make the recognition process complete. Feature volume details as extracted along with the user inputted surface finish and tolerance values are used for machining planning at the later stage.

References

1. Derli & Filiz, *A note on the use of STEP for interfacing design to process planning*, Computer Aided Design, 2002; 34: 1075-1085.
2. Dong & Vijayan, *Manufacturing feature determination and extraction* Part II: A heuristic approach, Computer aided Design, 1997; 6(29): 475-484.
3. Fu, Ong, Lu, Lee & Nee, *An approach to identify design & manufacturing features from a data exchange part model*, Computer Aided Design, 2003; 35: 979-993.
4. Han & Requicha, *Integration of feature based design and feature recognition*, Computer Aided Design, 1997; 5(29):393-403.
5. Huang & Hoi, *High-level feature recognition using feature relationship graphs*, Computer Aided Design, 2002; 34:561-582.
6. Jain & Sharavan Kumar, *Automatic feature extraction in PRIZCAPP*, International Journal of Computer Integrated Manufacturing, 1998; 6(11): 500-512.
7. Lee1 & Kim, *A feature-based approach to extracting machining features*, Computer-Aided Design, 1998; 13(30):1019–1035
8. Liu, Gonzalez & Chen, *Development of an automatic part feature extraction and classification system taking CAD data as input*, Computers in Industry, 1996; 29: 137-150.
9. Sandiford & Hinduja, *Construction of feature volumes using intersection of adjacent surfaces*, Computer Aided Design, 2001; 33: 455-473.
10. Woo & Sakurai, *Recognition of maximal features by volume decomposition*, Computer Aided Design, 2002; 34: 195-207.

MACHINING LINES WITH MULTI-SPINDLE STATIONS:
A NEW OPTIMIZATION PROBLEM

Dolgui A., Ignatenko I., Belmokhtar S.

Abstract: In this paper we examine a widespread industrial situation that is seldom researched, in effect, a new problem of designing transfer lines, with the potential of large cost savings. Compared to simple line balancing problem which deal with only elementary operations, in this case, operations are often combined into groups, called blocks, which are executed simultaneously by one spindle head. Clearly, this problem is more complex. The spindles are assigned such that we have a sequence of parallel blocks at workstations. It is supposed that the set of all available spindle heads is known. The goal is to find the optimal design by selecting a subset of spindle heads and assigning them to workstation such that all operations are executed, all constraints are respected and the line cost is minimal.

1. Introduction

Transfer lines are automated flow-oriented production systems which are typical in the mechanical industry (Hitomi, 1996). They increase the production rate and minimize the cost of machining parts. These lines are composed of linearly ordered multi-spindle machines (workstations), arranged along a conveyor belt (transfer system). At each workstation, there are several spindles heads and each is composed of several tools. Each tool executes one or several (combined tool) operations. These operations to be executed in the line are grouped into the so-called *blocks*. The operations of each block are executed simultaneously by one spindle head. When all blocks of a workstation have been accomplished, the workstation cycle time is terminated. The longest workstation cycle time defines the *line cycle time*. An example of such line is shown in Figure 1.

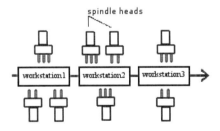

Figure 1. An example of transfer line

We are interested in the problem of designing modular mono-product transfer lines which are composed of standard spindle heads. At the preliminary design stage, the set of all available standard spindle heads is known. A subset of spindle heads should be selected and assigned to workstations while defining a fixed serial-parallel order of their activation. In hard automation lines the order cannot be modified during the exploitation period. The goal consists in finding the best configuration, i.e. an optimal ordered subset of spindle heads from the set of all available spindle heads and their best possible assignment to workstations.

A similar problem in literature is Simple Assembly Line Balancing Problem Type I (SALBP-I), where there is only a single product assembly line in which all operations and *precedence constraints* between operations are known. The SALBP-I consist of assigning the operations to workstations such that a given cycle time is not exceeded, the precedence constraints are respected and the total idle time is minimum. For SALBP-I, all operations are executed sequentially. The idle time is minimum iff (if and only if) the number of workstations is minimum. The comprehensive surveys on SALBP are presented in (Baybars 1986), (Scholl 1999), (Rekiek *et al.*, 2002), (Talbot *et al.*, 1986) and (Johnson *et al.*, 1988), where many heuristics, meta-heuristics and exact approaches are cited. Recent generalizations of SALBP tacking into account equipment selection are studied in (Graves and Redfield, 1988), (Bukchin and Tzur, 2000), (Bukchin and Rubinovitz, 2002). The goal of Line Balancing with Equipment Selection is defined as minimizing the total line cost which is composed of equipment and workstations costs.

The new problem is a generalization of line balancing with equipment selection problem. In previous works, some particular cases are studied, for instance, in (Dolgui *et al.*, 2002) and (Dolgui *et al.*, 2001) the set of all available spindle heads is not given beforehand (the structure of blocks is also a decision variable) and the objective function is formulated as minimizing the weighted sum of workstations and blocks number. For the case where the set of all available spindle heads is known, the papers (Dolgui *et al.*, 2005*b*) and (Dolgui and Ihnatsenka, 2004) show some interesting methods, but in (Dolgui *et al.*, 2005*b*) blocks at each workstation are executed simultaneously, and in (Dolgui and Ihnatsenka, 2004) blocks at each workstation are executed exclusively sequentially.

This paper deals with an unexplored transfer line design problem where the set of all available spindle heads is given, however spindle heads at each workstations can be activated in a mixed (serial-parallel) order. The objective is to minimize the total line cost while respecting precedence, inclusion, exclusion and parallelism constraints.

2. Problem statement and constraints

It is assumed that the set of all operations to be executed at the line is known. The set of all available spindle heads (blocks) is also known. The blocks costs and times are known. There are operations which can belong to different blocks (performed by many available spindle heads), in this case only one block (one spindle head) should be chosen.

The spindle heads of the same workstation are activated in a serial-parallel order, i.e. we have a sequence of parallel blocks. So, for each workstation an assignment of blocks can be represented by a family of sets of blocks. The blocks of each set are activated simultaneously. Different sets are executed sequentially. The processing time for a set is given by the maximal of their block times. The workstation time is obtained as the sum of the set times.

All admissible decisions should respect the following constraints:
- each operation is executed only once,
- the given line cycle time is not exceeded,
- precedence constraints should be satisfied.
  Note: There are some particularities of precedence constraints because of the grouping of operations into blocks. Let operation *i* directly precedes to operations *j*. There are two possibilities:
  - first, a block can contain both operation *i* and *j*. Which leads to a simultaneous execution of these operation by a combined tool,
  - second, these operations belong to different blocks. So, the block containing operations *j* must be assigned after the block with operation *i*.
- the constraints of the machining process itself can impose that some groups of operations be performed on the same workstation. This type of constraint is called *operations inclusions constraints*,
- there are blocks which cannot be assigned to the same workstation. This type of constraint is called *blocks exclusion constraint*,
- in this type of line, the blocks are activated in a serial-parallel order. The compatibility between blocks, i.e. the sets of blocks which can be executed simultaneously (in parallel) are given. This type of constraint is refereed as *blocks parallelism constraints*. Clearly, these constraints should be coordinated with the precedence and blocks exclusion constraints,
- for each workstation, the number of blocks which can be assigned is limited.

The aim consists in finding an optimal design, i.e. an ordered family of blocks subsets from the given set of all possible blocks. The optimal design minimizes the line investment cost by taking in account the above constraints. The line investment cost estimated as the sum of block costs and a cost depending on the number of workstations. This total cost should be as small as possible (cost minimization problem).

3. An optimization model

Let introduce the following notation:

$N=\{1,2,\ldots,n\}$	the given set of all operations which have to be performed on the line,
T_0	the given transfer line cycle time (1/target production rate),
n_0	the maximum authorized number of blocks in a workstation,
q_0	the number of all available blocks,
$\mathbf{B}=\{b_1, b_2, \ldots\}$	the given set of all available blocks,
$N(b) \subseteq \mathbf{N}$	the set of operations which belong to a block b,
$N(S) = \bigcup_{b \in S} N(b)$	the set of operations which belong to a block's set $S \subseteq \mathbf{B}$,
$t(b)$	the processing time of a block $b \in \mathbf{B}$,

$c(b)$ — the cost of a block $b \in \mathbf{B}$,

C_0 — the additional cost for each new workstation,

$P(b)$ — the set of operations which directly precede (in the ordinary sense) to each operation from $N(b)$. In other words, $P(b)$ is a set of operations which directly precede block b,

$P(S) = \bigcup_{b \in S} P(b)$ — the set of operations which directly precede (in the ordinary sense) $S \subseteq \mathbf{B}$,

$G^P = (N, D)$ — an acyclic digraph representing the precedence constraints for operations, $(i,j) \in D$ iff operation i directly precedes (in the ordinary sense) operation j,

$G^{BE} = (\mathbf{B}, E^{BE})$ — a graph representing the blocks exclusion constraints. $(b',b'') \in E^{BE}$, $b',b'' \in \mathbf{B}$ iff b',b'' cannot be assigned to the same workstation,

$G^O = (N, E^O)$ — a graph representing the operations inclusion constraints. $(i,j) \in E^O$, $i,j \in N$ iff i and j must be assigned to the same workstation. Also, the operations inclusion constraints can be defined by sets $I^O(i) = \{ j \mid j \in N$ iff $(i,j) \in E^O \}$ $i \in N$,

$G^{BP} = (\mathbf{B}, E^{BP})$ — a graph representing the blocks parallelism constraints. $(b', b'') \in E^{BP}$, $(b' b'') \in \mathbf{B}$ iff b' and b'' can be simultaneously (in parallel) executed at the same workstation,

Π — a binary relation. Let $G = (V, E)$ is a graph and $\mathbf{B} \subseteq V$. Notation $\mathbf{B} \, \Pi \, G$ means that all elements from $\mathbf{B}$ belong to the same clique in the graph G,

m — the number of workstations in a solution,

q_k — the number of the sets of simultaneously activated blocks assigned to workstation k in a solution, $k = 1, 2, \ldots, m$.

An assignment of blocks to workstation k is modeled by an ordered family $F_k = (S_1^k, \ldots, S_{q_k}^k)$, where: $S_u^k \subseteq \mathbf{B}$ is a set of simultaneously activated blocks. q_k is a number of such sets assigned to the workstation k, $k = 1, 2, \ldots, m$, $u = 1, 2, \ldots, q_k$. Index u indicates the order of the sequential executions of sets S_u^k.

We search an ordered family $L = (F_1, F_2, \ldots, F_m)$ taking into account all the following constraints (Dolgui and Ihnatsenka, 2005):

- all operations are performed:

$$\bigcup_{k=1}^{m} \bigcup_{S \in F_k} N(S) = N \tag{1}$$

- each operation is performed only once:

$$N(b') \cap N(b") = \emptyset, \ b' \in S_u^k, \ b" \in S_v^r, \quad (2)$$

$$b' = b", \ k, r = 1, 2, \ldots, m,$$

$$u = 1, 2, \ldots, q_k, v = 1, 2, \ldots, q_r,$$

- *The operation precedence constraints are respected by the following*

For all $k = 1, 2, \ldots, m$ and for all $S_u^k \in F_k$

$$P(S_v^r) \subseteq \left(\bigcup_{r=1}^{k-1} \bigcup_{v=1}^{qr} N(S_v^r) \right) \cup \left(\bigcup_{v=1}^{u} N(S_v^k) \right) \quad (3)$$

- *Block parallelism*

All blocks from set S_u^k are activated simultaneously $k = 1, 2, \ldots, m$; $u = 1, 2, \ldots, q_k$. Then, set E^{BP} must contain all edges $(b', b")$, $b' \neq b"$, $b', b" \in S_k$. In other words, S_u^k is a clique in G^{BP}. So, block parallelism conditions can be formulated as:

$$S_u^k \prod G^{BP}, \ k = 1, 2, \ldots, m, \ u = 1, 2, \ldots, q_k \quad (4)$$

- *Block exclusion*

$$(b', b") \notin E^{BE}, \text{ for all } \quad b', b" \in \bigcup_{u=1}^{q_k} S_u^k \quad (5)$$

$$k = 1, 2, \ldots, m, \ u = 1, 2, \ldots, q_k$$

- *Cycle time*

$$\sum_{S \in F_k} \max_{b \in S} t(b) \leq T_0, \ k = 1, 2, \ldots, m \quad (6)$$

- *Maximal number of blocks at workstation*

For each workstation, the number of blocks does not exceed the maximum authorized value:

$$\sum_{S \in F_k} |S| \leq n0, \ k = 1, 2, \ldots, m \quad (7)$$

- *Operations inclusion*

$$\bigcup_{S \in F_k} \bigcup_{b \in S} \bigcup_{i \in N(b)} I^0 \subseteq \bigcup_{S \in F_k} N(S) \quad (8)$$

$$k = 1, 2, \ldots, m$$

The objective: The line investment cost is as small as possible:

$$\text{Min } C(L) = \sum_{k=1}^{m} \sum_{S \in F_k} \sum_{b \in S} c(b) + C_0 m \qquad (9)$$

The decision variables are sets S_k grouped into families $F_k = 1,..., m$, $u = 1,..., q_k$.

A lower bound on the objective for this MIP model has been proposed in (Dolgui and Ihnatsenka, 2005). The computation of the lower bound is obtained by relaxing some constraints of the original problem leading to a special set partitioning problem. This lower bound can be used in a branch and bound procedure to solve the optimization problem which permits ILOG CPLEX to be more efficient in obtaining an optimal solution. The solutions provided by CPLEX can be satisfactory in industrial environment for a problem whose size does not exceed the sixteen operations.

4. A numerical example

Let us consider an example to illustrate the constraints and the input data of this transfer line design problem. The set of operations is $N = \{1, 2,...,12\}$. Table 1 gives the set of blocks. The cost of any workstation is equal to $C_0 = 380$ and the cycle time can not exceed $T_0 = 1$ minute.

Figures 2, 3, 4 and 5 provides precedence, inclusion, exclusion and parallel blocks constraints, respectively. Exclusion and parallel blocks constraints are expressed between pairs of blocks. For example: in G^{BE} both pairs (b_1,b_{12}) and (b_1,b_{11}) are incompatible, but blocks b_{11} and b_{12} are not so. The same relation is translated by the block parallelism constraints, for example: in G^{BP} blocks (b_1,b_4) can be in parallel, blocks (b_4,b_3) also can be in parallel, but not the blocks b_1 and b_3.

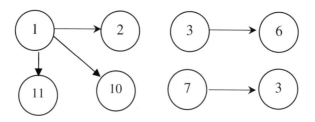

Figure 2. Precedence graph G^p

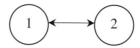

Figure 3. Inclusion graph G^0

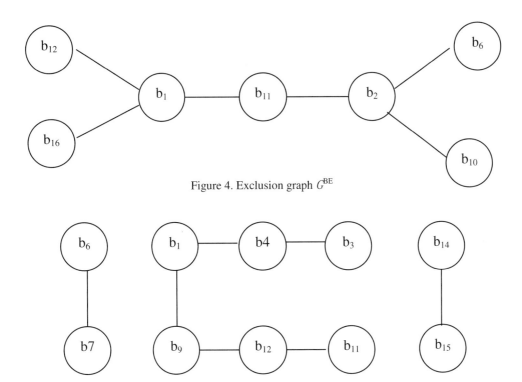

Figure 4. Exclusion graph G^{BE}

Figure 5. Parallel graph G^{BP}

Table 1. The blocks (spindle heads)

blocks	operations	$c(b)$	block time
b_1	{1,3,4,7}	240.004	55.12"
b_2	{1,4,7}	230.966	45.23"
b_3	{3}	210.74	49.45"
b_4	{5}	211.648	56.36"
b_5	{8,9,12}	230.612	58.25"
b_6	{12}	210.554	57.69"
b_7	{8,9}	221.55	59.69"
b_8	{8,12}	221.11	54.58"
b_9	{9}	210.762	57.32"
b_{10}	{2,10}	221.864	58.01"
b_{11}	{10}	210.212	54.24"
b_{12}	{2}	211.758	54.36"
b_{13}	{6,11}	221.688	57.85"
b_{14}	{11}	210.174	51.58"
b_{15}	{6}	210.924	57.47"
b_{16}	{11,12}	221.902	57.85"

The solution (see Figure 6) of this problem is described by the ordered family $L = (F_1, F_2)$, where $F_1 = \{S_1^1, S_2^1\}$ with $S_1^1 = \{b_1, b_4\}$, $S_2^1 = \{b_5\}$, and $F_2 = \{S_1^2\}$ with $S_1^2 = \{b_{10}, b_{13}\}$.

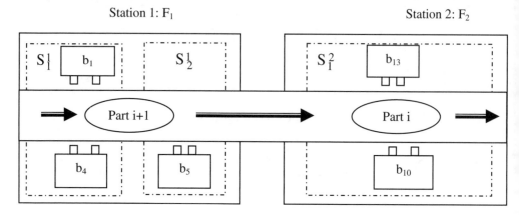

Figure 6. A feasible solution

Specifically, the solution is composed of 2 workstations. In the first workstation two sets of blocks are executed sequentially S_1^1 and S_2^1, where S_1^1 is composed of blocks b_1 and b_4 which are executed simultaneously. After the activation of b_1 and b_4, set S_2^1, which is composed of block b_5, is performed. For the second workstation, only one set S_2^1, which is composed of the blocks b_{10} and b_{13}, is executed by the two corresponding spindle heads.

5. Conclusion

In this paper we presented a model for a new optimization problem: machining transfer line design for the lines with mixed mode of spindle heads activation (a sequence of sets of simultaneous executed spindle heads at each workstations). The purpose is to find the optimal set of equipment and their logical layout for this line. The problem is both more general and complex than the simple assembly line balancing because of grouping the operations in blocks. Moreover, the specific constraints as exclusion, inclusion and parallel block constraints leads one to search for new approaches for this optimization problem. Because decisions concerning design have to be done only at the outset, we can devote relatively large computational resources to their resolution. The approach which use a Branch and Bound algorithm with a Lower Bound obtained by relaxation of this problem to a set partitioning problem seems promising.

Acknowledgements
Thanks to Christopher Yukna for his help in editing English of this paper.

References

1. Baybars I.: *A survey of exact algorithms for the simple assembly line balancing problem*. Management Science 1986; 32: 909-932.
2. Bukchin J., Rubinovitz J.: *A weighted approach for assembly line design with station paralleling and equipment selection*. IIE Transactions 2002; 35: 73-85.
3. Bukchin, J., Tzur M.: *Design of flexible assembly line to minimize equipment cost*. IIE Transactions 2000; 32: 585-598.
4. Dolgui A. Ihnatsenka I.: *Branch and bound algorithm for optimal design of transfer lines with multi-spindle stations*. Research Report. G2I Division, Ecole des Mines de Saint Etienne, France, N°2004-500-001, 33 pages.
5. Dolgui A. Ihnatsenka I.: *Balancing tranfert lines with mixed activation of spindle heads*. Research Report, G2I Division, Ecole des Mines de Saint Etienne, France, N°2005-500-002, 20 pages.
6. Dolgui A., Finel B., Guschinsky N., Levin G. Vernadat F.: *An heuristic approach for transfer lines balancing*. Journal of Intelligent Manufacturing 2005a; 16(2): 159-171.
7. Dolgui A., Guschinsky N. Levin G.: *A special case of transfer lines balancing by graph approach*. European Journal of Operational Research 2005b (In Press).
8. Dolgui A., Guschinsky N, Harrath Y Levin G. : *Une apprache de programmation linéaire pour la conception des lignes de transfert*. European Journal of Automated Systems (JESA) 2001; 36(1): 11–31.
9. Graves S.C., Redfield C.H.: *Equipment selection and task assignment for multi product assembly system design*. The International Journal of Flexible Manufacturing Systems 1988; 1: 31-50.
10. Johnson JR.: *Optimally balancing large assembly lines with FABLE*. Management Science, 1988; 34, 240.
11. Hitomi K.: *Manufacturing System Engineering*. Taylor&Francis; 1996, New York.
12. Rekiek B., Dolgui A., Delchambre A., Bratcu A.: *State of art of assembly lines design optimisation*. Annual Reviews in Control 2002; 26(2): 163–174.
13. Scholl A.: *Balancing and sequencing of assembly lines*. Physica-Varlag. Heidelberg, 1999.
14. Talbot F.B., Paterson J.H., Gehrlein W.V.: *A comparative evaluation of heuristic line balancing techniques*. Management Sciences, 1986; 32, 430-454.

CONTRIBUTIONS TO THE DESIGN FOR THE COMPUTER-AIDED MANUFACTURING OF CAR COMPONENTS

Duse D.M.

Abstract: The researches presented here aim to offer solutions, by means of computer-aided design and software modules for parametric modelling (CAD, DFMA, CAPP), for the restructuring and modernising of the technologies employed at S.C. COMPA S.A. Sibiu, Romania, in order to obtain products that are performance-optimised. The constructive-functional analysis of the car components uses modern design techniques based on the finite elements method. The static and dynamic loading state, the representative car components (shafts, bushings, gears, levers etc.) are analysed, as well as the dynamic behaviour of the assembly. The results are then implemented in the manufacturing of the assembly (hydraulic and mechanical systems, brakes, cardan transmissions etc.) as part of an integrated manufacturing system.

1. Introduction

In the design of manufacturing technologies for car parts, it can be noticed that there are differences between processing technologies for parts that are identical or very similar in geometry and dimensions. Certainly, technologists try to find always the "best" technology. However, their decision was based in most cases on their experience and intuition, so there is a pronounced empirical character.

The continuous accumulation of new data and information on technologies have made it impossible for them to be assimilated by technologists only by experience. The routine-based specialisation is now opposed by the design of manufacturing technologies using the scientific fundaments of the technical decisions theory. Therefore, information, reasoning and calculus are needed in the technology design. Manufacturing technologies must thus be conceived using optimal, typified processing technologies for representative parts, for part families, classes or groups. This optimal typified technology of the representative part is then used by the technologist, in a manual or computer-aided regime, for the development of the technology for a part.

The information is represented by the optimal typified processing technologies of the representative parts from all part families, classes or groups. Reasoning refers to determining the unfolding order of technologies in the stages $e_1, e_2, ...e_i, ..., e_n$. The calculus is applied in the stages $e_1, e_2, ...e_i, ..., e_n$, when the technological (technical and economic) optimum is determined with regard to the stages' contents.

The aim of these researches is the computer-aided design of technologies for manufacturing lines at S.C. COMPA S.A. Sibiu, for car components of the type of hydraulic and mechanical steering systems, braking systems, cardan transmissions, buffers etc., for the current production and for spare parts (Duse, 2003). Also, it is sought to create a pilot system for the computer-aided constructive and technological design for the increase of flexibility at the car components; manufacturing at S.C. COMPA S.A.

2. Computer-aided design for the manufacturing of car components

This research was determined by the need to adapt the industrial production of S.C. COMPA S.A. Sibiu to market economy conditions, through following components:
- diversification of the car components' manufacturing (for the internal and international market, for spare parts),
- alignment of the products and of the manufacturing technologies to the level of European standards and directives,
- improvement of the constructive and technological design activities,
- integration of human resources.

It was necessary to design and create modules for the parametric modelling through CAD (computer-aided design), for the product rehabilitation from the point of view of the cost/quality ratio through the DFMA method (design for manufacturing and assembling) and for the automated design of manufacturing technologies through CAPP (computer-aided process planning). The research also confers a higher quality and efficiency to car components redesigning and designing activities. The obtained data are used in the computer-aided design of manufacturing technologies in an integrated production system for representative parts of car components (see Figure 1).

A module for parametric design activities for part families that belong to car components, has been conceived and designed, that can later be generalised also for other products from the machine manufacturing domain. The parametrised part design module comprises:
- the constructive-functional designing or redesigning of representative parts, based on the generic part,
- the constructive design of tools and devices needed for the whole manufacturing process.

The module for computer-aided designing of manufacturing technologies comprises:
- technological sheets or operation plans obtained based on the generic technology,
- the computer-aided conceiving, modelling and simulation of the flexible manufacturing system,
- realising the computer-aided manufacturing line for the generic part,
- integrating the constructive and technological design activities in the production system by means of an adequate informational flow.

The two modules' integration will be done with the help of the informational flow (with adequate hardware and software interfaces).

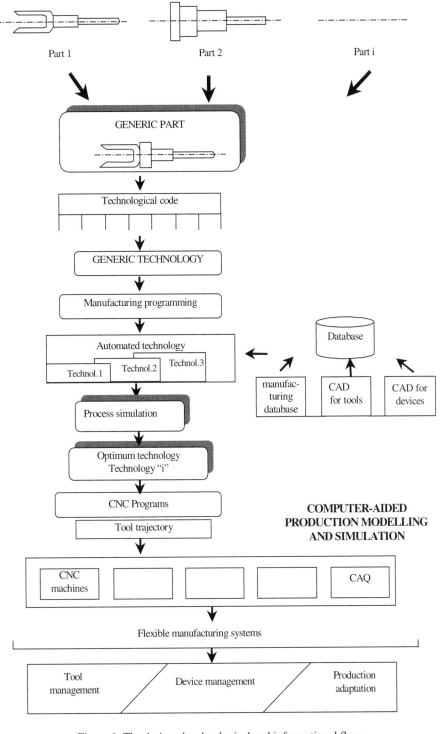

Figure 1. The designed technological and informational flow

The expected results refer to following aspects:
a) improving the efficiency of the constructive and technological design activities in the machine manufacturing – car components industry by:
    ❑ improving the quality of constructive and technological design activities through the informatisation of the design and manufacturing activities,
    ❑ reducing the manufacturing time of products and flexibilising the production,
    ❑ reducing the scrap by using CNC technologies,
    ❑ reducing the design costs for the manufacturing line;
b) improving the quality of resulted products by:
    ❑ redesigning the car components and representative parts,
    ❑ implementing the computer-aided design of tools and devices and an efficient control of the manufacturing technology;
c) aligning the products and manufacturing technologies to the level of European standards and directives by:
    ❑ designing the products and manufacturing technologies using specialised software (ProEngineer, MASTER CAM, CATIA etc.),
    ❑ manufacturing products by using NC technologies,
    ❑ assimilating into manufacturing products conceived under EU standards, norms and directives,
    ❑ improving the qualifying level of human resources;
d) integrating the human resources by:
    ❑ workplace offers for young specialists (with an adequate training in CAD/CAM/CIM) for diminishing immigration,
    ❑ restructuring the technical personnel;
e) diversifying the industrial production of car components from S.C. COMPA S.A. Sibiu, by:
    ❑ increasing the capability to assimilate into manufacturing any type of hydraulic and mechanic steering system, brake system, cardan transmission, buffer etc.,
    ❑ flexibilising the manufacturing technologies at S.C. COMPA S.A. Sibiu.

The expected results are:
❑ realising the modules for the computer-aided parametrised design of car components and for the computer-aided design of manufacturing technologies for representative parts (shaft, piston, cardan cross, lever etc.),
❑ creating a graphical database for the parts and the tools and devices corresponding to the manufacturing technology,
❑ realising of interfaces between the two modules and the production system,
❑ the management of the informational system needed for the integration of the two modules in a production system,
❑ designing computer-aided manufacturing lines at car components manufacturing companies, S.C. COMPA S.A. Sibiu, respectively,
❑ expanding the two modules (CAD and CAPP) to other activities within an integrated production system (CAM, CAQ, CAS),
❑ implementing the specific modules at car components manufacturing plants from S.C. COMPA S.A. Sibiu and at other companies (S.C. DACIA S.A. from Pitesti, S.C. ROMAN S.A. and S.C. TRACTORUL S.A. Brasov),
❑ based on this principle, manufacturing technologies will be realised for typified part families: shafts, bushings, complex bodies (casings, motors, pump bodies and valves,

discs and gears, levers and parts with crossed axes). As such, the specific modules can be extended to the majority of products manufactured in Romania which contain these part types.

3. Research goals, activities and methodology

The general scientific goal of the research is the informatisation of the car components manufacturing. By means of the computer-aided designing and realising of modules for the parametric modelling – CAD, rehabilitation products with regard to the cost/quality ratio through the DFMA method and for the automated design of manufacturing technologies CAPP it is sought to restructure the technologies at S.C. COMPA S.A. Sibiu. The constructive-functional analysis of car components and of the assembly envisions the constructive optimisation by means of modern methods like the finite elements method (FEM). This method is used to determine the static and dynamic loading state of the components and of the assembly. The emphasis will be on studying representative parts of car components (shafts, bushings, gears, levers, parts with crossed axes) and on the dynamic behaviour of the assembly. It will be sought to obtain an optimal product from the point of view of performance, and the results will be incorporated into the assembly manufacturing (hydraulic and mechanical steering systems, brake systems, buffers etc) as part of an integrated manufacturing system. Based on specific manufacturing requirements, an optimal variant of processing technology will be realised, using the computer-aided constructive design module for the generic part (Rehg, Kraebber, 2003).

Following operational objectives are pursued as part of the research:
- constructive-functional optimisation of car components and their parts,
- computer-aided parametrised design module for products: mechanical and hydraulic steering systems, brake systems, dampers, cardan transmissions,
- computer-aided design of manufacturing technologies for the product modules,
- design of the integrated manufacturing system.

The research in this domain is part of the Romanian national interest area regarding the re-technologising of manufacturing processes and the automation of production for achieving a higher flexibility (Duse (2), 2003). The research program comprises following activities:
A.0 – Project management
A.1 – Analysis of the state of the art regarding computer-aided design of products and manufacturing technologies, car components manufacturing
A.2 – Theoretical and experimental research regarding the constructive-functional optimisation of car components
A.3 – Realising a module for computer-aided constructive design based on the parametrised modelling of the generic part
A.4 – Elaboration of alternative manufacturing technologies for car parts
A.5 – Realising the module for computer-aided technological design for parts from car components, based on the generic technology
A.6 – Conceiving and realising the database system
A.7 – Evaluation of the possibilities for automating the production lines
A.8 – Computer-aided modelling and simulation of the manufacturing lines
A.9 – Implementing the research results at S.C. COMPA S.A. Sibiu

Table 1. The adopted research methodology

Name and contents of the research phases	Scientific goal
Phase 1 *Assessment of the state-of-the-art in computer-aided design and manufacturing of car parts* Assessment of the state of the art in Romania and abroad with regard to car components design and of the corresponding technologies. Assessment of the car components (new parts and spare parts) design at S.C. COMPA S.A., with regard to the realising conditions, design elements, existence of databases, possibilities for a computer-aided (re)design.	❑ synthesis of constructive-functional solutions; ❑ assessment of the design and manufacturing conditions existing at S.C. COMPA S.A. Sibiu; ❑ documentation on the manufacturing and assembling.
Phase 2 *CAD module for the computer-aided parametrised design of parts* Unfolding of experimental researches on car components by means of the FEM method. Conceiving a module for computer-aided parametrised part design. Generation of the representative parts, creation of the generic part. Automated generation of the models and of the part drawings. Creation of the geometrical elements database. Realising of the interfaces needed for the integration of the CAD module.	❑ constructive-functional optimisation of car components; ❑ design of the representative parts for the car components; ❑ forming of the database system; ❑ training of the speciality personnel.
Phase 3 *CAPP module for the computer-aided technology design* Design and realisation of the module for the computer-aided design of technologies for car components manufacturing, realising of the generic technology. Computer-aided modelling and simulation of the flexible manufacturing systems for part families, creation of the necessary interfaces.	❑ computer-aided design of the manufacturing technologies by realising the generic technology; ❑ design of the manufacturing lines.
Phase 4 *Data management system* Design of a relational data management system for designed or redesigned parts. Implementation of the relational data management system in an ORACLE database for the part and tool name folder. Design and management of the informational system for the integration of the realised modules in an integrated production system.	❑ management of the data regarding the parts and tools needed for car components; ❑ development of an informational system for the modules' integration in a production system.
Phase 5 *Implementation of the research results in production* Implementation of the computer-aided design and manufacturing program in production. Realising of the the necessary documentation and training of the personnel in the domain of computer-aided design on NC-machine-tools.	❑ implementation and homologation of the research results in production; ❑ training of the personnel in the domain of computer-aided design on NC-machine-tools.

A.10 – Quality control of the manufacturing process
A.11.a – Training course in applying CAD/CAE methods
A.11.b – Training course for database management
A.11.c – Training course for manufacturing on NC machine-tools.
The research methodology is presented in table 1.

4. Conclusions

The implementation of the two computer-aided constructive and technological design modules in the car components production at S.C. COMPA S.A. Sibiu implies:
- organising the constructive and technological design compartments after the realised CAD and CAPP modules,
- realising (or updating) and computer-aided management of the part database,
- implementation of the computer-aided technologies,
- realising the interfaces and connecting the NC machine-tools to the computer-aided module,
- training of the specialised personnel.

The main beneficiaries of this system are the specialty companies – S.C. COMPA. S.A. Sibiu, S.C. METALCAR S.A. Sibiu and other car components or machine manufacturers from Romania (S.C. ROMAN S.A. and S.C. TRACTORUL S.A. from Brasov, S.C. DACIA S.A. from Pitesti).

The foreseen economic effects for the beneficiary companies are:
- a decrease of the effective time needed for the constructive and technological design,
- an increase in the economic efficiency due to the informatisation of the constructive and technological design activities,
- the manufacturing of any type of car components at lower prices than those on the Western European market.

It is worth mentioning the results achieved by the company S.C. COMPA S.A. Sibiu in 2004, following the implementation of the presented program:
- realisation of the hydraulic steering box for the new car model Dacia Logan by applying the presented CAD and CAPP technologies,
- re-technologisation and modernisation of the manufacturing lines by acquiring CNC technologies,
- development, together with the "Lucian Blaga" University of Sibiu, of an Educational CNC Centre for training in the domain of computer-aided manufacturing technologies,
- forming and development of an IT centre for the informatisation of the design and manufacturing of car components.

The possibilities for expanding the application area of the research results are related to the realising of a pilot system for the computer-aided constructive and technological design and the integration of a production system that enables the expansion of the applications to any machine-manufacturing company. The application can be adapted to the actual conditions from a company by updating the realised databases.

The perspectives of continuing the research are related to following directions:
- the increase of the capability to assimilate in computer-aided manufacturing any part from the machine manufacturing area,
- the flexibilisation of the manufacturing technologies in the central part of Romania (Sibiu and Braşov),
- the modelling and simulation of the technological manufacturing lines from the machine manufacturing industry,
- the integration of the informational systems in the manufacturing systems from the machine manufacturing industry.

References

1. Duse D.M.: *Computer Integrated Manufacturing – CIM – of Cardan Transmissions* (in Romanian). Publishing House of the University of Sibiu, Sibiu, Romania, 2003.
2. Duse D.M.: *Informatisation of design and manufacturing of parts from the machine manufacturing industry – car components*. Research grant LBUS – CNCSIS, Romania, 2003.
3. Rehg I.A., Kraebber H.W.: *Computer-Integrated Manufacturing*. Prentice Hall, New Jersey, Columbus Ohio, USA, 2003.

A TOOL FOR CHARACTERISING MANUFACTURING STRATEGIES

Harrison D.K., Macdonald M.

Abstract: This paper describes the development of a modelling tool tailored to the task of manufacturing strategy evaluation. This modelling tool has been formed through an in-depth review of popular manufacturing system modelling techniques, and is based on an integration of System Dynamics, Activity Based Costing and Business Planning. This principle has then been verified through an industrially based experimentation that has tested a prototype modelling tool in the context of practical manufacturing strategy formulation. The results are encouraging and justify further development for evaluating manufacturing strategies.

1. Introduction

For a manufacturing company to consistently realise success invariably requires that organisation to achieve congruence between internal manufacturing capabilities and external market and financial environments. Actions that determine manufacturing capabilities can be referred to as a company's manufacturing strategy, and there is a close association between the existence of an explicit manufacturing strategy within a business, and prosperity.

A manufacturing strategy can be explicitly formed by a number of methods, but a particularly successful approach is that of practising managers being guided through strategy formulation by a formal planning process [1]. Usually, such a process is a sequence of activities that secure recognition of a company's existing manufacturing capabilities, structure an expression of the associated financial and market environments, and stimulate the evolution of a sequence of actions to overcome any deficits that may exist.

Evaluation is a principal stage in strategy formulation [2]. Such evaluation can be made through judgement of individual personnel, refined through bargaining between a number of personnel, and supported by analytical methods [3]. One such analytical method is building a model that emulates the behaviour of a system, which can then be used to gain insight into and make predictions about that system. Hence, a model can be created of a manufacturing system, a number of modifications can be made to the model to reflect the strategy under consideration, and the ensuing model behaviour treated as a prediction of future manufacturing capabilities. Such a model can then be used to support judgement and bargaining between industrialists in the strategy evaluation process.

Modelling is often used in detailed design of manufacturing systems [4]. However, manufacturing strategy formulation is different from detailed manufacturing system design.

Recent research has investigated how modelling can be used as an aid in manufacturing strategy evaluation [5]. Consequently, several modelling techniques have been reviewed against this role [6]. Hence, it is now known that no existing modelling approaches individually fulfil the demands of the manufacturing strategy evaluation task. Therefore, this research has formed the principles of a suitable modelling tool by integrating three modelling techniques: System Dynamics (SD), Activity Based Costing (ABC), and Business Planning (BP). This research has then verified these principles through industrially based experimentation.

2. Background to research

This research is concerned with models that are an abstract representation and emulation of a real world object or system. A model is a specific representation used in addressing a particular problem, whereas a modelling technique is a mechanism for constructing a collection of models that share distinguishing characterised, rules and/or properties [7]. A modelling technique may be applied in practice using a computer based modelling tool, and a number of tools are seen to exist for various modelling techniques. An example is Discrete Event Simulation which is a distinctive mechanism of model construction, and WITNESS and SIMON [4] are both computer based modelling tools that can be used to apply this technique and modelling techniques such as these are the governing principles that determine the capabilities of modelling tools.

Earlier published work has described how models can assist in strategy formulation [5]. This has led to establishing a set of criteria that are desired of a modelling technique by practising managers faced with formulating manufacturing strategies [6]. Summarising, a modelling technique should be flexible so that a variety of changes across the structure and infrastructure of a manufacturing system can be evaluated. Likewise, a modelling technique should assess strategic market, finance, and manufacturing based performance measures. Assessment of the dynamic behaviour of a manufacturing system is also important, as is serviceability in terms of application cost, application time, accuracy, and credibility.

To assess the capabilities of popular and diverse modelling techniques, an industrially based set of experiments has been conducted [6]. This assessment has reviewed $IDEF_0$, Enterprise Modelling (EM), Discrete Event Simulation (DES), System Dynamics (SD), Queuing Theory (QT), Business Planning (BP), and Activity Based Costing (ABC). The results showed that none of these modelling techniques provided all the capabilities required for manufacturing strategy evaluation. $IDEF_0$ [8] and Enterprise Modelling [9] are strong mechanisms for illustrating the activities in a system, and their interactions at an instance in time. However, these techniques lack any mathematically predictive capabilities.

Discrete Event Simulation [10] can evaluate a wide variety of issues, to a low level of detail, with relatively high model accuracy, and good model credibility. System Dynamics [11] has slightly less flexibility but exhibits a relatively rapid model build rate and execution time. However, because of the inherent approximation of treating a product as a flow, the depth of model detail, credibility, and absolute level of accuracy are less than for DES. Queuing Theory [12] enables an accurate model to be constructed and executed in a fraction of the time taken for comparable DES and SD models. The predominant concerns are that performance measures are given for conditions of steady state system behaviour, and that the inherent approximations restrict flexibility.

Activity Based Costing [13] and Business Planning [14] are both financial modelling techniques. ABC is focused at providing product cost, and has the flexibility to assess a range of strategic developments in terms of this measure. Models can be built relatively quickly if they are restricted to a strategic snapshot of manufacturing system performance, rather than operational costing systems. BP provides a broad financial perspective of manufacturing developments, but flexibility and credibility are limited because the manufacturing system characteristics are only superficially considered.

Although individually these popular modelling techniques do not provide the desired capabilities, some combinations do appear favourable. Particularly attractive combinations are SD, BP and ABC, or DES, BP and ABC. In each case ABC and BP can provide a financial perspective of strategic manufacturing proposals, whilst either SD or DES can provide the required predictions about manufacturing system performance. This research has taken these combinations as a foundation, and established from this the principles of a modelling tool for manufacturing strategy evaluation.

3. Formation of strategy evaluation tool

A logical approach to forming the principles of a modelling tool is to choose the most favourable combination of existing modelling techniques, either SD, BP and ABC, or DES, BP and ABC. ABC and BP are common to both combinations as they provide product cost information and an overall financial perspective of manufacturing developments. A choice can therefore be made by distinguishing between the suitability of SD and DES on the basis of the work described in [6] and [15].

DES and SD are distinctly different modelling techniques. DES models attempt to emulate the time dependent behaviour of a real system by acting out the activities that occur in the real system. Invariably, the number of activities will be rationalised to improve modelling efficiency. Likewise, the model execution time is reduced by approximating actual activities to discrete event, time is then scaled down, and skipped forward according to an event list.

SD is a modelling technique where the progression of information through a system is considered as a continuous flow. A system is described in terms of 'resources' which flow through a variety of 'states' according to 'rates'. An example of a resource is a product family, various states could be work in progress stores, and rates could be a representation of machines. Once the content of a system is defined a mathematical expression is constructed to link rates with states. This expression is then executed at various time intervals and performance of the system is recorded. As summarised in the previous section, there are five prominent distinctions in the behaviour of these two techniques: model execution time, credibility, accuracy, build time and detail. A choice between SD and DES can be made by considering these factors in the context of manufacturing strategy evaluation.

SD is initially appealing because it can provide shorter model execution times than a comparable DES model. However, this factor is dependent on the computer hardware on which a model is installed, and it is generally accepted that the performance of such hardware is being continually improved. Hence, improvements in the performance of computer hardware will themselves reduce the time taken for model execution. Whilst the execution time of both modelling approaches will be reduced, DES will benefit most significantly because of the associated larger model execution time. Therefore, the long term developments in computer hardware mean that the significance of this factor is reduced.

An advantage of DES is that the technique provides more credible models than SD since DES enables the construction of detailed and relatively life like models that include more detail than comparable SD models. For example, a DES model can represent individual products in a queue before a machine, whereas a SD model will be limited to showing an accumulation of product flow. Therefore, higher credibility can be considered to be roughly proportional to an increase in model detail. On this basis, the significance of the credibility issue depends on the level of model detail necessary for manufacturing strategy evaluation. This issue of model detail arises again later.

Model build time and accuracy present a dilemma; whether to choose an approach that provides faster model building rate, but to a lower level of accuracy, or a considerably slower model build, but eventually a better value of accuracy. A faster model build rate will mean that alternative strategies can be evaluated in less time. As found in [15], SD consistently required less time to build a model; approximately 30% of that taken for DES. SD is less accurate than DES, the experimental results show that the maximum accuracy achieved with SD is 7% less than the maximum value of accuracy attained for DES. Unlike model build time, this distinction only becomes apparent in the latter stages of model construction, where the detail included in the DES is greater than the SD model. Therefore, and yet again, a distinction between DES and SD is dependent on the level of model detail appropriate to manufacturing strategy evaluation.

The accuracy, credibility and flexibility associated with DES modelling depends on building a relatively detailed model. The benefits of DES can only be realised if during strategy formulation sufficiently detailed information is available about the system being considered. If the depth of information is insufficient to form a distinction between DES and SD, then SD is favoured because of shorter model build times. Therefore, choosing between SD and DES is pivotal on the level of model detail required in manufacturing strategy evaluation.

The literature holds little explicit guidance on the depth of detail to which a system should be considered during manufacturing strategy formulation. Notable exceptions are authors such as Parnaby [16] who cautions against becoming too overwhelmed in detail. In contrast, Schroeder and Lahr [17] reveal that companies which undertake discussions at a general level, partly in the belief that strategy is not detailed in nature, often have a superficial outcome. Unfortunately, these authors give no guidance as to what they consider an acceptable level of detail to be.

Some resolution to this situation is given by Mintzberg [18] who sees that organisations often pursue what may be called umbrella strategies; the broad outlines are deliberate while the details are allowed to evolve within them. Likewise, Quinn [19] argues that a strategy must deal with some unknowable factors. Dealing with strategy in an umbrella fashion reduces the amount of detail considered by strategy formulators in practice, and hence strategy formulation is likely to be more manageable and the result easier to communicate within a company. On this basis, manufacturing strategy can be seen as a sequence and framework of deliberate actions that should be sufficiently robust to accommodate for some variance in detail, and indeed the concept of tactical actions can be encouraged to assist in the management of strategy implementation.

This view of strategy favours the SD approach to modelling. The level of detail at which modelling is appropriate to manufacturing strategy evaluation is considered here to be such that a distinction between SD and DES on accuracy, credibility and flexibility is not significant. Indeed, a benefit of SD may be that it prevents the strategy formulator becoming overwhelmed in unnecessary detail. Invariably, this will lead to a criticism of inflexibility with SD, but this criticism itself may well reflect a practitioner who is taking an inefficient approach to strategy formulation. It has been shown that SD is a more appropriate foundation on which to build a modelling tool for manufacturing strategy evaluation than DES. On this basis the complete principles of such a tool will be formed by integrating SD, ABC and BP.

4. Operation of modelling tool

It has been argued that the principles of a modelling tool for manufacturing strategy evaluation should be based on SD, ABC and BP. Using these principles, a model can be created of a manufacturing system modified to reflect the strategy under consideration, and the ensuing model behaviour treated as a prediction of future manufacturing capabilities. Such modelling should support judgement and bargaining between industrialists in the evaluation of a manufacturing strategy. Figure 1 shows how the chosen modelling techniques can be integrated to enable such evaluation.

The specific role of SD is to provide the measurements of manufacturing capability, such as delivery reliability, delivery lead time, volume flexibility [1], along with information about the internal state of the manufacturing system in operation. To provide this information, an SD model will need to be configured to represent the system being studied. This configuration will typically need to contain data about product families, manufacturing processes, information flows within the company, estimated product demands, etc. An input will also be necessary to describe the manufacturing strategy under evaluation. For example, how the flow of product families through a factory is intended to be transformed as a reorganisation of facilities or an investment in processes takes place.

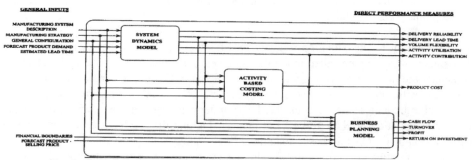

Figure 1. The structure of the proposed manufacturing strategy evaluation tool

The role of ABC in the manufacturing strategy evaluation tool is primarily to give a value of cost for each product family included in the SD model. It should be noted that this is a relatively restricted use of an ABC model, which can also be used to analyse such issues as overhead cost structures. This restrictive approach is necessary to avoid excessive model build times. Configuration of the ABC model will also require a description of the host company's manufacturing system and strategy. Principally, this description will be in terms of cost pools and cost drivers that are representative of the organisation being modelled. For example, a company may contain a large production control department, where the primary task is to operate a computer based manufacturing planning and control system. If alternative forms of manufacturing control strategies are under consideration, it may be pertinent to group production control costs as a cost pool, and to link this cost pool to manufacture through a cost driver of computer generated production control information. The manufacturing strategy may then, through changes to the cost driver, describe how the role of the production control department will be changed as different strategies are considered.

The role of BP in the manufacturing strategy evaluation tool is to directly provide the financial measures of profit, turnover, cash flow and return on investment. However, to perform these calculations, values for product cost, volume, and lead-time are required from the ABC and SD models, along with estimates for product selling price to enable calculations of profit. Descriptions of the manufacturing system and strategy will again be required to configure the BP model. The manufacturing system description may contain information about company assets, and the manufacturing strategy will indicate how such assets are intended to change as a strategy is implemented. It is now necessary to assess whether this modelling approach does indeed assist practising managers who are faced with the task of evaluating a manufacturing strategy.

5. The design of experimentation programme

Formal experimentation as reported by Baines [15], has been chosen to assess the utility of the manufacturing strategy evaluation tool described above. The principal question that experimentation must address is whether strategy evaluation will, in practice, benefit from the application of the modelling tool? The intended purpose of the strategy evaluation tool is to improve the understanding and prediction of practising strategy formulators. Therefore, experimentation must assess whether the understanding and prediction that formulators hold about a manufacturing strategy improves through the use of modelling.

On this basis, the chosen experimental approach has been to ask a group of industrial managers, who are themselves practising strategy formulators, to provide considered estimates on the effects of a specific set of strategic developments to a manufacturing system. Simultaneously and independently, a series of validated models have then been constructed that incorporate the same set of strategic manufacturing scenarios. Then, having made their estimates based on judgement and bargaining, the industrial managers have been exposed to the valid models and any changes that have occurred in their estimates measured as illustrated in Figure 2.

To operationally apply a modelling technique a modelling tool is required. This presents a dilemma; it is intended that the modelling principles are applied as a tailored computer modelling tool, yet investment in a purpose built software package is dependent on the results of this study. Therefore, integration is required between the modelling techniques at a minimum cost. The selected solution has been to manually integrate individual computer tools by instructing the model builders to apply the information flows described in Figure 1.

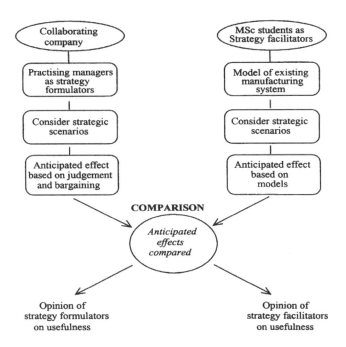

Figure 2. Measuring the utility of the modelling tool

Finally, there are many controls that have been incorporated into the experiment design. For example, the influence of the research team has been isolated as far as possible from the experimental results. This has been achieved by using a group of seven post-graduate students, studying computer aided engineering, to construct and validate the necessary models. The overall structure of the experimentation programme is summarised in Figure 3.

6. Execution of experimentation programme

The collaborating company used in this study is relatively large and typical of a multi-product batch manufacturing environment. The first activity was to identify a group of practising managers from this company who were currently acting as strategy formulators. Through discussions, five personnel became involved in this study, and they will be subsequently termed the strategy formulators.

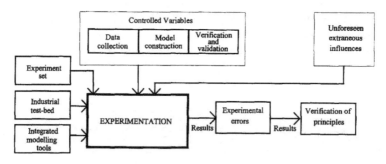

Figure 3. Overview of the experimentation programme

The model builders were then introduced to the study. This introduction commenced with dividing the group of post graduates into three teams. Training was then given on the individual modelling tools, data collection, model verification and validation methods. Finally, the teams were instructed to define the necessary data to construct a model of the manufacturing system at the collaborating company, and a mechanism to provide the information flows between models was set up. Throughout the study, the research team ensured the adoption of common assumptions, product families, etc. The research team also policed the model building process by requiring the model builders to give formal weekly presentations of their progress. A formal introduction of the strategy formulators to the model builders, and vice versa, was then performed at the collaborating company.

The model builders initially focused on developing models of the company's existing manufacturing system. Simultaneously, the research team carried out interviews with the strategy formulators to define a representative set of strategic development scenarios as shown in Table 1.

Once this set of nine scenarios had been developed the strategy formulators were asked to judge what the likely effect of each of these scenarios would be on manufacturing system performance. This was done on a one-to-one interview basis to capture each individual's judgement. These interviews themselves revealed that the strategy formulators were unwilling to give precise numerical estimates of the effect of each development. Hence, the process had to be simplified by requesting only the direction and order of magnitude of change. The strategy formulators were then questioned as a group as to whether any amendments were necessary to encourage bargaining between individuals. The outcome of this process was a set of scenarios and anticipated effects from the strategy formulators.

Table 1. Summary of manufacturing scenarios evaluated with the modelling tool

Decision represented	Overview of manufacturing scenario evaluated
Facilities	Storage of components on assembly line. Removal of storage areas, stocking policy changed.
Capacity	A linear increase in sales volume over 2 years, so that after this period overall production has increased by 40%.
Span of process	Subcontracting some sub-assembly operations so that components arrive ready for assembly onto Product A.
Processes	Integrating sub-assembly operations into main assembly. Setting up sub-assembly cells which are spurs off the main assembly line.
Human Resources	Double day shift in assembly and test areas.
Quality	100% inspection of some product's at delivery to ensure acceptable quality.
Control policy	Shop floor scheduling programme for Product A manufacture.
Suppliers	All Product A supplied 'right first time'.
New products	Product mix changing over 1 year so that Product C accounts for 40% of volume.

Notes:
1. A fuller description of these manufacturing scenarios and their role in experimentation - Baines [15].
2. The chosen manufacturing scenarios represent actions being considered by the collaborating company.
3. Explicit reference to actual products at the collaborating company has been avoided.

In the fifth week the teams of model builders were given the set of manufacturing scenarios to evaluate. Again, the research team only intervened to ensure that all assumptions were as far as possible common, and to police the transfer of information between the teams of facilitators.

On the tenth week of experimentation the work was drawn to a conclusion. A presentation was given to the strategy formulators, by the model builders, on the models constructed and the model based predictions about the effects of the strategic scenarios on manufacturing system performance. On conclusion of the presentation the research team gave the predictions made by the strategy formulators on the effects of the strategic scenarios. A debate was then encouraged between the two groups in order to ensure completeness and validity of results, and the research team observed how understanding and predictions of the groups changed. Subsequently, each team of model builders were interviewed and their opinions sought on problems encountered in the study. A similar meeting was then held with the strategy formulators and their opinions sought as to the usefulness of the models constructed.

7. Discussion of results

The views of the model builders were mainly concerned with problems that arose when applying the modelling tools which included carrying out data collection. When the necessary information about the company had been defined, it was either difficult to find, often in an inappropriate form, or totally inaccurate. For example, financial accounts information was not in a form that suited the models, or even suited the management of manufacture. Likewise, the standard times for component manufacture were both inaccurate and inconsistent. Indeed, much information on the company's manufacturing planning and control system was inaccurate and misleading. To overcome this particular problem the model builders had to resort to gathering the opinions of employees directly involved in product manufacture.

A second concern that arose was with training. Whilst basic training on each of the modelling techniques and modelling tools had been conducted, the research team purposely did not influence the form of the actual models constructed. However, the result was that the model builders complained that their progress was hampered by the lack of previous modelling experience. This was particularly the case with SD, as detailed information about applying this technique to manufacturing system modelling is scarce. Nevertheless, model build times were reasonable. This was felt to provide a case for some form of education and training guide for personnel faced with manufacturing system modelling in the context of strategy evaluation.

A third concern also arose as a consequence of the limited focused training given on the modelling tools. The issue was with the use of discrete elements in the SD model [11], which subsequently caused problems with the level of detail that a model contained. For example, a machining cell was modelled as individual machines, materials handling was modelled as an individual activity, etc. In this particular example it would have been adequate to consider the whole machining cell as a production rate. It appears that once the principle of being concerned with an aggregate flow is discounted, the amount of detail quickly proliferates.

A further concern of the model builders was that no attempt was made to integrate the three modelling tools at a computer level, rather formal communications were set up between the student groups. In practice it was found that in spite of these communications being very formal and policed by the research team, there were occasions where inconsistencies arose in the three models constructed. In one example, the ABC model contained departments within the factory that the SD model did not consider. In a second instance, the BP model included some financial information for a second manufacturing site that was outside the scope of the study. It can be argued that in many cases the inconsistencies between the models may not have occurred if the company's own personnel, with their inherent knowledge of the manufacturing system, had been used. However, as the model builders demonstrated, models have to be consistent to a low level of detail and this is impractical to ensure manually. Likewise, manual integration effectively triplicates the data entry requirements. Hence, a strong case was observed to exist for automated integration of the modelling tool.

Generally the strategy formulators felt that the models offered a greater insight into the behaviour of their company's manufacturing system. Two issues in particular were raised during execution of the experimentation programme. First, the strategy formulators felt that the models enhanced their understanding of their company's manufacturing system because a holistic view of developments could be seen. Furthermore, the models were observed to stimulate debate and discussion between the strategy formulators. However, they also felt that this understanding would have been enhanced if they themselves had constructed the models.

Second, the strategy formulators did appreciate the predictive capabilities of the models, and although the numerical values provided by the models were often treated with some scepticism, a major advance was considered to be made by such values existing. Prior to the study, the strategy formulators were unwilling to make such numerical predictions themselves. Again, the strategy formulators felt that the predictive power of the models would have improved if they had been directly responsible for model construction. For example, on one particular occasion the BP model predicted a considerable financial loss during one month of the company's operation. It transpired during the final presentation that the model did not account for a factory 'shut down' period. This validation error would probably have been avoided if the strategy formulators had been more closely involved in model building.

8. Concluding remarks

The principles of a modelling tool tailored to the task of manufacturing strategy evaluation has been formed by choosing the most suitable combination of modelling techniques. This choice has been based on the comprehensive knowledge of the capabilities of modelling techniques established in preceding work. On the basis of this analysis a combination of System Dynamics, Activity Based Costing and Business Planning have been chosen to form the modelling tool principles. By executing an experimentation programme at an industrial collaborating company, practising managers proved the usefulness of the modelling tool. Three important issues have been highlighted that should be considered for future work:

1. Develop an integrated computer tool on the basis of ABC, SD and BP.
2. Strategy formulators to be model builders. This should be reflected in the design of a modelling tool and creation of a supporting methodology.
3. The need for an education guide on how to approach data collection and model building for manufacturing system modelling when addressing strategy evaluation.

The research activity described in this paper combines with the previous contributions made by the authors. Preceding work has provided knowledge of strategy formulation processes and popular modelling techniques [5, 6]. The research described in this paper has then followed a complete sequence of conjecture, deduction and verification. It is however important to relinquish any claim that the modelling solution formed has been proven to be forever more suited, or always the best solution, to manufacturing strategy evaluation. The research process undertaken has not attempted such a proof, rather the modelling tool principles have been verified as contributing to manufacturing strategy formulation and are worthy of further development.

References

1. Platts K.W.: *Manufacturing Audit in the Process of Strategy Formulation*. Ph.D thesis, University of Cambridge, 1990.
2. Hofer C.W., Schendel D.: *Strategy Formulation: Analytical Concepts*. West. 1978.
3. Mintzberg H., Raisinghani D., Theoret A.: *The structure of unstructured decision processes*. Administrative Science Quarterly, vol.21, June, 1976, 246-275.
4. Carrie A.: *Simulation of Manufacturing Systems*. Wiley, 1988.
5. Baines T.S., Hamblin D.J., Harrison D.K.: *A unified classification of manufacturing strategies and design processes*. IEE Engineering Management Journal, 1993, vol. 3, no. 6, 281-286.
6. Baines T.S., Kay J.M., Hamblin D.J.: *Modelling in the evaluation of a manufacturing strategy*. The 1st Int. Conf of the European Operations Management Association, June, University of Cambridge, 1994, 213-218.
7. Banerjee S., Basu, A.: *Model Type Selection in an Integrated DSS Environment*, Decision Support Systems, vol. 9, 1993, 75-89.
8. Brovoco R.R., Yadav S.B.: *A Methodology to Model the Functional Structure of an Organisation*. Computers in Industry, 1985, Oct., no. 6, 345-361.
9. Johansson H.J., McHugh P., Pendlebury A.J., Wheeler W.A.: *Business Process Reengineering*. John Wiley & Sons, 1993.
10. Law A.M., Kelton W.D.: *Simulation Modeling and Analysis*. McGraw-Hill, 1991.
11. Wolstenholme E.F.: *System Enquiry: A System Dynamics Approach*. John Wiley, 1990.
12. Suri R., Diehl G.W.: *A Precursor to simulation for complex manufacturing systems*. Proceedings of the Winter Simulation Conference, 1985, 411-420.
13. Gray P.: *Using the Interactive Financial Planning*. 1984, December, 286-292.
14. Kaplan R.S.: *Yesterday's Accounting Undermines Production*. Harvard Business Review, 1984, July/August, 95-101.
15. Baines T.S.: *Modelling in the evaluation of a manufacturing strategy*. Ph.D. Thesis, Cranfield Uni., 1995.
16. Parnaby J.: *The Design of Competitive Manufacturing Systems*. Int. Journal of Technology Management, 1986, vol.1, no 3/4, 385-396.
17. Schroeder R.G., Lahr T.N.: *Development of Manufacturing Strategy: A Proven Process*. Proc of the Joint Industry University Conference on Manufacturing Strategy, Michigan, 1990, 8th-9th Jan., 3-14.
18. Mintzberg H.: *The Rise and Fall of Strategic Planning*. Prentice Hall, England, 1994.
19. Quinn J.B.: *Strategic Change: Logical Incrementalism*. Sloan Management Review, 1978, Fall, 7-21.

CRITICAL CHAIN SCHEDULING: APPLICATION TO PRODUCTION JOB ORDER MANAGEMENT

Morandi V., Aggogeri F., Baroni S.

Abstract: Companies, that supply custom-made products or services to their customers, have to deal with typical project management problems such as the difficulty in respecting due dates, budgets and customer requirements. So the implementation of project management techniques in this field covers a reduction in the project duration and an increase in the reliability of time estimation and project development. Among all the techniques, the "Critical Chain" algorithm, that was developed at the end of the last century by Goldratt, stands out because it provides a structured, logical approach to project management problems that seems to be more applicable than others to the production job order scheduling. This paper describes its application to the job order management of a special machine tool manufacturer showing its merits and pitfalls.

1. Introduction

In the last years product concept development teams which focused upon the market found themselves in a virtuous cycle which tended to improve performance. This is an example of the very positive association of market orientation and business performance that many economists found in their researches. As a consequence of these findings, customer driven manufacturing (CDM) emerges to meet the latest trends such as the increase of variety of products, small batches sizes, shorter lead times and globally oriented production. In particular, many companies operating in mechanical field began to adopt the One-of-a-Kind Production (OKP) as production philosophy. The OKP, an extreme case of CDM, means a particular production method that converts the customer's development ideas or requirements into a product by a "once" successful approach under constraints of a critical delivery date, cost and quality. The characteristics of OKP are: High Customisation (as a consequence the products change frequently); Continuous customer influence through the production process (product specifications may change partially after the production has started); Prototype-based evolutionary and concurrent approach of product development and production; Capacity-oriented production resources and different organisation for producing different products; Adaptive production planning and control (production often starts without finalised product design and consequently without the complete determination of the production process)

Moreover companies making this kind of products, that may be variations of a basic design or a totally new design, are essentially project driven and are typically involved in several concurrent projects at any one time. They may therefore be seen as project and systems integrators. In particular, planning and scheduling in OKP system means the co-ordination of a number of activities that have to be processed in one of several modes by selected members of the principle company functions such as sale, design, manufacturing, control. So the management of these operations implies typical problems of project management.

For this reason, recently it can be observed the development of the application of project management techniques to the OKP companies.

This paper presents an analysis of a scheduling approach, the Critical Chain, that buds from the combination of the Theory of Constraints (TOC) and project management philosophy. The Authors applied this technique to an industrial case evaluating its potentialities in managing the last and more complex stage of a job order in a special machine tool producer.

2. Project management problems

Every project is made up of a set of tasks each of which has to be allocated to a given resource, has a duration and a possibly completion cost and may have a set of other tasks without which it cannot start. The tasks may be assigned to a number of different internal resources or to several sub-contractors and component suppliers. As the number of tasks, sites, contractors and suppliers increase, the management of the project involves the planning, organisation and control of a even more large number of complex factors, activities and their interrelations.

The judgement criteria of the project management effectiveness are relevant to quality, time and costs. Although the attention to this aspects, often projects are late, don't respect the budget and quality specification. To solve this problems, in the last years, many methodologies-techniques emerge (i.e W.B.S., Risk Management, Gannt Chart, P.E.R.T., C.P.M., C.S.C.S.). Unfortunately all of them have critical aspects that don't let achieve the best project management, so their application isn't enough to guarantee the fulfilment of project specification respecting the due date.

Project scheduling and coordination imply the management of uncertainty and changes related to project parameters. In particular, in the OKP system, uncertainty means that when a production of a product has to start some production resources, operation sequences and part routings cannot be determined while changes mean that after the production has started, changes and modifications to production resources, operation sequences and part routings may be required. These changes result from that previous uncertainty become certain, modifications to the products are required by the customer, new product are scheduled to be produced and the actual production may not keep up with the production schedule.

The uncertainty, caused by many factors (i.e. lack of a clear specification of what is required, novelty lack or experience of this particular activity, limited analysis of the process involved in the activity and possible occurrence of particular events or conditions, etc.) can be reduced but not removed. So the way the uncertainty is managed in projects is the core of improvement of project performance that depends on two conflicting aspects: the increasingly important need for speed in project delivery and the equally important need for reliability in delivering the project as promised.

3. Critical chain scheduling

Developed through the application of the Theory of Constraints (TOC) to the subject of projects, Critical Chain is a common-sense combination of techniques and philosophy that can enable dramatic, rapid and ongoing improvements by helping identify the core problems in a project system and by providing an algorithm that gives a systematic approach in scheduling.

As in traditional TOC, in Critical Chain the key concept is the constraint, or the "factor" that restricts the ability of the system to do work and thereby earn revenue. To improve the project management it is fundamental that the focus of project manager should be the constraint. As a consequence, it is the protagonist of the five steps of this methodology that suggests to identify the constraint(s), decide how to exploit it (them), subordinate everything else to the previous decision elevate the system constraint and to avoid the inertia in case the constraint has been broken. The project manager should also pay a particular attention to the variability management that, according to Critical Chain approach, is a main cause of scheduling failure. To solve the matters due to a wrong variability management, Critical Chain scheduling rests upon a different management of activities according to their influence on the project lead time, it distinguishes between critical and non-critical activities that is to say between tasks that could delay the end of project and tasks that haven't a great impact on it. In particular, the objective is to identify the critical chain, a sequence of critical activities that determines the project lead time, taking into account the interdependencies between activities and the availability of the resources. This identification allows to highlight the priorities in management decision. Furthermore it has to be underlined that the critical chain normally is also the constraint of the project in single-project environment. So looking up for the critical chain means to make the first step of the methodology.

Another peculiarity of Critical Chain is that it attempts to account for certain typical human behaviour patterns during project planning and execution. For example, it takes into account that people do considerable provision for contingencies when estimating activity duration. Rational people responsible for project activities therefore attempt to make commitments that they could meet with a level of certainty so they add "safety" time to the amount of working time. Critical Chain don't challenge the habit to build too much contingency into schedules but rather that it is built in at the wrong place, to be more precise, at activity level rather than at the project level. This approach applies the principle of aggregation to project schedule risks: contingency reserves for individual activities are reduced so that activity durations are realistic but challenging. The provision for contingencies that are removed from the individual activity durations are replaced by a contingency reserve or "buffer" at project level. As a result of the effect of aggregation this buffer is smaller than the sum of the individual reserves. Thus project duration is reduced and the project control is simplified. In fact, a delay could be seen only monitoring the use of project buffer. The most serious implication of aggregating reserves is that all due dates on individual activities and sub-project have to be eliminated. In Critical Chain scheduling only the project manager makes a commitment on the project delivery date, all levels below him limit themselves to make estimates and communicate expectations. So "Critical Chain Scheduling" suggests the shifting of focus from assuring the achievement of task estimates and intermediate milestones to assuring the only date that matters- the final promised due date of a project.

Moreover, the lack of intermediate due dates helps in hindering the negative effects of the "Parkinson Law" and of the "Student Syndrome" that contribute in wasting the safety time.

So this approach tries to protect the project from the attitude that work expands so as to fill the time available for its completion and from the inclination to postpone the activity start until it becomes urgent.

Critical Chain introduces several new measures to manage the uncertainty and, as consequence, to respect the final due date complying with the budget and the customer requirements (for a widening study of this approach see [2]).

4. The case study

4.1. The objective

The main purpose of this study is an evaluation of the Critical Chain applicability to industrial context in order to understand not only the advantages from it but also the troubles in putting on this methodology. In particular, this application was used to verify the following hypothesis:
- H1: Critical Chain Scheduling helps in reducing project lead time,
- H2: Critical Chain Scheduling allows to exploit the positive fluctuation (in other words, it undoes the effect of Parkinson Law and it exploits the earlier finish of the activities),
- H3: Critical Chain simplifies the project control.

4.2. The project context

The Authors applied the Critical Chain Scheduling to the job order management of an Italian machine tool manufacturer who produces special machine tools that are marketed all over the world. The range of its production, that belongs to the OKP typology, meets the requirements of different sectors such as the automotive, the transport, the energy industry in which it has gained a leadership position. The strength point of this company is the ability to design and build machine tools according to the customer requirements while the weak point is the high time to market that restricts the number of job order and that implies a large WIP and its consequences. Moreover, it must deal with adaptive production planning and control, frequent rescheduling and typical project management problems as late completion, over spending, the need to cut specifications. For these reasons, the company would like to experiment a new management method in only a phase of the job order in order to verify the potential improvement. Between all the job order phases (design, production, assembly, setting up), which management could be considered single project management for the complexity of their activities network, the Authors chose the last one because of its high level of variability and uncertainty. In fact, the setting up of different machine can involve enormously different activities and it cannot be completely defined in planning stage.

The Authors, therefore, focused on the setting up of a machine tool for locomotive engines that, after the machine assemblage, implies CNN initiation, setting up of different machine groups, software inspection, geometric inspection, machine testing and production test. It should be noted that the activities that made up these stages sometimes could be overlapped and that management difficulties depend also on the need of coordination between different company departments.

4.3. The planning phase: Critical Chain Scheduling application

At first, the Authors required separate activities planning to the different company areas involved in the setting up stage. In particular, every functional foreman was asked to identify the activities network of its area starting form the last one (the production test at the customer) and proceeding backwards until the machine tool assemblage. Then, in order to identify the global activities network, the different plans were integrated in a general scheme thanks to the collaboration of the several areas. According to the consequent plan, the setting up was composed by 44 activities that required the contribution of mechanical, electronic, software and quality functions. Besides the operations description, in this outline, the activities precedence relationships have been represented.

During this phase, some contradictions caused by a separate planning approach emerged: some activities weren't univocally identified that is to say that different areas named the same activity in different ways or pointed it out with different detail levels; the activities precedence of different plans didn't correspond. As a consequence, the first consideration of this case study is related to the importance of an integrated approach in the activities planning.

Starting from the integrated plan, the operators were asked to estimate the time duration of the activities. These estimates were analysed and, as the critical chain suggests, were reduced, with the exception for automated activities. This reduction is a consequence of the human disposition to overestimate time duration in order to have safety time.

After this first step of the critical chain scheduling algorithm, the Authors implemented all the other steps following the scheme represented in figure 1. In particular, in the second step, analysing the resources contention, the machine (object of the setting up) availability proved to be a critical factor while the availability of the factory equipments used in the activities was not a problem. So the Authors did a distinction between activities that required the machine availability and activities that could be processed without the machine presence. With these distinctions and durations determined, the next task that had to be tackled was the identification of the key activities that form the Critical Chain. On the contrary of the traditional critical path, the critical chain is defined considering the resources contention. In this case it was composed by a sequence of 18 activities that required the machine tool availability and that determined the overall project duration (more or less 60 working days).

In order to provide protection against statistical variation, in the following steps, the authors put in a project buffer of 26 days at the end of the Critical Chain, feeding buffers whenever a non critical chain activity joined the critical chain and resources buffers whenever a resource had a job on the critical chain and the previous critical chain activity was done by a different resource. The project buffer, was a reserve of safety time that is necessary because the time estimations of the activities was reduced in the first step and as a consequence they didn't considered the possibility of unexpected events. The feeding buffers, that are reserves of safety time, and resources buffers, that can be treated as just a wake-up call which alerts resources to be ready to work on their critical chain tasks when needed, were inserted in order to protect the critical chain from the delay of no-critical activities and from the resources unavailability.

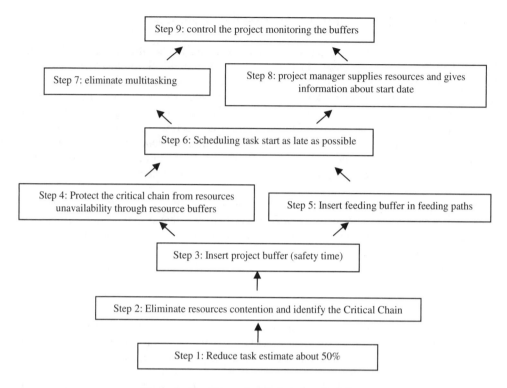

Figure 1. Algorithm of critical chain scheduling

After these phases, in accordance with the due date and the principle of scheduling as late as possible, building a Gantt diagram, the Authors defined the start date of the setting up and of the single activities of it in order to respect the established delivery date. When the setting up began, the Authors monitored it and intervened in it modifying the activities scheduling when it was necessary, that is when the risk of delay in due date is high or when the milestones (customer required that particular tasks would finish by a particular date) could be not respected.

4.4. The control phase: the buffer management

The monitoring phase is as important as the planning phase because it allows to single scheduling disruptions out and to embark corrective actions in order to respect the due date. At this point, a clarification is needed: the Critical Chain approach has, among other things, as purpose the reduction of the number of rescheduling during the job order management. Therefore a planning activity is considered good if its result is the definitive activities scheduling that is when it is "robust" in respect to the uncertain events. Although this consideration, sometimes the changes become fundamental for the goal fulfilment so identifying these special instances, through the job order monitoring, is extremely important.

Against the other approaches, in Critical Chain, controlling the work is simple because it means monitoring the buffers consumption focusing mainly on the project buffer that is meant to be sufficient to cover all the uncertain events which may be encountered.

Moreover, for pragmatic management reasons, the project buffer is divided into three parts: the safety zone (green zone), in which the project can be permitted to go without any problems, the planning zone (yellow zone), that arises the need to draw some plans up to put the project back on schedule, and the implementation zone (red zone), in which begins the implementation of the previous plans. The aim of this method is to have a total contingency with warning, get ready and do something.

To implement this procedure, the Authors asked the operators to make report of their progress work every two working days. The information of these reports was useful not only to make decisions during the job order management (to control the project buffer consumption) but also to evaluate the influence of Critical Chain at the end of the setting up. In the control stage, it could be observed that the beginning of the setting up activities has been postponed because of a delay of the other job order phases. In particular, according to the planning these activities should start at the 4th August while in reality they began the 10th September with a delay of 20 working days. So, already at the beginning the project buffer turned out to be wasted and it could be highlighted a penetration in the "red sector" of it. In other words, the piled delay had wasted all the planned safety time. Moreover, it removed also the usefulness of feeding buffers. As a consequence, the need to take corrective actions emerged. Thinking about a solution that could permit to respect the established delivery, the opportunity to work on two shifts seemed to be the best one because it permitted to make parallel activities in the same day. Although the great delay, the Authors decided to implement this solution not at the beginning but only when the project buffer seemed to be more compromised (20 working days after the setting up start). This is only an example of the acting way followed during this phase and that allows to respect the delivery date, although the great initial delay. This good result could be achieved also thanks to the exploitation of positive fluctuations that are the completion in advance of some activities.

It should be underlined that the application of Critical Chain Buffer Management, simplified enormously the control activity because it requires only to focus on just few points.

4.6. The approach evaluation

Critical Chain satisfies the company requirements such as the reduction of the time to market and facilities in project management and control.

First, the application of Critical Chain Scheduling reduces the lead time of setting up. This is a consequence of the aggregation principle of contingency and the reduction of time estimates, that is based upon the assumption that people usually make pessimistic forecasting. The fact that the due date has been respected although this reduction and that the effective activity duration was lower than operators estimate (generally, the ratio between effective duration and estimate was less than 0,5) shows the effectiveness of the Critical Chain theoretical foundations. So, through this case study, the first hypothesis has been verified. Moreover, this experience allows the Authors to experiment that the elimination of all due dates on individual activities and subprojects, that is a Critical Chain principle, helps to exploit the positive fluctuation consenting the trade-off between earlier and later activities finish.

Regarding the project control, in the operative context of this company, because of the intrinsic complexity of its production, a steady management and control of resources (human resources and materials) interdependencies seems to be fundamental. The fact that Critical Chain focuses on global optimum (due date) instead of local ones, enabled a systematic job order management, that allowed to point out beforehand potential delays in the delivery date and to have a long time to attempt corrective actions (double shifts, activities order permutation, etc.). The protection at the end of the project (project buffer) and in no-critical branches (feeding buffer) was enough to absorb the ordinary variability of activities. The buffers demonstrated to be very useful not only to protect the project from delays but also to make the job order control easier. Therefore, it seems to be evident that the third hypothesis was verified.

The main trouble in the implementation of Critical Chain is people resistance to change. In fact the Critical Chain approach requires to modify the way people think about project scheduling and it implies a paradigm that is radically different from the prevailing one. So it is very difficult to acquire necessary cooperation and collaboration.

5. Conclusion

The application of Critical Chain principles to the job order management shows that this method fits with the industrial context, especially in OKP companies, where it implies a lot of advantages in the scheduling of single job order. The main ones are the reduction of project lead time and the simplification of project control.
In the future, the Authors will be engaged in the evaluation of the validity of Critical Chain approach in the management of parallel job order.

References

1. Burchill G., Fine C.: *Time versus market orientation in product concept development: empirically based theory generation.* Management Science (43),465-478.
2. Goldratt E.M.: *Critical Chain.* Great Barrington (MA), The North River Press, 1997.
3. Newbold R.C.: *Project Management in the Fast Lane.* Boca Raton (FL), St Lucie Press, 2000.
4. Rand G.K.: *Critical Chain: the theory of constraints applied to project management.* International Journal of Project Management (18), 2000, 173-177.
5. Steyn H.: *An investigation into fundamentals of critical chain project scheduling.* International Journal of Project Management (19), 2000, 363-369.
6. Wortmann J.C.: *Production management for One-of-a-Kind.* Computer in Industry (19), 1992, 79-88.

APPLICATION OF SIMULATION AND ARTIFICIAL NEURAL NETWORK TO THE PROBLEM OF SCHEDULING OF FLEXIBLE MANUFACTURING SYSTEM

Noor S., Khan M. K., Hussain I., Ullah I.

Abstract: A hypothetical Flexible Manufacturing System (FMS) consisting of four CNC machines, one loading machine, two Automated Guided Vehicles (AGV) with loop layout is simulated with ARENA software package. Various performance measures of the system such as Work In Progress (WIP), utilization of machining centres, average part cycle time, and numbers of parts leaving the FMS are measured under different scheduling scenarios, eight different dispatching rules, for different part types and mixed mode of product-sequence. A series of one hidden layer Artificial Neural Networks (ANNs) 11-S1-2, and two hidden layers neural networks 11-S1-S2-2, with varying number of neurons in the hidden layers are trained and tested for the selection of a suitable scheduling scenario for the desired performance measures. A maximum of 58% testing performance of product-mix sequence alternative that matched the expected output data was achieved. Similarly 36% and 23.33% was recorded for dispatching rules and scheduling scenarios, respectively. This low testing performance of the ANNs for scheduling scenario is due to possible trapping of the ANNs into local minima and thus cannot be recommended alone for implementation. Further research is being conducted to improve this performance.

1. Introduction

The rapid growth in demand of quantity, quality and variety of products has resulted in the design of Flexible Manufacturing System (FMS), which is considered as one of the key ways to achieve organizational flexibility. FMS combines the efficiency of high production line and the flexibility of a job shop to best suit the batch production of mid-volume and mid-variety of products (Chan and Chan, 2004).

Scheduling of FMS is a complex problem, known to be NP-Hard combinatorial problem (Meeran, 2003., Jain and Meeran, 1998., Hitomi, 1996). Analytical methods have been used as the traditional approach for the solution of scheduling problems which give exact solutions for static and simplified problems. However these fail to deliver for real time and dynamic scheduling of manufacturing system, especially FMS. The dynamic nature of such systems demands a scheduling procedure, which is reactive and sensitive to the system's status. It is not yet known whether policies and procedures designed to schedule and control traditional manufacturing processes are appropriate for an FMS.

An efficient scheduling system, which is very important for FMS environment, depends on the scheduling scenario (Chryssolouris et al, 1990). A scheduling scenario is a combination of job sequences, mix routing alternatives and dispatching rules for machining centres, AGVs, robots and other resources. An efficient scheduling system would be that which can select the best possible combination of job sequences and routing alternatives and dispatching rules for resources for realization of the best performance measures.

In this paper, the possibility of ANN-based scheduling system (for multi-objectives) in conjunction with simulation software ARENA is explored, which can suggest the best possible scheduling scenario for the desired performance measures of the FMS.

2. Simulation of FMS

Simulation has been widely used for the design of operation and control strategies of manufacturing systems (Chan and Chan, 2004). Simulation models are useful to evaluate various operational strategies through computer based experiments in a realistic environment. This type of modeling is a bridge to the artificial intelligence approach. Simulation can also handle stochastic data which cannot be handled in analytical models without major simplifications.

Consider a hypothetical FMS consisting of loading/unloading unit (L/UL), two AGVs and four CNC machining centres (MC), as shown in Figure 1. The working of this FMS was simulated under a variety of scheduling scenarios to determine the influence of scheduling scenario on the performance of the FMS. A scheduling scenario is defined as the combination of a product-mix sequencing alternative and a dispatching rule. The Table 1 lists the three product-mix sequencing alternatives used and Figure 2 shows how these are combined with eight dispatching rules to form 24 scheduling scenarios. Eleven different performance measures were calculated as a result of the simulation, as shown in Figure 2.

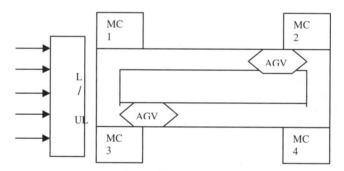

Figure 1. Configuration of the hypothetical FMS

This FMS was simulated in ARENA software using *Arrive, Assign, Transport, Server, Sequence, Distance, Simulate, Statistics* modules. Exponential distribution was used to randomly generate the time between part arrival, and normal distribution for cycle time, loading time and due date (Kelton et al, 1998). Priority 2, 3, 1, 4, 5 and part ratios 0.3, 0.2, 0.1, 0.1, 0.3 were assumed for part types 1, 2, 3, 4 and 5 respectively. A *Server* module was used for loading unit with a capacity of one unit. Four server modules were used for four machining centers (MC) with a general processing time attribute and the processing time for each part on each machining centre was given in *Sequence* module from Table 1. The parts are carried through free transporters (two AGVs), and distances for AGVs to different possible points are given in *Distance* module.

Table 1. Operation sequences (Processing Time) for part types of different product-mix sequences alternatives

Product-Mix Sequences(Routes)	Part Type	Operation Sequences(Processing Time)					
		1	2	3	4	5	6
Alternative 1	1	Loading	MC3(20)	MC1(10)	MC4(15)	MC2(22)	Unloading
	2	Loading	MC2(10)	MC3(8)	MC4(10)	Unloading	--------
	3	Loading	MC1(13)	MC2(14)	MC4(20)	Unloading	--------
	4	Loading	MC3(15)	MC4(10)	MC1(16)	Unloading	--------
	5	Loading	MC1(20)	MC2(12)	MC3(10)	Unloading	--------
Alternative 2	1	Loading	MC1(10)	MC2(22)	MC4(15)	MC3(20)	Unloading
	2	Loading	MC1(10)	MC4(10)	MC2(8)	Unloading	--------
	3	Loading	MC3(15)	MC4(12)	MC2(20)	Unloading	--------
	4	Loading	MC3(10)	MC1(16)	MC4(15)	Unloading	--------
	5	Loading	MC1(20)	MC2(12)	MC3(10)	Unloading	--------
Alternative 3	1	Loading	MC2(22)	MC1(10)	MC3(20)	Unloading	--------
	2	Loading	MC1(10)	MC4(10)	MC3(8)	Unloading	--------
	3	Loading	MC2(14)	MC4(13)	MC3(20)	Unloading	--------
	4	Loading	MC1(16)	MC2(12)	MC4(13)	Unloading	--------
	5	Loading	MC1(20)	MC2(12)	MC3(10)	Unloading	--------

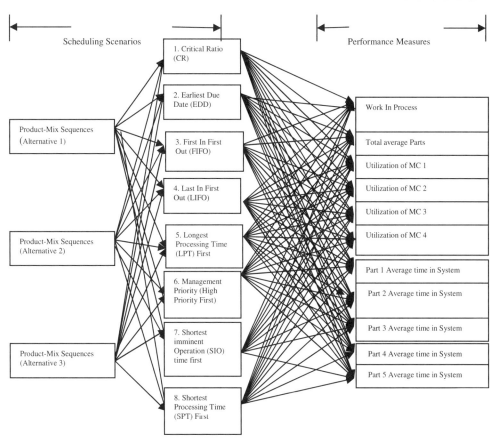

Figure 2. Performance measures under different scheduling scenarios

3. Data generation

The *Simulate* module governs the whole simulation and running condition. A shift of 1000 minutes is defined in this module. The *Statistics* module is used for gathering all the data of performance measures shown in Figure 2. The data of performance measures is extracted from simulation run under each scheduling scenario and scaled according to *(X-Min)/ (Max – Min)* between 0 and 1. Scheduling scenarios are coded with two digits where the first digit represents the product-mix alternative routes and the second represents the dispatching rule. For example in scheduling scenario 71 , 1 represents product-mix sequences alternative 1 and 7 represents dispatching rule , i.e. shortest imminent operation time first, (SIO). Each scheduling scenario was replicated 30 times to cover possible variation of data in the same class and thus 720 data sets of input-output pairs were obtained. The results from the simulation exercise provided the data to the ANN model.

4. Architecture of the ANN model

The eleven performance measures form the input and the two components of a scheduling scenario-dispatching rule and product-mix sequencing alternative form the output of the ANN. Accordingly 11 input neurons and 2 output neurons were used in the feed-forward back-propagation error neural network (Hagan et al,`2002) as shown in Figure 3. To search for the optimal ANN architecture, 55 different neural networks including one hidden layer and two hidden layers networks with varying number of neurons from 6 to 40 and 6 to 25 are used respectively. The other parameters i.e. training function (TrainLm),Transfer function (Logsig), adaptation learning function (TrainGDM) and transfer functions for hidden layers (logsig) and output layer (purelin), performance function (MSE) remain fixed. A typical ANN topology is shown in Figure 3.

5. Training and testing of ANN

Both input and output data of simulation model are used to train the ANNs as per training and testing scheme shown in Figure 4. The data is divided into two groups, training and testing data. 120 data sets (approx 17%of the total data) are selected randomly for testing and the remaining data is used for training. Each of the above networks is initiated, trained and tested 20 times in search of the best weights of the architecture of the ANN model. Figure 4 shows the training and testing scheme of ANN model.

6. Results and discussion

Given a set of performance measures values, the ANN predicts the scheduling scenarios that will produce the performance. During training, the objective is to minimise the Mean Square Error (MSE), which is the square of the difference of ANN output and the target. After training, the performance of the ANN was tested by calculating the percentage of cases in which the ANN correctly predicted the scheduling scenario, dispatching rule and product-mix sequences alternative. For example, a set of 20 data, as shown in Table 2, was simulated (tested) on a trained ANN. In only three out of 20 cases, the ANN matched (√`) the expected output for scheduling scenario Thus the testing performance for scheduling scenario (TPSS) for this ANN is 15%. Similarly, 7 and 10 cases (marked '• & o ') were correctly predicted for dispatching rule and product-mix sequences alternative, thus the testing performance for dispatching rule (TPDR) and Testing performance for product-mix sequences alternative (TPPMA) are 35% and 50% respectively.

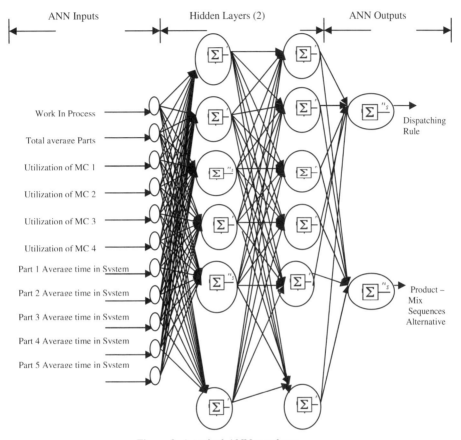

Figure 3. A typical ANN topology

A series of ANNs, 11-S1-2 (one hidden layer), 11-S1-S2-2 (two hidden layers) were trained for the 600 data sets and tested for 120 data sets. The results are shown in Figures 5 to 12. Figures 5 to 8 show the results of 11-S1-2 networks (one hidden layer S1) for MSE, TPSS, TPDR and TPPMA respectively. Similarly, Figures 9 to 12 show the MSE, TPSS, TPDR and TPPMA of 11-S1-S2-2 (two hidden layers S1, S2) networks.

Figure 5 shows that there is a decreasing trend of MSE, better training performance with the increasing number of neurons in the hidden layer S1 but this better training performance does not reflect in the testing of the networks (Figure 6). A maximum of 20 % TPSS has been achieved for the series of 11-S1-2 networks. Where as a maximum of 32% and 57% were recorded for TPDR and TPPMA, shown in Figures 7 & 8 respectively. Similarly Figures 9 and 10-12 explain the training and testing performance of 11-S1-S2-2 networks where a maximum of 23.33 %, 36% and 58% have been achieved for TPSS, TPDR and TPPMA respectively.

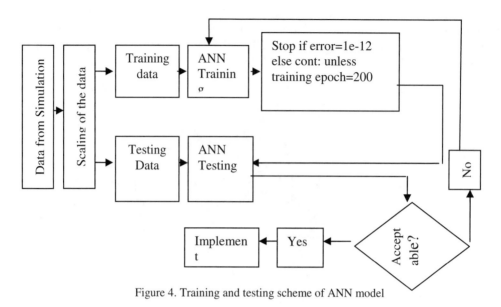

Figure 4. Training and testing scheme of ANN model

Table 2. An Example for calculation of Testing Performances of the trained ANNs

* dispatching rule (DR), ** production mix sequences alternate

No of ANN output (scheduling scenario) matching the expected output of the testing data (√) SS=3

No of ANN output (dispatching rule) matching the expected output of the testing data (•) DR=7

No of ANN output (production mix sequences alternative) matching the expected output of the testing data (O), PMSA=10

Total Testing Data=20

Testing performance for scheduling scenario (TPSS) of the trained ANN=3*100/20=15%

Testing performance for dispatching rule (TP DR) of the trained ANN=7*100/20=35%

Testing performance for production mix sequences alternative (TPPMSA) of the trained ANN=10*100/20=50%

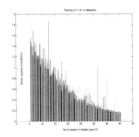

Figure 5. MSE for 11-S1-2 networks

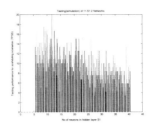

Figure 6. TPSS for 11-S1-2 networks

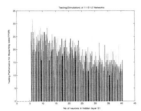

Figure 7. TPDR for 11-S1-2 networks

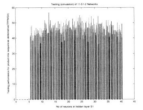

Figure 8. TPPMSA for 11-S1-2 networks

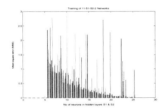

Figure 9: MSE for 11-S1-S2-2 networks

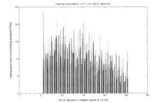

Figure 10. TPSS for 11-S1-S2-2 networks

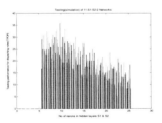

Figure 11. TPDR for 11-S1-S2-2 networks

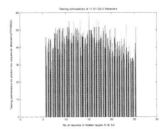

Figure 12. TPPMSA for 11-S1-S2-2 networks

In both series of networks, the testing performance needs further improvements and cannot be recommended alone for implementation. This unsatisfactory performance of the networks can be attributed either to lack of exploration of optimized architecture or vulnerability of back error propagation to trapping into local minima (Feng et al, 2003, Negnevitsky, 2002).

To improve the testing performance of the networks so that it can be applied to a real scheduling system, a suitable procedure is needed to take the network training out of the local minima and try to find the global minimum. This can be possibly achieved through optimization of the network layers weights with genetic algorithm which is well known for solving difficult optimization problems (Negnevitsky, 2002). This kind of research is being presently investigated.

7. Conclusions and future work

ANN can be easily applied to multi-output problems and thus can be very effective tools for multi-objective scheduling problems in complex manufacturing systems such as FMS. However, ANN has an inherent problem of being trapped into local minima which needs to be addressed. This problem can be tackled by a hybrid approach of evolutionary neural network. Genetic algorithm (GA) for optimizations of the ANN weights is planned for future work. The output of the new hybrid system of evolutionary neural networks can be modified through a suitable heuristic when necessary. It is expected that the new hybrid system will improve the quality of the solutions so that they can be implemented in real environments.

References

1. Chan S.T.F., Chan H.K.: *A Comprehensive survey and future trend of simulation study on FMS scheduling*. Journal of Intelligent Manufacturing, 15, 2004, pp 87-102.
2. Chryssolouris G., Lee M., Domroese M.: *The use of Neural Networks in Determining Operational Policies for Manufacturing Systems*. Journal of Manufacturing Systems, Volume 10/No.2, 1990, pp. 166-175.
3. Feng Shan I. Ling., Cen Ling, Huang Jingping: *Using MLP networks to design a production scheduling system*. Computer & Operation Research, 30, 2003, pp. 821--832.
4. Hagan M. T., Demuth H. B., Beale M.: *Neural Network Design*. Vikas Publishing House Pvt. Ltd., New Delhi, 2002.
5. Hitomi Katsundo: *Manufacturing Systems engineering – A unified approach to manufacturing technology, production management, and industrial economics*. Taylor and Francis Ltd, London, 1996.
6. Jain A.S., Meeran S.: *Job-shop scheduling using neural networks*. International Journal of Production Research, 1998, Vol. 36, No. 5, PP 1249-1272.
7. Jain A.S., Meeran S.: *Deterministic job-shop scheduling: Past, Present and Future*. European Journal of Operational research, 113, 1999, pp. 390-434.
8. Kelton W. D., Sadowski R. P., Sadowski D. A.: *Simulation with ARENA*. McGraw-Hill, 1998.
9. Meeran S.: *A History of Encounters with Intelligent Search Job Shop Scheduling- Failures, Successes and Lessons Learnt*. Proc. of 19th International Conference on CAD/CAM Robotics and Factories of the Future, Kuala Lumpur, Volume 1, 22-24 July 2003, pp K3-1-K3-23
10. Negnevitsky M.: *Artificial Intelligence-A guide to intelligent Systems*. Addison-Wesley, Essex CM20 2JE, 2002.

EXPLORING WEB-BASED COLLABORATIVE PARADIGMS FOR MANUFACTURING COMPETITIVENESS

Pun K.F., Ellis R., Lewis W.G.

Abstract: Traditional mass production is becoming a thing of the past. The shifting paradigm has already begun leaning towards customerisation of varying demand. Web-based manufacturing or e-manufacturing fosters collaboration among businesses and innovation in integrated design and lean/flexible operations. Coupled with enterprise resource management systems, it can help manufacturers to enhance greater flexibility and contribute potentially to cost reductions and sustainable growth. This paper explores the determinants of web-based collaborative paradigms and discusses the importance of their developments for manufacturing competitivess.

1. Introduction

It has been widely agreed that the business success of manufacturing companies depend on their ability to identify the needs of customers and to quickly create products that meet these needs and can be produced at low costs, and with the shortest delivery time (Pun and Rao, 2002). Achieving these goals is not solely a marketing problem, nor is it solely a design problem or a manufacturing problem involving all functions. These are the watchwords for success in this new business environment, as they act on 1) bringing products to market faster; 2) responding faster to customers; 3) avoiding production problems related to inventory and materials; and 4) reducing inventory and production cost (Toussaint and Cheng, 2002). Therefore, communication and collaboration across companies and their partners become vital. Understanding market drivers based on mutually beneficial collaborative problem solving processes enables companies to move beyond traditional design, manufacturing and trading mechanisms to new and different ways of solving challenges and capturing new markets. The challenges are enormous because most of the old models are changing at a rapid pace. Internet/ intranet/extranet is the means for enhancing their design agility and manufacturing responsiveness, and thus their competitiveness (Toussaint and Cheng, 2002).

With the emerging applications of Internet and Web-enabled technologies, many organisations have shifted from traditional manufacturing operations towards an e-manufacturing paradigm and e-business enterprise integration (Pun and Rao, 2002; Rooks, 2000). While e-business focuses on cost savings, customer satisfaction and revenue growth, collaborative manufacturing fosters innovation in integrated design and agile manufacturing on a real-time basis. Thus, while researchers and designers are making continuous strides in developing e-manufacturing systems for manufacturing companies, these systems have not been employed to the fullest potential. As is with any new technology, there is always some degree of skepticism involved on the users' part. Some of the most important tasks facing engineers, managers, researchers and IT professionals include the creation of secure internet-based financial transactions, collaborative design and manufacturing, supply chain and system integration and sustainable growth models (Pun and Rao, 2002; Yang and Xue, 2003). This paper discusses the determinants of web-based collaborative paradigms, and presents with examples their applications in various areas of manufacturing domains. The need for the development of web-based collaborative paradigms for manufacturing competitiveness is explained.

2. Determinants of web-based paradigms

The ability to re-module and reconfigure the manufacturing systems from a set of exchangeable modules or reusable components can improve business processes, information systems, application systems and knowledge systems. In this context, the collaborative paradigms require that the technology of Computer Integrated Manufacturing (CIM) systems be successfully carried out. Enterprise Modelling and Integration (EMI) technology provides the necessary modelling tools to understand, analyse, reengineer and optimise the manufacturing system and reusable components while enterprise integration provides the necessary integrating infrastructure (Pun and Rao, 2002).

Web-based/-enabled manufacturing or e-manufacturing is a collaborative paradigm that incorporates the use of Internet technologies that allow for total integration from Product Life Cycle Management (PLM) to Enterprise Resource Planning (ERP) and from shop floors to Customer Relationship Management (CRM) (Yang and Xue, 2003). This is an idea of cooperation, mutual interest and trust translated onto the web. It encompasses the exchange and sharing of information as the foundation for business decisions and relations among departments within a company, same companies in various geological locations and among rival companies (McClellan, 2002). This practice could be extended to suppliers and customers alike to enhance the ultimate goal of customer satisfaction through the Internet.

Table 1. Determinants of collaborative manufacturing

Determinants	Impacts on collaborative manufacturing
1. Collaboration among Partners	❑ Increase responsiveness from both manufacturing and the design side ❑ Increase information availability ❑ Enhance mass communication
2. Communication Methods	❑ Reduce transfer lines ❑ Reduce data navigation iterations
3. Content Management and Tools	❑ Provide consistent and accurate product definition ❑ Provide helpful product viewing from 2D drawings to 3D virtual models
4. System Implementation	❑ Encourage trusts via the creation of virtual cooperation ❑ Recognise the need to customer service and prepare for the technology-driven changes. ❑ Appreciate the use of necessary infrastructure and technological software to provide results

Varying customers' demands and expectations for specific products and quick delivery at competitive prices have led to the need for improved global responsiveness. The use of the Internet or Internet technology is changing the way of product development, ranging from information gathering, product managing and commerce to product development, and maintenance (Huang and Mak, 2001; McClellan, 2002). This could allow companies to cut product development and manufacturing costs, and greatly increase collaboration among partners internally and externally (Peng, 2002). There is an increasing need for effective collaboration, communication, content management and system implementation that determine the efficacy of web-based manufacturing paradigms. These determinants are depicted in Table 1 and elaborated as follows.

2.1. Collaboration among partners

Collaboration focuses on expanding information sharing throughout internal company-wide operations and external supply-chain partners (McClellan, 2002). Many web-technologies (e.g. HTML, ASP, Java, and Open Database Connectivity protocol) have been used for improving the capabilities and efficiency of collaboration (Yang and Xue, 2003). However, trust is a critical issue when it comes to information sharing. Information across the Web has to be closely monitored on the client side from viruses, poor programming and unauthorised utilisation. Issues of piracy and awareness are sited as reasons that should discourage the use of Open-source Software. Policies must also be enforced to ensure intellectual property rights.

2.2. Communication methods

There are generally two methods of communication for implementing collaborative functions. Synchronous communication refers to the method by which designers can access the same product base simultaneously and make changes together (Qiang *et al.*, 2001; Chang *et al.*, 2001). Asynchronous communication is the method whereby a designer can access the product database simultaneously but act independently so that other team designers do not readily observe any changes made but later come together to identify the optimal design process and object (Cutkosky *et al.*, 1996). The main difference between these two methods is that the former allows for designers to make changes to the product simultaneously while the latter isolates the changes by a particular designer and then later team designers come together to decide upon the product of optimal choice. Other technical aspects involve data modelling such as 3D geometry and XML; and system architecture design of which many examples include QFD modelling system by Rezayat (2000) and a Web-based patent database by Roy and Kodkani (2000).

2.3. Content management and tools

Many generic Web-based tools demonstrate their usefulness to improve the capability and efficiency of a collaborative design (Roy and Kodkani; 2000; Huang *et al.*, 2001a). Data modelling incorporates the development of various database system for modelling products, manufacturing resources (including materials, machine and personnel) in implementing advanced manufacturing systems (Yang and Xue, 2003). The use of client-server communication methods allows the modules and databases to be distributed at both server and client ends. In Web-based applications, the client's programs are usually downloaded automatically from the server sites to local computers through Web-browsers. Moreover, many electronic design library modelling systems have been developed to preserve knowledge in the Web servers that can be accessed by designers through Web-browsers (Yang and Xue, 2003). Researchers are able to share, access, focus and identify areas of research need. End users on the Web have the capability to select particular designs and create their own unique design suitable to their needs. They can combine existing design primitives into different configurations and order their customise design using the Web-browser.

2.4. System implementation

Collaborative manufacturing requires a cross-functional, outcome-oriented type of organisational structure. The fundamental rethinking and radical design of business processes is to achieve dramatic improvements in critical measures of performance including cost, quality, capital, service and speed (Cheng et al., 2001; McClellan, 2002; Pun and Rao, 2002). The traditional way of setting up an organisational structure has to be re-defined. Teams of individuals in various departments must work together to foster creativity and productivity. The desire to be creative and meet and surpass customer expectations and demands must be fuelled by providing supporting infrastructure and highlighted by the knowing customers satisfaction. This requires an attitude adjustment and employees must be able to relate to the changes and feel the importance of their contribution (Hormozi, 2001). Moreover, security management is always a major implementation issue to ensure against piracy, viruses, tampering of information and copyright infringements (Zhuang et al., 2000; Camarinha-Matos et al., 2001).

3. Applications of web-based paradigms

Web-based technology provides the capability for the collaboration and integration of various departments within an organisation and among business partners (McClellan, 2002). One important feature that has been made possible on the Web is the ability to customise a design unique and suitable to the needs of the user. Many web-based systems have been designed and developed for optimal product design using CAD/CAM software (Adamczyk and Kociolek, 2001). Applications of web-based paradigms also extend to other areas in production planning and control (Cheng et al., 2001), simulation (Adapalli and Addrpalu, 1997), information management (Song and Nagi, 1997), enterprise integration (Trappey and Trappey, 1998), and supply chain management (Li et al., 2001). A list of reported applications of these systems is depicted in Table 2.

3.1. Product design and development

Product development life cycle includes the areas of design, process planning, production, etc. it has therefore become necessary to integrate these areas in order for product development teams to effectively collaborate from various locations (Adamczyk and Kociolek, 2001). An important feature, apart from being able to integrate the various aspects of a product development life-cycle, is the capacity to provide feedback to improve the design. This concurrent engineering approach incorporates the downstream product development life-cycle considerations into the design phase (Yang and Xue, 2003).

Table 2. Examples of web-based collaborative paradigms

Areas	Systems	Descriptions	Authors
Product Design and Development	CyberEye system	Incorporates the technologies of HTML, ASP, Java, ActiveX, OOM, and Open Database Connectivity protocol in the development of a platform for collaborative product design	Zhaung et al. (2000)
	Web Product Design Support System (WPDSS)	Used for synchronising group co-modification operations and maintaining consistency of the shared CAD models in collaborative design review	Qiang et al. (2001)
Production Planning and Control	TeleDSS (Decision Support Systems)	Use workflow management as a mechanism to facilitate the teamwork in a collaborative product development environment where team members who are geographically distributed	Huang et al. (2000a)
	TeleRP System (Rapid Prototyping)	Use RP tele-manufacturing Web server for remote part submitting, queuing, and monitoring	Jiang and Fuyuda (2001)
Simulation	UMAST (Uninhabited Aerial Vehicles Modeling and Analysis Simulator Testbed)	Uses Web-based modelling and simulation system to emulate characteristics of uninhabited aerial vehicles for studying human/system interactions	Narayan et al. (1999)
Information Management	Integrated Tele-Cooperative Manufacturing (ITCM)	Handles heterogeneous data and integrates software applications using Web servers, databases, document conferencing software, operating systems, and integration utilities	Nee and Ong (2001)
	Engineering Change Management (ECM) Framework	Supports ECM procedures and activities, clarifies responsibilities, and provides associated documents	Huang et al. (2001a, b)
Enterprises Integration	Persistent Distributed Store (PerDiS) Platform	Offers persistence, concurrent access, coherence, and security on a distributed data-store based on the distributed shared memory paradigm	Sandakly et al. (2001)
	Web-based knowledge management (WKM) System	Enables a content-based search that can improve search effectiveness, and supports automatic translation and reuse of product data among different application systems	Yoo and Kim (2002)
	Web-based Electronic Product Cataloging (EPC Web)	Uses database, multimedia, and Internet in compiling the product information, and access the resulting electronic product catalogues through Web-browsers.	Huang and Mak (1999)
Supply Chain Management	EXPRESS (a Web-based system for managing product and market information)	Used for automated shelf layout, continuous sales analysis, and real-time logistic management are incorporated into the Marketing Information System (MIS), which can be used by store managers through an easy-accessed Web-based interface	Trappey and Trappey (1998)
	WeBid System	Provides a suit of tools for establishing and managing the customer-supplier relationships, and for supporting early supplier involvement in new product development on the Internet.	Huang et al., (2000b)

3.2. Production planning and control

Many Web-based manufacturing systems allow users at different locations to have their designs manufactured at remote locations by expensive equipment such as CNC centres, robots, and rapid prototyping machines (e.g. Hu *et al.*, 2001; Luo and Chang, 2001). Others are geared towards allowing and improving collaborative planning and control of production (e.g. Cheng *et al.*, 2001; Leong *et al.*, 2001)

3.3. Simulation

Development of Web-based manufacturing systems using multimedia and virtual reality has attracted the attention of many researchers. Simulation in production and business processes allows for virtual characteristics to mimicked real situations of a system without the actual full financial cost of the physical problem or situation. Adapalli and Addepalu (1997) discussed the different methods for integrating manufacturing process simulations with the Web and implemented a Web-based metalworking process simulation system for defect prediction.

3.4. Information management

In manufacturing enterprises, information needs to be linked from various niches of the extended enterprise. Management of this information is critical to empower and better support manufacturing related processes (McClellan, 2002). This can help communicate with various departments within a company on different geological location or even with competing companies. The design and implementation of a Web-based information system integrating manufacturing databases can be dispersed at various partner sites (Song and Nagi, 1997). Information and data exchange can be modelled in a hierarchical fashion, and can be controlled by interacting with partners' knowledge bases.

3.5. Enterprise integration

As industries become consolidated and/or grow, the need for integration becomes more apparent (McClellan, 2002). Virtual enterprising is becoming more common since companies are recognising the opportunities and benefits of joining forces and bringing together the greatest source of strength of each party involved to produce the optimal product. A collaborative approach is based on an exchange of requests and information between collaborative autonomous agents that support the design, manufacturing planning, and facility formation activities (Yoo and Kim, 2002).

3.6. Supply chain management

Supply chain management involves conceptualisation of a product to production, to quality control, to sales and then to customer satisfaction. Many Web-based systems have been developed to obtain requirements from customers, analyse these requirements, and get feedback from customers (Yang and Xue, 2003). Frohlich and Westbrook (2002) contend that there is a close relationship between Internet enabled supply chain integration strategies and performance in manufacturing and services. The Web/Internet technologies provide an excellent environment for communication among partners in a supply chain. This also enables customers to interactively communicate their needs directly. More services are made available to the customer on-line and in remote locations (Li *et al.*, 2001; McClellan, 2002).

4. Conclusion

The Web-based collaborative paradigms are of great use in the integration of various aspects of the product development lifecycle, integration of management and enterprises. Collaboration can expand information sharing among companies both internally and externally. This paper discussed the concepts associated with the paradigms for attaining manufacturing competitiveness. The development of these paradigms rely significantly on several determinants, including 1) collaboration among partners, 2) communication methods, 4) content management and tools, and 4) system implementation.

Trust is a cornerstone for the success of Web-based manufacturing or e-manufacturing. This paper concludes that adopting Web-based collaborative paradigms is an emerging way for manufacturing companies to work collectively to produce products conforming to customers' needs and expectation in a timely manner with quality intact and at affordable prices. In order to help build and sustain manufacturing competitiveness in industry, further research will 1) investigate the critical factors that affect the development of web-based collaborative paradigms and 2) explores their potential applications in different manufacturing sectors in a regional and global context.

References

1. Adamczyk Z., Kociolek, K. (2001): *CAD/CAM technological environment creation as an interactive application on the Web*. Journal of Materials Processing Technology, Vol.109, 222-228.
2. Adapalli S., Addrpalu K. (1997): *World Wide Web integration of manufacturing process simulations*. Concurrency-Practice and Experience, Vol. 9, pp. 1341-1350.
3. Camarinha-Matos L. M., Afsarmanesh H., Osorio A. L. (2001): *Flexibility and safety in a Web-based infrastructure for virtual enterprises*. Internat. Journal of Computer Integrated Manufacturing, Vol. 14, pp. 66-82.
4. Cheng K., Pan P.Y., Harrison D.K. (2001): *Web-based design and manufacturing support systems: implementation perspectives*. International Journal of Computer Integrated Manufacturing, Vol. 14, pp.14-27.
5. Cutkosy M.R., Tenenbaun J.M., Glicksman J. (1996): *Madefast: collaborative engineering over the Internet*. Communications of the ACM, Vol. 39, pp. 78-87.
6. Frohlich M.T., Westbrook R. (2002): *Demand chain management in manufacturing and services: web-based integration, drivers and performance*. Journal of Operations Management, Vol.20, pp. 729-745.
7. Hormozi A.M. (2001): *Agile Manufacturing: the next step to logical manufacturing*. Benchmarking: An International Journal, Vol.8, No.2, pp. 132-143.
8. Hu H.S., Yu L.X, Tsui P.W., Zhou Q. (2001): *Internet-based robotic systems for teleoperation*. Assembly Automation, Vol. 1, pp. 143-151.
9. Huang G.Q., Mak K.L. (2001): *Issues in the development and implementation of Web applications for product design and manufacture*. International Journal of Computer Integrated Manufacturing, Vol.14, pp. 125-135.
10. Huang G.Q., Huang J., Mak, K.L. (2000a): *Agent-based workflow management in collaborative product development on the Internet*. Computer-Aided Design, Vol.32, pp. 133-144.
11. Huang G.Q., Huang J., Mak K.L. (2000b): *Early supplier involvement in new product development on the Internet: implementation perspectives*. Concurrent Engineering: Research and Applications, Vol.8, pp.40-49.

12. Huang G.Q., Nee W.Y., Mak K.L. (2001a): *Engineering change management on the Web*. International Journal of Computer Applications in Technology, Vol.14, pp. 17-25.
13. Huang G.Q., Nee W.Y., Mak, K.L. (2001b): *Development of a web-based system for engineering change management*. Robotics and Computer-integrated Manufacturing, Vol.17, pp.255-267
14. Jiang P., Fukuda S. (2001): *TeleRP: an Internet Web-based solution for remote rapid prototyping service and maintenance*. International Journal of Computer Integrated Manufacturing, Vol.14, pp.83-94.
15. Leong H.V., Ho K.S., Lam W. (2001): *Web-based workflow framework with CORBA*. Concurrent Engineering: Research and Applications, Vol.9, pp.120-129.
16. Li D., McKay A., Depennington A, Barnes C. (2001): *A Web-based tool and a heuristic method for cooperation of manufacturing supply chain decisions*. Journal of Intelligent Manufacturing, Vol.12, pp.433-453.
17. Luo R.C., Tzou J.H., Chang Y.C. (2001): *Control and monitoring through Internet*. IEEE-ASME Transactions on Mechatronics, Vol.6, pp.399-409.
18. McClellan M. (2002): *Collaborative Manufacturing: Using Real-Time Information to Support the Supply Chain*, St. Lucie press.
19. Narayanan S., Edala N.R., Geist J., Kumar P.K., Ruff H.A., Draper M., Haas, M.W. (1999): *Umast: a Web-based architecture for modeling future uninhabited aerial vehicles*. Simulation, Vol.73, pp.29-39.
20. Nee A.Y.C., Ong S.K. (2001): *Philosophies for integrated product development*. International Journal of Technology Management, Vol.21, pp.221-239.
21. Peng Q. (2002): *A survey and implementation framework for industrial-oriented Web-based applications*. Integrated and Manufacturing Systems, Vol. 13, No.5, pp 319-327.
22. Pun K.F., Rao U.R.K. (2002): *A proposed e-manufacturing systems framework for e-business enterprise integration*. CD-ROM Proc. of the Int. Conference on E-Manufacturing, The Institution of Engineers, Bhopal.
23. Qiang L., Zhang Y.F., Nee, A.Y.C. (2001): *A distributive and collaborative concurrent product design system through the WWW/Internet*. International Journal of Advanced Manufacturing Technology, Vol.17, pp.315-322.
24. Rezayat M. (2000): *The enterprise-Web portal for life-cycle support*. Computer-Aided Design, Vol.32, pp.85-96.
25. Rooks B. (2000): *Technomatrix weaves its manufacturing web*. Assembly Automation, Vol.20, No.3, pp.213-217.
26. Roy U., Kodkani S.S. (2000): *Collaborative product conceptualisation tool using Web technology*. Computers in Industry, Vol. 41, pp.195-209.
27. Sandakly F., Garcia J., Ferreira P., Poyet, P. (2001): *Distributed shared memory infrastructure for virtual enterprise in building and construction*. Journal of Intelligent Manufacturing, Vol.12, pp.199-212.
28. Song L.G., Nagi, R. (1997): *Design and implementation of a virtual information system for agile manufacturing*. IIE Transactions, Vol.29, pp. 839-857.
29. Toussaint, J. and Cheng, K. (2002), "Design agility and manufacturing responsiveness on the Web", *International Journal of Production Research*, Vol.38, No.12, pp. 2743-59.
30. Trappey C.V., Trappey A.J.C. (1998): *A chain store marketing information system: realising Internet-based enterprise integration and electronic commerce*. Industrial Management and Data Systems, Vol.98, pp. 205-213.
31. Yang H., Xue, D. (2003): *Recent Research on Developing Web-based Manufacturing Systems: A Review*. International Journal of Production Research, Vol. 41, No. 15, pp. 601-3629.
32. Yoo S.B., Kim, Y. (2002): *Web-based knowledge management for sharing product data in virtual enterprises*. International Journal of Production Economics, Vol. 75, 173-183.
33. Zhaung Y., Chen L., Venter R. (2000): *CyberEye: an Internet-enabled environment to support collaborative design*. Concurrent Engineering: Research and Applications, Vol. 8, pp. 213-229.

PARALLEL KINEMATIC MACHINES (PKM) AND MULTITASKING IN AIRCRAFT COMPONENTS MANUFACTURE

Selvaraj P., Radhakrishnan P.

Abstract: Even though PLM solutions play a major role in time compression in design and manufacturing activities, technologies for realizing the hardware have to be improved further in terms of realizing the products better, faster and cheaper. Evolution of NC/CNC machines replaced the conventional milling and copy milling machines and also reduced the fixture requirements. Five axis CNC machines reduced the number of work piece setups to machine the aerodynamic sculptured surface components and also the polishing effort involved. The recent development of multitasking CNC machines integrate various heterogeneous operations like milling, turning, drilling, reaming, and boring etc. and it brings down the number of work piece setups, machines required, work in process inventory, and increased accuracy. The future trend indicates the evolution of Parallel Kinematic Machines for machining of aerospace components through multitasking approach. This paper brings out the benefits which could be reaped out and the issues that are to be sorted out in PKM to make it available for aircraft industry.

1. Introduction

Design and development of aircraft needs the best of the technologies available in any domain. PLM solutions enable to accelerate the various design activities starting from conceptual design to detail design and assembly, work flow, configuration management, change management, design in context methodologies, visualization in digital mock up, multi-directional communication on clearance of design and production queries etc in the development of any complex product like aircraft. In aircraft industry majority of the machined parts are made of aluminium, and more than 90 percent of the material is machined out to meet the one of the prime objectives of 'light weight design'. The concept of multitasking operations [1] reduces the number of machines required, work in process inventory, manufacturing throughput time and increased efficiency. The following chapters bring out the complexity involved in machining of aircraft parts, benefits of multitasking approach and relative merits of PKM machines comparing to Serial Kinematic Machine (SKM) for utilizing this concept.

1.1. Serial and parallel kinematic machines (SKM and PKM)

Conventional 3 axis and 5 axis CNC machines (SKM), have perpendicular connection between adjacent links from the base to the spindle unit requiring larger work space and simple control operating systems. In contrast a parallel link machine tool has a completely different structure and it is driven by parallel links and rotary joints. Relative advantages of PKM machines are listed below:
1. Parallel Mechanism – speed receives more attention due to their high stiffness to mass ratio.

2. PKM is five times as stiff as conventional machining centers to take heavy cuts without chatter and gives the highest precision [2].
3. Higher speeds and accuracy is possible in PKM than SKM.
4. In parallel links, their lengths can be changed independently and also the 6 degrees of freedom of the movement of the spindle can be achieved.
5. Increased velocity and acceleration of the machine is possible in PKM due to its decreased inertia mass.
6. Stiffness of each link of the PKM can be increased substantially, since only tension and compression is acting on the links, not bending.
7. Accuracy of the PKM system can be increased, since each link error does not accumulate while in SKM it accumulates.
8. In PKMs, multiple ball screws (expanding or contracting supporting ball screws) are used and the machining loads are shared.
9. PKM has mechanical simplicity and high volumetric accuracy.
10. Less installation requirements for PKM, and needs limited work space [4].
11. The stiffness of the mechanism in PKM against the cutting force is not as high as SKM. However for machining of soft materials like aluminium alloy, this problem is not remarkable.
12. In SKM machines there are guide ways, whereas in PKM there are no guide ways and can control six axes with very simple structure.
13. PKM machines have energy saving capability since those machines have a light movable body and can move or accelerate it at high speed with low energy.

The figures below [3] brings out the basic structural differences of SKM and PKM.

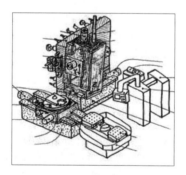

Figure 1. CNC machining center (SKM)

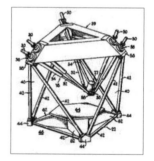

Figure 2. Parallel kinematic machine

1.2. PKM mechanism

One of the most widely used parallel mechanism structures is hexapod type machine tool of 'Stewart Platform Link Mechanism' [5]. It has six telescoping struts, each of which is connected to the base by a 2 degrees of freedom (DOF) joint. The other end of each strut is connected by a 3 degrees of freedom joint to the platform, wherein the machine spindle is mounted. Thus (3 x 2) six degrees of freedom control of the spindle position is possible. The schematic diagram of such mechanism is shown in Figure 3.

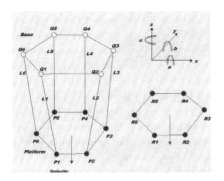
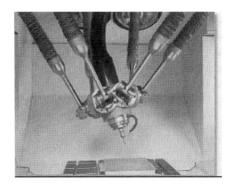

Figure 3. A Stewart platform parallel link mechanism

Figure 4. PKM Okuma Cosmo Center PM-600

One of the widely used OKUMA COSMO CENTER PM-600 [3] PKM based multitasking machine is show n schematically in Figure 4 [6].

2. Aircraft components machining

Aircraft components are of basically two different types viz. monolithic parts (frames, stringers, brackets, longerons etc.) and prismatic parts (housings, covers, pumps, filters etc.).

2.1. Monolithic parts

The monolithic parts require mainly surface milling operations followed by drilling, boring (in-line boring) etc. The surface milling calls for either three axis CNC machines or 5 axis machines. Long cycle time (due to ball nose milling), increased number of set ups, ridges (scallop heights) formation , requirements of additional tooling for inclined hole processing operations etc. are the factors that we encounter in 3 axis milling. Higher machine hour rate, availability of reliable post processor, avoidance of tool-work piece, tool-fixture, tool-machine bed due to over rotary motions etc. are the factors that restrict the application of 5 axis surface milling.

2.2. Prismatic Parts

This category of parts require heterogeneous operations like milling, turning, grinding, hole processing, and super finishing operations such as lapping. Conventional CNC machining route need more number of different variety of machines which lead to multiple setups, work in process inventory, reduced accuracy and long cycle time. Such components require anything between 10-20 setups to make. Sample aircraft prismatic components are shown in Figures 5 and 6.

Figure 5. Component requiring turn-mill-drill-bore-tap

Figure 6. Component requiring mill-turn-drill-bore-contour grinding etc.

3. Requirements of CNC aircraft components machining

Due to the complexity involved in machining of surface dependent aircraft parts, following factors need attention on priority.

3.1. High Speed Machining

As already mentioned most of the monolithic aircraft components require scooping out of almost 90 percent of the raw material. The concept of High Speed Machining is a must, where in the low depth of cut and higher speed and feed that the chips are removed faster, there by surface finish of the component is improved, built-up-edge formation is minimized and also the warpage of the component due to heavy cuts is reduced. Both SKM as well as PKM machines will meet the requirements. But in many of the cases these components call for angular drilling and in line boring for which additional tooling is required in case of five axis SKM where as due to the independent of movement (6 degrees of freedom) of spindle of PKM, the additional tooling is eliminated. Comparing to SKM is, higher speed and feed is possible in PKM. (cutting velocity is of the order of 100 to 240 m/min) and the rpm 20000 to 30000 [6, 7, 8] for machining aluminium material.

3.2. Machining of closed bevel angle surfaces

The components cross sections shown above are designed keeping the principles of integrated design and manufacture of components [9] which lead to parts reduction. Availability of extensive CAD/CAM systems facilitate such advanced design where as the same benefits could be obtained only through multitasking approach using PKM machines. In case of SKM CNC machines special tools with additional setups are required even if we go for a five axis machining center, where as in PKM, the additional set ups could be eliminated due to the independent sixth axis movement of spindle.

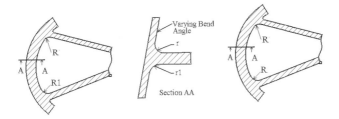

Figure 7. Intricacy of aircraft monolithic parts contour

3.3. Advanced CNC control systems

The advanced high speed milling controls can impact aircraft component surface milling productivity manifold. The most significant benefits of high speed machining for aircraft components come by way of 3D contour milling where in complex surface milling programs having millions of line segments is not uncommon. This benefit is still enhanced when the work piece material is aluminium. The desirable features of the CNC control for high speed milling is as follows [10]:
1. Look ahead feature (built in intelligence to analyze geometries to prevent gouges and overshoots).
2. Accurate control at higher feed rates PKM comparing to SKM.
3. Direct CNC Networking
4. Open System Architectures (the design of choice that empowers the customer with choices).
5. Multi-processor strategies to take the CNC beyond just CNC control.

This high speed machining benefits in terms of improved accuracy, fit, finish and cutter life, less polishing effort. The features mentioned above are more easily attainable in PKM machines due to its light weight, low inertia mass and advanced CNC control systems [11].

3.4. CNC machining through multitasking

The prismatic components such as jet engine cases, pump bodies, filters and various system components of aircraft require various heterogeneous operations like milling, turning, hole processing, contour grinding, lapping etc. The ideas of 'from start to part' and 'done in one' of multitasking machines is well suited for aircraft parts, an example of which is shown in Table 2. To cite an example Mori Seiki MT2000 SZ turning center having twin spindle, and two turrets (lower turrets have 12 tools, can accommodate a turning tool or live tool). Instead of upper turret, there is an upper tool spindle with four axis movement capability (X,Y,Z and B+/- 120 degrees to present its tool to a work piece in either spindle. An automatic tool changer with 120 tool magazine of turning and live tools [12]. These features are well suited for the machining of certain category of aircraft components. Another example is Okuma's Cosmo Center PM-600 as already brought out in the earlier section 1.2.

3.5. CNC software for multitasking operations

Tool Path generation and post processing the Cutter Location (CL) data for multitasking machines poses a real challenge. Controlling multiple tools across one or more spindles at one time introduces a whole new level of complexity to programming, especially if one wants to create efficient programs. Synchronizing multiple tools cutting simultaneously is important to reduce the overall non-cutting time. The software should generate the code and insert waits in the appropriate places to avoid collisions. The programmer need not worry about the top turret colliding with the bottom turret because they both are trying to machine something on the same spindle at the same time. The option of having dynamic solid simulation provides the dynamic and realistic views of the entire machine set up, including part, fixtures, and the machine tool. The simulation should take care of collision detection between the part and tool; the part and tool with holder etc. Gibbs CAM [13], ESPRIT 2005 [14] are a few CAM software available for multi tasking machines programming suiting PKM.

3.6. Linear motor servo drives

Linear motors account for part of the reason to nearly double the productivity of many manufacturing facilities. The high feed rate and high acceleration of the linear motors make possible [15] in reducing the cycle times. Linear motors allow the cutting velocity of 100 to 240 m/min speeds above 20,000 rpm with the acceleration of 1.5 g (14.5 m/s2) in the rapid motion in Z axis, and 1 g in X and Y axes to 76 m/min [8] (g- acceleration due to gravity). Even of these high speeds, the linear motors position the spindles in the entire area to a degree of accuracy of about 5 microns [8,11]. Deployment of linear motors in PKM is well received.

4. Analysis of aircraft component geometry

A sample aircraft 'bracketary' component with billet shapes and different work piece setups with operations and tools involved are shown in Figure 8 and in Tables 1 and 2. These operations are done in a 3 axis CNC milling machines. There are totally six setups including sine-table requirements and also billet preparation. Figure 8 and Table 2 are illustrated only to highlight the complexities involved in machining a typical aircraft part, which can be handled easily using multitasking approach of PKM. The details of component geometry are not given, only the billet size (400x100x90 mm) is given to get the idea of the part size.

Brief of the operations for Table 1, Figure 8 and Figure 9.

Using a 3 axis CNC machine, a minimum of six setups are required excluding the billet preparation. Total number of machining operations involved are 35 (end milling, surface milling, line drill and ream etc.) and the number of cutting tools used are 15 (including drill and ream) . If we use 5 axis machining center (SKM), the number of setups could be reduced, however the inline drill and ream of the holes in the fork is not possible directly. In PKM due to the independent movement of the spindle in the sixth axis, in line boring operations could be carried out. This operation is very critical in terms of assembly point of view.

Table 1. Sequence of operations with tools for a component in Figure 8 using 3 axis CNC

Sl. No.	Setup Details	Operations Name		Cutting Tools Used (diameter x corner radius x length) in mm
1.	Billet preparation	i.	vertical setup -sine table reference machining	20 x 0 x 100
		ii.	inclined face machining for setup 5	20 x 5 x 80
		iii.	roughing (removal of extra material)	20 x 0 x 100
2.	Setup1	i.	roughing	20 x 0 x 100
		ii.	machining inside channel	8 x 0 x 100
		iii.	semi-finish (surface mill)	8 x 4 x 80
		iv.	finishing (surface milling)	10 x 5 x 80
		v.	inclined face (surface) milling	6 x 3 x 80
3.	Setup2	i.	roughing	20 x 0 x 100
		ii.	scooping - inside the channel	8 x 0 x 100
		iii.	semi-finish (surface mill)	8 x 4 x 80
		iv.	finishing (surface mill)	10 x 5 x 80
		v.	roughing to have 3 mm radius	10 x 5 x 80
		vi.	finishing (surface mill	6 x 3 x 80
		vii.	finishing of inclined face (surface mill)	6 x 3 x 80
		viii.	reference face (setup 5) rough	8 x 0 x 50
		ix.	reference face (setup 5) finish	8 x 4 x 50
4.	Setup3	i.	contour facing (semi finish)	16 x 5 x 80
		ii.	contour finishing (surface milling)	16 x 8 x 80
		iii.	grooves for (sine table setup) reference	16 x 0 x 80
		iv.	inclined face for 'setup6' (sine table setup2) surface milling	10 x 5 x 80
		v.	reference face for 'setup6' roughing	8 x 0 x 80
		vi.	reference face for 'setup6' finish milling	8 x 4 x 80
5	Setup4	i.	surface machining (semi-finishing)	16 x 5 x 80
		ii.	surface machining (finishing)	16 x 8 x 80
		iii.	merging area machining (surface milling)	10 x 5 x 80
		iv.	3 mm corner radius maintaining	6 x 3 x 80
6	Setup5 (sine table setup1)	i.	machining of fork	16 x 0 x 100
		ii.	trimming fork outside (rough)	16 x 6 x 100
		iii.	trimming fork outside (finish)	16 x 6 x 100
	Setup6 (sine table setup2)	i.	finishing top face of the fork	10 x 5 x 80
		ii.	trimming (roughing)	16 x 0 x 80
		iii.	trimming (finishing)	16 x 6 x 80

The above component also calls for line drill and ream operations of $14.1^{+0.018/-0.000}$ mm diameter holes.

Figure 8. Shape of billet after preparation; billet envelope size is 400x100x90 mm

Figure 9. Sequence of setups involved in 3 axis CNC machining of a part

5. Issues in PKM for multitasking applications

Even though there are distinct advantages of using PKM in place of SKM, still the issues which are given below have to be sorted out so as to see the PKM multitasking technology is available to aircraft industry:

1. On a PKM, the position of the spindle head is indirectly estimated from the rotational position of servo motors. It is critical to identify accurately the length of struts and the position of joints.
2. Deformation due to gravity is another critical factor.
3. The stiffness of the mechanism in PKM against the cutting force is not as high as SKM. This has to be addressed to machine materials like steel.
4. Position error due to the thermal expansion of ball screws (Albert, 1998).
5. Accurate identification of geometrical (kinematic) parameters and thermally insensitive strut length measurement for performance assessment of PKM (Albert, 1998).
6. Difficulty in setting appropriate feed rates for motions that involved change in orientation as well as translation (Albert, 1998).
7. The CNC control units of PKMs are complex.
8. On a PKM, the position of the spindle head is indirectly estimated from the rotational position of servo motors. It is critical to identify accurately the length of struts and the position of joints.

6. Conclusion

This paper brought out the requirements of machining aircraft components, the shape and geometry of each part is unique in design. The advantages of going for multitasking approach using CNC machines (SKM) as well as PKM were also highlighted. The additional benefits of PKM machines for multitasking of aircraft component machining was brought out in terms of high velocity and feed rates due to light mass inertia, simultaneous six axis of motions for the spindle, non-accumulation of errors, look ahead feature of CNC control systems etc. However the technical issues such as deformation due to gravity, stiffness mechanism to machine variety of steel materials, positional error due to thermal expansion of ball screws etc. were also mentioned. The authors also brought out with a suitable example showing the difficulty involved in machining an aircraft component and how the evolution of a full pledged PKM will increase the productivity.

Acknowledgements

The authors wish to thank the Managements of Aeronautical Development Agency, Bangalore, India and Vellore Institute of Technology (Deemed University), Vellore, India for encouraging and allowing the authors to publish this paper.

References

1. http://www.johnhart.com.au/news/articles/the_evolution_of_multitasking/
2. http://www.ingersoll.com/ind/hexapod.htm
3. Selvaraj P., Radhakrishnan P.: *Innovations in the Design of CNC Machine Tools to Maximize Productivity and Cost Reduction – A Review*. National Aerospace Manufacturing Seminar NAMS 2003, Thiruvanthapuram, India, June 2003.
4. Albert J., Wavering K: *Parallel Kinematic Machine Research at NIST: Past, Present, and Future*. First European-Americal Forum on Parallel Kinematic Machines Theoretical Aspects and Industrial Requirements, 31st August-1st September, Milan 1998.
5. http://precnt.prec.kyoto-u.ac.jp/mmc/research/para/pa_overview_jp.html
6. http://www.americanmachinist.com/printout.php?EID=7418
7. http://www.geindustrial.com/cwc/products?
8. http://www.ipnews.com/archives/moldmaking/may03/okuma_cosmo.htm
9. Selvaraj P., Radhakrishnan P.: *Technologies on Integrated Design and Manufacture of Aircraft Components*. Proceedings of National Conference on Advanced Manufacturing & Robots, CMERI, Durgapur, India, January 2004, p:105.
10. Todd A., Schuett J.: *Advanced Controls for High Speed Milling*. High Speed Machining Conference, Chicago, IL, May 7-8, 1996.
11. http://www2.automation.siemens.com/mc/mc-sol/en/
12. http://www.production-machining.com/articles/1104cip1.html
13. http://www.production-machining.com/articles/110202.html
14. http://www.dptechnology.com/Press_Releases/04-10_JIMTOF_Japan.htm
15. http://www.mmsonline.com/articles/0201bp1.html

HOLONIC THEORY AND MODELING APPLIED TO MES FOR METALWORKING SMEs

Baqueiro L.G., Cadena M.R., Molina A.

Abstract: This document describes the modeling process and the laboratory based demonstration of a Holonic Manufacturing Execution System (MES) for Metalworking SMEs, starting from the environment in which is embedded to the crumbling of its composing elements. Decomposition of information components and their relationships represent the Static View (comprised by different information interrelation panes) for the recursive Holarchies of which Holonic MES is constituted. MES-Holarchy level dynamic behavior is modeled through Petri Nets, as well as this Holarchy interaction with other Manufacturing element, both during MES normal operation as well as when facing a disturbance. The results of this model implementation are deployed and future research lines are stated.

1. Introduction

An initiative for a Holonic Manufacturing Execution System for Mexican Metalworking SMEs –Small and Medium Enterprises- was presented in 1st International Conference on Applications of Holonic and Multi-Agent Systems -HoloMAS 2003- (Gaxiola et al., 2003). The essential point of that work consisted in a novel approach for both the modeling and design a Holonic Manufacturing Execution System for Mexican SMEs (Small and Medium Enterprises) in the metalworking area, focused to MTO (Make to Order) Environments. Basic developed points presented in this initial work were:
- a conceptualization of MES as a Holon,
- the role performed by a Holon-MES inside a HMS through its autonomous and collaborative features,
- the definition of a Static Information Model, in order to conceptualize the information flow through a Holonic Information System (HIS),
- MES-Holon Functionalities Definition,
- proposed Implementation Tools, mainly through Web Services.

It is important to mention that one of the major challenges of this initial work was the modeling of disturbances in a Manufacturing Environment. Leitão's (Leitão's et al., 2003) and Fan-Tien Cheng's (Fan-Tien Cheng et al., 2001) work (among some others) were a strong base reference for the continuation of present work. Leitão modeled the dynamic behavior that occurs in a Manufacturing Environment through Petri Nets. Fan-Tien Cheng, on the other hand, presented an elements' decomposition analysis of a MES functional holon. This paper shows the methodology exploited to model both the static and dynamic behavioral aspects of a Holonic MES.

2. Basic concepts

2.1. Manufacturing Execution Systems

The concept of MES (Manufacturing Execution System) has acquired a huge importance in today's modern manufacturing industries. MES Systems are an information bridge between Planning Systems used in Strategic Production Management (such as ERP) and Manufacturing Floor Control (such as SCADA, Supervisory Control and Data Acquisition). It links the Manufacturing Information System's layers (Strategic Planning and Direct Execution) through the adequate on – line managing and control of updated information related with the basic enterprise resources: people, inventory and equipment (McClellan, 1997) (See Figure 1).

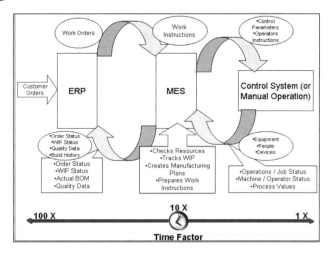

Figure 1. Enterprise's Information Flow Model (adapted from MESA International, Pittsburgh, PA, USA, 2000, and from TATA Consultancy Services; Mumbai, India, 2002)

In an effort to create a common information structure base for such systems (MES and ERP mainly), organizations like ISA have developed standards like ANSI/ISA – 95.00.01 and 02, which can be applied regardless of the degree of the environmental automation degree (ANSI / ISA, 2001).

2.2. Holonic Theory

The origin of Holonic Theory took place in 1967, when the Hungarian philosopher Arthur Koestler edited his book "The Ghost in the Machine" (Koestler, 1967). According to Koestler, these kinds of organizations, during their evolution, reach some intermediate stable and auto configurable states. Koestler was motivated to mint the word "Holon", which is the combination of the suffix "Holos" ("the whole") and "on" ("the part") in order to define a system comprised by entities with Autonomous –the inherent Self Reliance property- and Cooperative properties. Other important concepts concerning with Holons are the *Holarchy* and the *Interlocking Holarchies* (Wyns, 1999). A Holarchy is a structure in which Holons are organized (even in an inter-hierarchically manner) and it has an inherent

Recursive property (this is, a holon in a higher Holarchy can be by itself another Holarchy, comprised by lower-level holon). Interlocking Hierarchies can be defined as the Holons' trend to group in a non neither centralized nor hierarchical manner, even in an interlocking structure among holon branches.

3. Holonic theory applied to manufacturing: holonic manufacturing systems

A formal definition of the term Holon was proposed by Holonic Manufacturing Systems - HMS- as "an autonomous and cooperative building block inside a manufacturing system to transform, transport, store and/or validate information and physical objects" (Manufacturing Systems/ HMS, 2000). Under the manufacturing context, a Holarchy can be conceptualized –as HMS does- as a ...*Holarchy that integrates the all complete domain of manufacturing activities, from Product Orders' reservation to Design, Production and Marketing in order to achieve an agile manufacturing enterprise* (Botti et al., 2002).

Important developments and theoretical models have been developed to achieve an Agile Manufacturing Environment. One of these models has been developed by Rahimifard (Rahimifard, 2002). Extrapolation of this model is done in the present work, with the following four key elements comprising it: a Holonic Information System (HIS) supported by two databases – an Order database and a Manufacturing database -, an Operation Floor Holon, comprising all manufacturing direct operations on the product, a MES Holon, in charge of all direct manufacturing operations administrative tasks and, finally, an ERP or Planning Holon, in charge of all ultimate decision making activities.

3.1. Holonic manufacturing system and sanufacturing execution system Holarchies: static aspects

Present work's designed MES model can be seen, based on Rahimifard's concept focused on SMEs, as a Holonic element inside a Holonic information system (HIS), as depicted in Figure 2.

Being supported on the recursive property possessed by Holarchies (Wyns, 1999), and, after the fact that MES is proposed as a Holarchy embedded on another (a Holonic Manufacturing System), this Manufacturing Execution System can be seen as a Holarchy comprised by other holons (with their comprising elements not necessarily located physically together) (Gaxiola, 2004); this elements are *Planning Management and Shop Floor Interfacing Holon*, *Machines and Maintenance Management Holon*, *Work Order Management* (WOM) *Holon*, *Simulation Holon*, *Scheduling Holon* and, finally, *Production Management and Tracking Holon* (see Figure 3).

3.1.2. Databases Model

On his original work, Rahimifard (Rahimifard, 2002) established the necessity of having databases supporting SME's manufacturing activities. Information stored in that databases should be focused on Manufacturing Resources, Processes and Orders. Adapting Rahimifard's concept, it was determined that the connection among Holarchy elements depicted in Figure 2 could be possible only if all of them were fully interconnected through the Holonic Information System (HIS). Thus, the next step was oriented to model that elements belonging to HIS. On this respect, Toh (Toh, 1997), on his Doctoral Thesis, proposed a three-interconnected-views model in order to determine information requirements and specifications for metalworking SMEs. This Model is comprised of two views for the static aspects (Facility and Information Views) of a Metalworking SME Manufacturing Environment and one view for the behavioral aspects of the system (Organization / Behavior View).

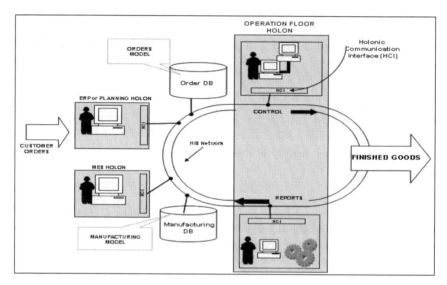

Figure 2. MES integration with a Holonic Information System –HIS (Adapted from Rahimifard, S.: *A Practical Representation of Holonic Manufacturing Systems*, 2002)

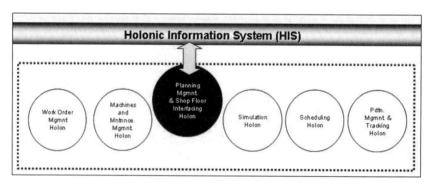

Figure 3. MES Holon components deployment

Analysing Toh's proposal, the relationships among all sublevels contained in the first two mentioned Toh's model views were determined. In present work, a UML modeling notation, Class Diagrams, was employed in order to have a more powerful and updated modeling tool for this project's databases. Figure 4 depicts an example of this modeling through Class Diagrams, in which the conceptual relationship among Resources, Processes and Facility levels is established. After that, in a lower level (programming level), the required Manufacturing and Order Databases are created (*e.g.* Order's Model). This Class Diagrams Modeling permits to build the two required basic databases for the HIS in an easier way.

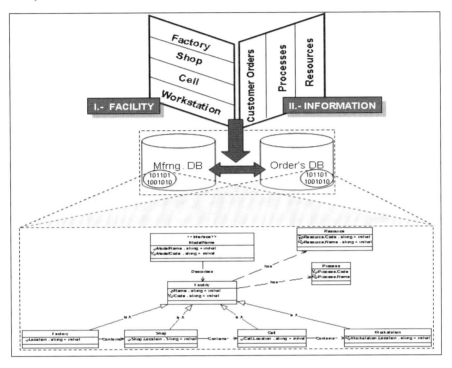

Figure 4. Order's model (example)

3.2. Holarchy behavior model and holarchy configurations under disturbances: dynamic aspects

Toh's three-interconnected-views model is pivoted precisely in one of these views: the Behavioral View. This behavioral view is comprised of three sub views, namely Cooperation, Control and Autonomy (Holonic features). These behavioral aspects are inherent to each Holon depending on the Holarchy in which they are immersed. Reaching this point of the modeling process of the MES Holon its surrounding Holonic Manufacturing Environment and its comprising Holarchies under study, the use of some of the modeling tools offered by UML wouldn't make clear at all the interactions that exist during normal manufacturing system operation or, moreover, when an exception occurs.

To model the Holonic behavior aspect, with its inherent characteristics of *Autonomy*, *Cooperation* and *Control*, Petri Nets are a good alternative. Some advantages offered by Petri Nets, especially in Manufacturing Environments Modeling are their ability to handle operations' sequence concurrency and non synchronous events, conflict and mutual exclusion in systems (University of Skövde, 2003). According to McClellan, some of the most common disturbances experienced by a Manufacturing System are: *Resource*

Availability Loss, Modifications in Sequencing or in Prioritizing, Quantity Changes, Making Routing Revisions, Machine Adjustments for Quality Assurance, and *Marking of Orders for Material Shortage*. Figure 5 depicts the concept in how MES Holon Holarchy acts under a disturbing circumstance and its representation, taking in count its Cooperation, Control and Autonomy features.

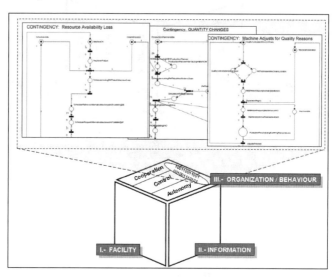

Figure 5. Organizational / behavioral MES-Holon Holarchy Feature. Different adopted configurations for the same Holarchy under different circumstances

These issues belong to organizational/behavioral part of the MES Holon conceptual Model. It is important to remember that, in Petri Nets notation, each Place (denoted by a circle) represents the state adopted by a system's element let represent the dynamic part of the MES software system.

3.3. MES Holon comprising elements

At this point, is necessary to define which the basic structure elements are for a Holon inside the MES-Holon Holarchy. Fan-Tien Cheng et al. determined that a Holon being part of a Manufacturing Execution System, in order to execute its tasks, ensure communication with the Holonic Manufacturing System and have a certain degree of resistance to environmental disturbances must have the following basic elements: a *Holon Kernel*, which manages and/or uses the *Holon Configuration* (comprised by a *Remote Communication Interface*, a *Security Mechanism*, a *Local Database* and a *Knowledge Base*). All these elements and their internal hierarchy are named the *"Generic Holon"*. Besides the Generic Holon, there exists the *Specific Configuration* inherent for each kind of Holon. MES' elementary holon common structure, adopted and adapted for present's research is depicted in Figure 6.

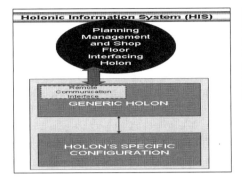
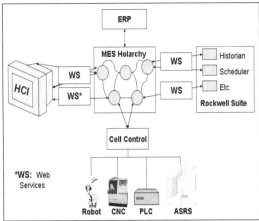

Figure 6. MES' elementary holon common structure

Figure 7. MES-Holarchy interaction with outside elements through Web Services

4. Implementation tools

In order to reach the implementation of the MES Holon and its components inside of a Holonic environment, it is necessary to have an informatics' architecture with the required characteristics for an easy and low-cost remote communication interface. This can be done through Web Services. In a Web service, message description specifies the operations that can be requested, the necessary parameters and the type of results produced. Such platform independent specification provides a mechanism that can be implemented in different commons programming languages. The Web Services can provide the infrastructure to support whole MES Holon Functionalities. Under a Web Services approach, these functionalities become a collection of Web Services that work together to control the information flow between different Holons and applications. When information is required, the application client invokes remote functions on the server to send and receive data (Canché, 2004).

5. Preliminary validation of the proposal

A laboratory validation of this proposal was done in a Manufacturing Cell, through a first instantiation of the WOM and Scheduling Holons and their interaction. The decision on the first Holons to instantiate was done based in the fact the scheduling tasks are considered the most important in a MES System. Inherently, a Scheduling Holon requires the direct interaction with the WOM Holon in order to manage adequately the orders to be scheduled. This first controlled environment case was adopted due to the fact that the Manufacturing Cell operations are analogous to a SME metalworking environment, in which machining, assemble and inspection operations occur. Comprised resources in this Manufacturing Cell can be classified in Human Resources (a Managing and Operation Responsible), Material Handling Resources (Automatic Storage and Retrieval System -ASRS- Station, Containers and Pallets), Production Resources (Machining Station, Assemble Station and Robot), Measuring and Testing Resources (Camera Inspection Station). RS BizWare software suite, granted by Rockwell Automation, as well as some other control and Human-Machine Interface (HMI) software were used in the project.

MES-Holarchy compounding elements interact with some of the mentioned applications through Web Services mechanisms in order to consume the required and specific services provided by these applications located outside MES-Holarchy. The required services are necessary in order to achieve the designed MES-Holons functionalities (Figure 7).

WOM tasks (Gaxiola, 2004) can be carried out instantiating the proposed components in the following:
- ❏ WOM Generic Holon:
  - Remote Communication Interface: ASWM,
  - Security Mechanism: an installed component on the ASWM Light Client (for this specific case),
  - Holon Kernel: an installed component on the ASWM Light Client (for this specific case),
  - Local Database: Access '97 database (CELMEX, created by CSIM-ITESM),
  - Knowledge Base: User's expertise provided through Rockwell's Software RS Scheduler ® Interface.
- ❏ WOM Holon Specific Configuration:
  - Manager: Microsoft Excel ® comma separated worksheet (*.csv file),
  - Managed Information Objects: the information objects that comprise the Microsoft Excel ® comma separated worksheet,
  - User Interface: those provided by Rockwell's Software RS Scheduler ® Interface, Microsoft Excel ®, worksheets and INTERMON ® didactic work cell system.

Scheduler Holon tasks (Gaxiola, 2004) can be carried out instantiating the proposed components in the following:
- ❏ Scheduling Generic Holon:
  - Remote Communication Interface: ASWM,
  - Security Mechanism: the punctual security mechanism embedded in the Rockwell Software RS Scheduler ® application,
  - Holon Kernel: an installed component on the ASWM Light Client (for this specific case),
  - Local Database: Microsoft SQL ® Database (RS Factory Data Model ®),
  - Knowledge Base: User's expertise provided through Rockwell's Software RS Scheduler ® Interface.
- ❏ Scheduling Holon Specific Configuration:
  - Manager: Rockwell Software RS Scheduler ® application,
  - Managed Information Objects: the information objects that comprise the specific tables of the Rockwell's Software RS Scheduler ® application in the context of the RS Factory Data Model ®,
  - User Interface: the user's interface provided by Rockwell's Software RS Scheduler ® Interface.

5.1. Uploading and Managing the Work Orders in the context of SME Manufacturing Systems

Firstly, the authentication procedure to interact with the WOM Holon is done through the previously described. The definition of the SME Manufacturing System Holarchy components and the uploading of the work orders (that can constitute also a mechanism for modifications of Work Orders Management) begins with an authentication process through a software component on the ASWM.

After that, the information uploading process (work orders uploading, which originally were deployed in Microsoft Excel® comma separated worksheet) is carried out through another information processing object or component of the ASWM (acting as an interface), which is depicted in Figure 8.

After the work orders uploading process (a part of the pool of digitalized unallocated orders can be seen in Figure 9) has been carried out, it takes the turn to schedule such orders (the scheduling process is omnipresent both in normal operation and disturbances). Before to schedule, a detailed digitalized work order is depicted for a specific order the general data involved with it (customer, due date, product requested, etc.), the involved operations, resources (principal and additional) as well as the involved material actions (the involved materials to accomplish the order, their attributes, and the assigned level in the context of the finished product).

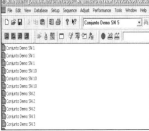

Figure 8. SME Manufacturing Information System

Figure 9. Pool of unallocated Work Orders

Figure 10. Case Study's Specific Schedule for sample work order number

Finally, the resulting schedule (focused on Study Case's work order number 5, but constrained by the rest of the other product's comprising operations) is shown in figure 10. The scheduling approach followed for this specific case was a Forward Scheduling, based on Due Dates.

6. Conclusions and further research

Through the iterative process of knowledge acquisition inherent to the project, the major contributions of present work can be considered in the following issues:

- an ampliation of Toh's proposal of the Metalworking SME Manufacturing Information System through the identification and proposal of the specific elements and information objects that would support the Holonic behavior of the entire Metalworking SME Enterprise under the conceptualization of a *Meta-Holarchy*,
- a tasks-logical-affinity segmentation is proposed, in order to determine the specific functional Holons to be included in the MES Holarchy.

A new architecture for both a MES System under the Holonic Concept and its comprising elements is proposed. The new architecture's proposal of the comprising elements of the MES Holarchy as the building elements of the Holonic MES System is relevant from the point of view that it provides a novel approach to conceptualize and implement a MES System (normally restrained to stable industrial environments, like in the commodities fabrication or massive production) in SMEs environments with high instability conditions.

- it is proposed the use of cheap punctual software solutions with the intention of just to consume the necessary and specific required services (conceptualizing them as information processing objects), the communication among Holons based in variegated software platforms is ensured through a novel technology that nowadays is increasingly invigorating: Web Services. The use of easily acquirable technology facilitates in an outstanding manner all the process, becoming it reachable to the majority of Metalworking SMEs,

❏ the implementation of two Holons from the proposed architecture (the Work Order Management and the Scheduling Holons) constitute a good departure base for a further complete concept's implementation and future work on this research line.

It is expected that present work, despite the fact that it is just a proposed concept, if instantiated, would be useful to ameliorate the nowadays huge Information Technology gaps and scarcities that prevail in the Small and Medium Enterprises (the majority of the Industrial Base) in industrial developing countries, like Mexico, helping to improve their competitiveness.

Acknowledgements
The research reported in this paper is part of a Research Chair in Manufacturing of ITESM titled "Design, Manufacturing and Integration of Reconfigurable and Intelligent Machines". The authors wish to acknowledge the support of this grant in the preparation of the manuscript. The authors also want to manifest their special gratitude to Rockwell Automation Mexico donator and technical advisor of the complementary software used in this research.

References

1. Gaxiola L., Ramírez M.J., Jiménez G., Molina A.: *Proposal of Holonic Manufacturing Execution Systems Based on Web Service Technologies for Mexican SMEs*. HoloMAS 2003: 156-166.
2. Leitão P., Colombo A.W., Restivo F.: *An Approach to the Formal Specification of Holonic Control Systems*. HoloMAS 2003: 59-70.
3. Cheng F., Wu S., Chang C.: *Systematic Approach for Developing Holonic Manufacturing Execution Systems*. IECON'01: The 27 th Annual Conference of the IEEE Industrial Electronics Society. Inst. of Manuf. Eng., National Cheng Kung Univ. Taiwan R.O.C. (2001) 261 – 266.
4. McClellan M.: *Applying Manufacturing Execution Systems*. ISBN 1-57444-135-3, 1997.
5. American National Standard & ISA – The Instrumentation, Systems, and Automation Society: Standard ANSI/ISA–95.00.02–2001, app. October 2001, USA, ISBN: 1-55617-773-9, Pg. 13
6. Koestler A.: *The ghost in the machine*. London, Arkana, 1990; ISBN 0140191925.
7. MESA International, A High Level Vision, White Paper, Pittsburgh, PA, USA.
8. TATA Consultancy Services: Manufacturing Execution Systems - A Concept Note; Air India Building, 11th Floor, Nariman Point, Mumbai 400021, India; February. Pg. 2 – 3. Feb. 2002.
9. Wyns J.: *Reference Architecture for Holonic Manufacturing Systems – the key to support evolution and reconfiguration*. Doctoral Thesis; Katholieke Universiteit Leuven.
10. Manufacturing Systems (HMS) Internat. Symposium. Kitakyushu, Japan, Oct. 19th-20th,2000
11. Botti N., Vicente J., Giret B.A.: Aplicaciones Industriales de los Sistemas Multiagente. Departamento de Sistemas Informáticos y Computación Universidad Politécnica de Valencia, Camino de Vera s/n, 46022. Valencia {vbotti, agiret}@dsic.upv.es.
12. Rahimifard S.: *A Practical Representation of Holonic Manufacturing Systems*. Fifth IFIP/IEEE (BASYS'02), September 25-27, 2002. Cancun, Mexico (2002), Pg. 323 –330.
13. Gaxiola L.: *Holonic environment integration methodology for metalworking SMEs manufacturing systems: MES Holon*. MSc Thesis. ITESM, Campus Monterrey. N.L., 2004.
14. Toh, K.T.K.: *A Reference Model for Information Specification for Metalworking SMEs*; Ph.D. Thesis, Loughborough University, 1997.
15. University of Skövde: *An Informal Introduction to Petri Nets*; Sweden, 2003; Last access on January, 2003, http://www.ite.his.se/ite/utbildning/kurser/auc111/pdf/pn_informal.pdf.
16. Canché L., Ramírez M., Jiménez G., Molina A.: *Manufacturing Execution Systems (MES) based Web Services Technology*. 7th IFAC Symp. on Cost Oriented Automation, Gatineau/Ottawa, Canada, June 7 - 9, 2004.

SHOP FLOOR CAPACITY EVALUATION

Haddad R.B.B., Carvalho M.C.H., Bera H.

Abstract: This work analyses the integration of mathematical programming with ERP to solve capacitated production scheduling problems. The limits of critical resources on the shop floor are recognised and modelled or as a capacitated network flow problem with additional constraints or as a graph. The proposed formulation add to the standard ERP the "temporal view" of the scheduling problem that overcomes overload situations.

1. Introduction

Production scheduling in large enterprises is a multi period, multi stage, multi product decision problem. As a consequence, in order to determine a good solution for this problem, a manager needs to be supported by some software tools. Enterprise Resource Planning (ERP) is one of the industrial practices employed for supporting decisions on finance, quality control, sales forecast and manufacturing resource planning apart from other fields (Gershiwin, 1986; Clark, 2002).

The ERP composed of modules, proposes an integrated solution for the whole enterprise. It covers, besides Production Areas, also Account, Financial, Commercial, Human Resources, Engineering, Project Management, etc. Inside the Production area, there is the Manufacturing Resource Planning module, which is the focus of this paper. It establishes the quantities and due date for the item to be manufactured or assembled. When it generates the production orders to be implemented on the shop floor, it does not consider the actual installed or production capacity. It works as if the shop floor had infinite capacity and can generate production orders that overload critical resources, hiding good shop floor planning implementation (Carvalho, 1999; Rom, 2002) When overload, the manager has to decide, based on his experience, the orders to be anticipated, postponed and those not to be implemented. As it is a large problem that evolves with time, it is impossible for the manager to reach a good solution, without any supporting tool.

Considering that ERP is an industrial practice for manufacturing production planning and scheduling, encountered in most of the medium size enterprises (Clark, 2002; Rom, 2002), this paper proposes to incorporate a capacity module in order to overcome the major disadvantage of this approach, that is infinite capacity consideration.

This module can be implemented by Integer Programming, which is an Np hard problem (Papadimitiou and Steiglitz, 1982) and has the drawback of dimensionality even for small problems. An approximated approach that escapes the dimensionality problem and generates reasonable solutions for actual systems, is Linear Programming. To take advantage of the very special structure, the problem can be modelled using network flow with side constraints algorithms. Other consideration to simplify the model is to consider only the production bottlenecks optimisation. In some cases it is also possible to transform the network flow problem into a graph.

This paper is organised as following: section two presents conceptual aspects of discrete production systems; section three presents the mathematical model; section four is dedicated to the integration between ERP and Mathematical Programming; section five presents an application example and finally, in section six, are the conclusions.

2. Conceptual aspects of manufacturing production systems

In a manufacturing system, items flow through the production resources suffering transformation in the form, constitution or location, following a predefined sequence of operations to become a final product. The production-planning problem would be easy to solve based on known demand and infinite raw material availability. But actual manufacturing systems have to process many products, and share limited resources to supply an unknown demand.

In this complex environment, the production capacity will depend on several factors, such as: production mix, production sequence, sharing resources rate, set-up times and temporal demand behaviour. Therefore the production-planning problem evolves to management of temporal resources availability according to demand flotation (Carvalho at al., 1999). The challenge is to establish a production planning that determines the temporal allocation of production resources (raw material and machines), that meet temporal demand requirements and maximise the long term enterprise objective, the revenue. The ERP is a software that tries to help the decision maker in achieving the above objectives. It is composed of modules that work on the same database in such a way to normalise all enterprise activities. One of its modules, the manufacturing module, is responsible for production planning management, and its approach is start the production on the latest date that reaches the demand due date. To meet demand due date is a crucial point for the enterprise since it results in client satisfaction and consequently increases enterprise competitiveness (Haddad, et al 2002).

Benton and Shin (1998), highlight that the manufacturing module is also responsible for establishing an efficient production program, promoting fast response to marketing changes. Although this module is considered to be a production organiser", since it specifies the task to be implemented, it does not consider properly the actual production capacity of each resource. One consequence of this simplification is presented in the Figure 1. Production planning allocates for the first week enough production to meet the specified demand and has the same procedure for the second week. As it did not consider the production capacity, the second week is overloaded. Probably, the best policy would be to anticipate part of the production needed in week (i+1) for week (i), to avoid outsourcing, duty or demand cutting. But, in this case, the information of the temporal behaviour of the demand and the production system capacity, during the planning horizon, have to be considered in the decision making process.

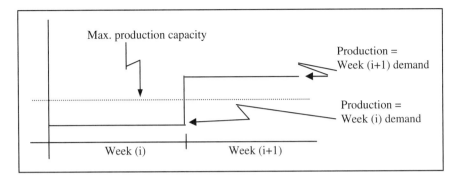

Figure 1. Example of Two Weeks ERP Production Scheduling

The proposal of this paper is to restructure a existing ERP by the introduction of a new Module. According to this approach, the decisions are made in sequence and each hierarchical level has its own characteristics, including length of the planning horizon, level of detail of the required information and forecasts of demand and resources. It includes a new planning level that generates a limited number of optimum scenarios through the PRONET algorithm to be evaluated in detail by an ERP Module, them this approach suggests solving the production scheduling planning by integrating optimisation method (Linear Programming) with industrial practices (ERP).

3. Mathematical model

3.1. Mathematical model for the production system

The manufacturing process is composed of a sequence of production activities over the production network following a pre-defined technological sequence. In their route through the production system, the items, raw material, semi-manufactured and products wait in queues for release conditions, and are subjected to fabrication or assembly operations until they reach the last production stage. These flow characteristics allow to model the production problem as a Network Flow model (Carvalho et al., 1999), as shown in the Figure 2. In that figure A_i is the incidence matrix of product **i**, **x** is the decision variable representing the material flow in the production resource, limited to the lower bound **l** and upper bound **u**; **b** are the demands and raw material availability vector, S_i describes the mutual capacity and mutual inventory constraints, called side constraints, and **d** is the production capacity vector. This problem exhibits a special structure that is exploited in the solution considered by PRONET algorithm (Yamakami et al, 2000).

i = 1,...,n index for products	u = upper bound
x_i = amount of product i produced	b = demands and raw material availability vector
Ai = incidence matrix of product i	Si = side constraints
l = lower bound	d = production capacity vector.
c = unit cost for production of product i	t = periods considered in the model

Figure 2. Block angular structure of Matrix A

$$\text{Min} \sum_{i=1}^{n} c_i^t x_i$$

$$S.A.: \begin{bmatrix} A_1 & 0 & \cdots & 0 & 0 \\ 0 & A_2 & \cdots & 0 & 0 \\ \vdots & \vdots & \ddots & \vdots & \vdots \\ 0 & 0 & \cdots & A_n & 0 \\ S_1 & S_2 & \cdots & S_n & I \end{bmatrix} \begin{bmatrix} x_1 \\ x_2 \\ \vdots \\ x_n \end{bmatrix} = \begin{bmatrix} b_1 \\ b_2 \\ \vdots \\ b_n \\ d \end{bmatrix}$$

$$l \leq x \leq u$$

If the production lines are decoupled the Si matrices are equal to zero and the optimal manufacturing scheduling is reduced to a simple network flow problem.

3.2. Solution algorithm

The large dimension of Linear Program models associated to the production planning, incentives the development of algorithms that explores the special structure of the problem. One approach should be the Netside Algorithm (Kennington, 1990), but its efficiency is limited to the size of the problems (Carvalho et al., 1999). Interior Points Methods (IPM), has been grown importance since the good results obtained in Adler et al. (1989). Nowadays the IPM for Linear Programming are well established for practical applications (Gondzio, 1996), and good algorithms are available (Wright, 1996).

Although IPM should be indicated to large scale problems, faster solutions can be achieved applying some practical knowledge of actual problems together with mathematical transformation. The first one is the bottleneck management defended by the Theory of Constraints (TOC) (Goldratt and Cox, 1986). According to this theory, a productive system can be divided into two kind of resources: the bottlenecks and the others. Bottlenecks are those resources with limited production capacity and therefore need an special treatment. Planning and scheduling these resources must be managed carefully, looking for better production dates. The decisions regarding to other resources are submitted to bottleneck decisions.

The second assumption is the use of mathematical transformation over the constraint matrix. According to Zahorik et al. (1984), "the immense size of these problems and the imprecise nature of many of their costs and demands further suggest that good heuristics may be as desirable as (presumably) more costly optimisation algorithms". The authors presented an heuristic to transform a special case of a network flow with side constraints in a pure network flow. This special case assumes multi-level, multi-item, a three periods long problem, no backorders, and one of the possibilities below:
- inventory side constraints at any level, or
- production and inventory side constraints only at the finished level.

4. MRP and linear programming integration

Although ERP is an industrial practice largely employed in manufacturing production planning, it is not enough to completely solve the complex production-scheduling problem mainly because:
- the basic form of an MRP assumes that there are no capacity constraints,
- as a consequence it does not consider the time co-ordination of production capacity and raw material with demand requirements.

The temporal co-ordination can be reached by the integration of a Linear Programming algorithm with the MRP module of an ERP. MPS generates long term planning for product types as shown in Fig 3. a). PRONET assesses the ERP Data Base, and starting from MPS targets, considering production and raw material constraints, product priorities, and aiming at the maximisation of the revenue, generates feasible production scheduling. It suggests, when necessary, the anticipation or postponing of demand attainment, according to a pre defined criteria. An alternative scheme is shown in Figure 3.b) that suggests to run the MRP module, and verify capacity or raw material constraint violation. When violation occurs, PRONET is runed to generate a feasible solution. In both approaches, the objective is to maximise customer service level by co-ordinating the temporal distribution of the production according to production availability of machines, storage and raw material

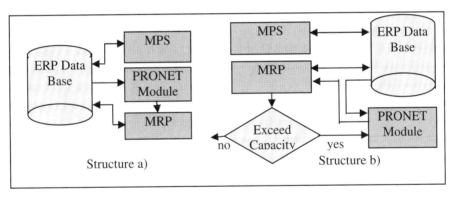

Figure 3. Structures for linear programming and ERP integration

The Module PRONET was developed in two versions. In the first one the scheduling is generated by a Network Flow Algorithm with Side Constraints. When Zahorik et al. (1984) conditions are satisfied, the problem is transformed into a graph and the second version can be runned. Linear Programming generates the optimal result to the production scheduling, defined as temporal production organisation of the products, considering capacity availability The results need to be refined and MRP starts from the targets established by the PRONET algorithm, considering explicitly purchase, storage, production orders liberation and set-up times and determines the scheduling without observe the actual production capacity. Therefore, Linear Programming and MRP are complementary planning tools since the first one has the temporal visibility of a capacitated production problem and the second one has the visibility of the items and the set-up time allowing a more accurate representation of the problem.

5. Example

Consider a spring factory with a bottleneck work centre "B", like showed in Figure 4. Work centre B is composed of two machines – M15 and M19. Consider two products – PR1 and PR2- that have one of their operations in this restricted work centre. The available production time of B is 40 hours per week and it has different production rates for PR1 and PR2. B can process 144000 pieces of PR1 or 128000 pieces of PR2 in this period.

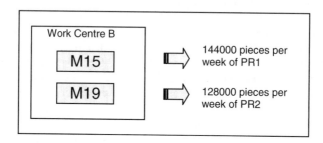

Figure 4. Restricted work Centre in a Spring factory

Suppose PR1 and PR2 have demands like presented in table 1 for weeks 35 and 36. In this situation if the ERP was adopted it should provide a scheduling with an overload of 8.41hours in week 35.

Table 1. PR1 and PR2 demands for weeks 35 and 36

	PR1	PR2
Week 35	118000 pieces	50000 pieces
Week 36	40000 pieces	40000 pieces
total	158000 pieces	90000 pieces

In this situation the manager must decide the orders to be postponed or, possibly, those not to be implemented. The PRONET Module helps the manager in these decisions. Applying, then, the PRONET to Table 1 data, will generate the optimal scheduling as showed in Figure 5, for these two weeks demands.

The upper part of this figure is related to PR1 and the lower to PR2. The part related to each product is divided in two branches: one for week 35 and the other one for week 36. The PRONET solution suggests to make 98582 pieces of PR1 and 41730 pieces of PR2 in the week 35 and 59418 pieces of PR1 and 48270 pieces of PR2 in the week 36. 19418 pieces of PR1 and 8270 pieces of PR2 will be made in backorder. The need of backorders were expected because week 35 was overloaded and week 36 was idle.

Table 2 presents the capacity utilisation of machines M15 and M19 in weeks 35 and 36. M15 has total load in both weeks and M19 is idle in second week. The PRONET scheduling is the data entry for the MRP. MRP must provide production and purchase orders and consider set-up times.

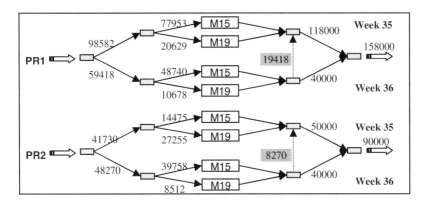

Figure 5. Minimum cost scheduling for table 2 demands in work Centre B

Table 2. Percent utilisation of M15 and M19

Machine	Week 35	Week 36
M15	99,30%	100,00%
M19	99,76%	39,98%

The opposite situation was also studied. That means, first week idle and second one overloaded. In this situation the best solution was to advance part of the second week production. That happened because according to the parameters entered in the system it was more expensive pay extra hours than to storage products for one week.

6. Conclusion

Production Planning are complex problems that need different approaches to balance, at the same time, resource availability, enterprise interest and client needs. Taking into account these aspects, this work associated Mathematical Programming (Linear Program) with Industrial Practice (ERP). The combination of these two approaches showed to be adequate for the solution of production planning problems in actual environments. The production optimisation balances the capacity with temporal demand oscillation and induces, when necessary, production transfer from overloaded to idle periods, avoiding in this way extra hours and outsider services. It was presented two different approaches for the Production Optimisation Module. In the first one the problem is structured like a network flow and in the second one problem is solved using a network flow with constraints algorithm. For the first approach some conditions must be considered:
- inventory side constraints are considered only in first three periods, or
- production and inventory side constraints only at the finished level.

Mathematical Program results will be a "near optimum" value, since calculations does not consider set-up times. To better represent set-up times, the module MRP is included. The advantage of this integration is clear. It helps with multiple approaches the manager in decisions by adding enterprise competences, overcoming the usual disruption by radical changes in enterprise procedures. The optimisation software was developed independent of the ERP structure. In this way it can be integrated in any ERP package available in the market.

Aknowledgements

We would like to thank LOGOCENTER for the ERP LOGIX installed at CenPRA and CNPq for the traineship support

References

1. Adler I., Resende M. G. C., Veiga G., Karmakar N. (1989): *Implementation of Karmakar Algorithm for Linear Programming*. Mathematical Programming, 44; 297-335.
2. Benton W. C., Shin H. (1998): *Manufacturing Planning and Control: The Evolution of MRP and JIT Integration*. European Journal of Operational Research; 110; 411-440.
3. Carvalho M.F.H., Fernandes C.A.O., Ferreira P.A.V. (1999): *Multiproduct Multistage Production Scheduling (MMPS) for Manufacturing Systems*. Production Planning & Control, 10; 7; 671-681.
4. Clark A. R. (2002): *Optimization Aproximations for Capacity Constrained Material Requirement Planning*. Production Economics; 84; 115-131.
5. Gershwin S.B., Hildebrant R. R., Suri R., Mitter S. K.: *A Control Perspective on Recent Trends in Manufacturing Systems*. IEEE Control System Magazine, vol 6 n. 2, (1986); 3-15.
6. Goldratt E., Cox J. (1986): *The Goal: A Process of Ongoing Improvement*. New York: North River Press.
7. Gondzio J. (1996): *Multiple entrality Corrections in a Primal-Dual Method for Linear Programming*. Computacional Optimization and Applications; 6; 137-157.
8. Haddad R.B.B., Cravalho M. F. H. (2002): *Complementaridade entre Programação Matemática e ERP*. Congresso Brasileiro de Automática, Natal; 245-250.
9. Kennington J.L., Whisman A. (1990): *Netside User's Guide*. Department of Computer Science and Engineering, Southern Methodist University, Dallas.
10. Papadimitriou C. H., Steiglitz K. (1982): *Combinatorial Optimization- Algorithms and Complexity*. Prentice-Hall, Inc; New Jersey.
11. Rom W.O., Tukel O.I., Muscatello J.R. (2002): *MRP in a Job Shop Environment Using a Resource Constrained Project Scheduling Model*. The International Journal of Management Science; 30; 275-286
12. Wright S. J. (1996): *Primal-Dual Interior Point Methods SIAM Publications*. SIAM; Philadelphia, PA.
13. Yamakami A., Takahashi M.T., Carvalho M.F.H. (2000): *Comparison of Some Algorithms for manufacturing Production Planning*. IFAC-MIM 2000 Symposium on Manufacturing, Modeling, management and Control, Rio Patras; 280-284.
14. Zahorik A., Thomas L.J., Trigeiro W.W.: *Network Programming Models for Production Scheduling in Multi-Stage*. Multi-Item Capacitated Systems Management Science, Vol. 30, no 3, Março 1984.

IMPROVING SURFACE QUALITY USING MQL MACHINING

Jung J.Y., Lee C.M., Cho H.C., Cui H., Hwang Y.K.

Abstract: This paper presents an approach to improve the quality of machined surfaces by machining with MQL (Minimum Quantity of Lubrication). This research sets up an experiment with experimental factors and three levels for each factor. The factors are cutting speed, feed rate, depth of cut, and lubrication. The surface quality in this experiment means surface roughness. This experiment is carried on a lathe to cut aluminum. This research finds the optimal levels of each cutting parameter that produces good surface roughness.

1. Introduction

Precision machining is a primary manufacturing process in metal working industries. It is well known that metal cutting takes a great important part in metal working industries. Machining requires coolant and lubricant to improve surface roughness. Coolant is effective to decrease heat generated in machining. Lubricant gives good dampness for lubrication. Thus, coolant and lubricant are necessary for machining. However, they produce serious circumstantial problems in machining since it consumes a lot of them. Very small particles of lubricant and coolant float in air in machining shops. This aerosol of cutting fluids causes many diseases in the respiratory organ of workers. Cutting fluids are necessary to improve the quality of machined surface, but they give havoc on both workers and circumstance. Thus, many developed countries limit its usage to keep clean circumstance.

Many literatures (Klocke, 1997, Diniz, 2002, Baraga, 2002, Kelly 2002, Vieira, 2001) suggest approaches to reduce the consumption of cutting fluids. Dry cutting, semi-dry cutting, and chilled air cutting are alternatives arose to resolve the problems in machining. Dry cutting does not consume a cutting fluid. Since this has poor cooling and dampness effect, a cutting tool wears out or chips in a short time. Chilled air cutting gives cooling effect, but less effect on lubrication. It makes rust on the machined surface of ferrous material. Also, a nozzle linked to the supply tank of chilled air is frozen in about a couple of ten minutes after use, which chokes the nozzle. This shuts off the supply of chilled air. Thus, these two methods do not give acceptable solution to reduce the consumption of cutting fluids.

Semi-dry cutting consumes minimum quantity of lubricants. It gives fairly good cooling and dampness effect for machining. Even it uses cutting fluids, but only small amount that does not give havoc to workers and circumstances in machining shop. This can be a good solution to replace wet cutting that consumes a lot of cutting fluids. However, the characteristics of the semi-dry cutting are not well known in detail. This research investigates some effects to improve surface quality in MQL machining for turned material on a CNC lathe.

2. Experimental design for MQL machining

The purpose of the experiment for MQL machining is to find any effect of experimental factors to surface quality. The quality of machined surface is evaluated in several ways. However, the surface quality set limit to surface roughness in this paper.

This research deals with surface quality of turned parts on a lathe. The experimental factors for this problem are to find effects on surface roughness. This research confines experimental factors to feed rate, cutting speed, and amount of MQL. Since depth of cut does not have great effect on a surface roughness (Shaw 2005), this research excludes the depth of cut from the experimental factors. The levels of each experimental factor are set to be three levels each. The best levels for each factor that are obtained from the previous literature (Lee, 2004) are the mid levels in the experiment. One level higher or lower than the mid level is added for each factor.

3. Apparatus for an experiment of MQL machining

A CNC turning center manufactured by D machine tool Company in Korea is utilized for this experiment. The following Figure 1 shows the CNC lathe that is adopted for this experiment. The machine is MQL only purposed.

Figure 1. CNC lathe for MQL machining

The cutting tool insert is coated tungsten carbide with TiN-TiCN-Al2O3-TiN by the method of Chemical Vaporized Deposit. Tool holder is PCLNR 2020 K12. MQL supplier utilized is Vario UFV10-001 manufactured by VOGEL Company in Germany. The following Figure 2 is MQL supplier that is utilized in this research. The Table 1 is a detail specification of the supplier.

From the previous experience (Lee, 2004), machining feature is fully dependent on the distance from a tip of MQL supplier to work-piece. This research set the length 19 mm. A copper horse is linked from the MQL supplier to the tip of a nozzle. Its diameter is 5 mm and tip's is 2 mm. Lubricant is supplied with jet in about 45 degrees angle to the machining surface. The pressure of the compressed air is about 45 pounds per square inch and LubriFluid F100 is used in the experiment. The following Figure 3 shows nozzle set-up on a machine. The following Table 2 is the specification of the lubricant utilized in the experiment.

Figure 2. MQL supplier   Figure 3. Nozzle set-up

Table 1. Specification of the MQL supplier

Oil drop-let size (μm)	Aerosol quantity (Mℓ/h.)	Air consumption (Mℓ/min)
0.5	5~150	140~300

The turned surfaces are measured on the Surftest SV-624 made by Mitutoyo in Japan. This machine measures the surface roughness of turned parts in the centerline average method. The Figure 4 shows measuring machine that is used in this experiment.

4. Experiments and measurements

The work-piece for this experiment is SM45C carbon steel and its length is 99 mm, diameter 53 mm. Its hardness is measured in BHN 170~190. The following Table 3 shows the chemical composition of the material and the Figure 5 shows shape and dimension of the test specimen. The experiment is planned by an experimental design. It is designed with three factors and three levels for each factor. It is repeated two times to reduce potential errors. Experimental data are collected from the apparatus shown in the previous section. The surfaces of the turned specimen are measured on the measuring machine. The measuring direction is shown in the following Figure 6. One surface is measured three times to reduce potential errors. The measuring points are located on a surface with 120 degrees on a cylindrical circumference of specimens. The measured data are listed in the following Table 4.

Figure 4. Measuring machine

Table 2. Specification of the lubricant

Type	Base	Density at +20°C (g/cm^3)	Viscosity at +40°C (mm^2/s)	Flash point (°C)
LubriFluid F100	higher alcohol	0.84	25	184

Table 3. Chemical composition of the work-piece material

Composition Metal	C	Mn	Si	P	S
SM45C	0.45	0.64	0.18	0.021	0.03

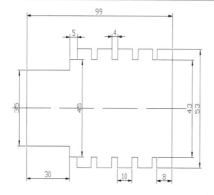

Figure 5. Dimensions of the specimen

Figure 6. Measuring points and direction

5. Analysis and result of the experiment

The collected data from the experiment are analyzed with MINITAB to identify variances. From the analysis, it is shown that the amount of MQL, feed rate, and interaction between them greatly affect to the surface roughness. Cutting speed does not considerably have effect to the roughness, which is not true in wet cutting. The following Table shows ANOVA (analysis of variances) for this experiment.

Table 4. Collected data from the experiment

Repeat 1					Repeat 2				
Order	MQL	Cutting speed	Feed rate	Surface roughness, Ra(µm)	Order	MQL	Cutting speed	Feed rate	Surface roughness, Ra(µm)
1	150	180	0.15	1.374	1	80	180	0.15	1.146
2	150	250	0.10	1.045	2	80	250	0.20	1.402
3	150	180	0.20	1.737	3	80	220	0.20	1.449
4	150	180	0.10	1.067	4	80	220	0.15	1.340
5	150	220	0.15	1.425	5	80	250	0.10	0.962
6	150	220	0.20	1.785	6	80	180	0.10	0.948
7	150	250	0.20	1.801	7	80	180	0.20	1.588
8	150	250	0.15	1.441	8	80	220	0.10	1.013
9	150	220	0.10	1.100	9	80	250	0.15	1.388
10	10	220	0.20	0.794	10	10	180	0.20	0.985
11	10	180	0.15	0.447	11	10	220	0.20	1.055
12	10	250	0.20	0.819	12	10	180	0.10	0.580
13	10	220	0.15	0.505	13	10	180	0.15	0.733
14	10	250	0.15	0.551	14	10	250	0.15	0.685
15	10	180	0.20	0.826	15	10	220	0.10	0.630
16	10	250	0.10	0.696	16	10	220	0.15	0.859
17	10	180	0.10	0.580	17	10	250	0.20	1.006
18	10	220	0.10	0.430	18	10	250	0.10	0.635
19	80	180	0.20	1.037	19	150	220	0.10	1.044
20	80	220	0.20	1.140	20	150	250	0.20	1.833
21	80	180	0.15	0.928	21	150	220	0.15	1.405
22	80	250	0.15	1.006	22	150	180	0.20	1.857
23	80	220	0.10	0.735	23	150	180	0.15	1.409
24	80	250	0.10	0.817	24	150	220	0.20	1.895
25	80	180	0.10	0.875	25	150	180	0.10	0.919
26	80	220	0.15	1.129	26	150	250	0.10	1.051
27	80	250	0.20	1.261	27	150	250	0.15	1.452

Table 5. ANOVA table

Factor	DF	SS	MS	F	P
A	2	4.5399	2.26997	101.30	0.000
B	2	0.0199	0.00996	0.44	0.646
C	2	2.3749	1.18744	52.99	0.000
A*B	4	0.0011	0.00028	0.01	1.000
A*C	4	0.4388	0.1097	4.90	0.004
B*C	4	0.0273	0.00682	0.30	0.872
A*B*C	8	0.0345	0.00431	0.19	0.990

The factors A, B, are C in the ANOVA table mean the amount of MQL, cutting speed, and feed rate respectively. A*C is the interaction of MQL and feed rate. From the Table, 1-P means reliable range. The amount of MQL affects to surface roughness with 100% reliable range.

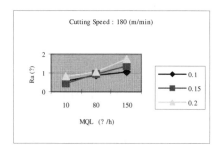

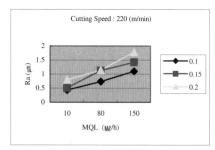

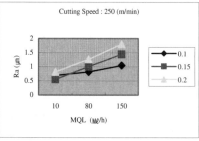

Figure 7. Amount of MQL and surface roughness

From the experiment, the amount of MQL has great effect on the surface roughness. MQL has good wetting and cooling effect since it increases the contacting area of lubricant to work-piece. In MQL machining, lubricant is highly pressed and jetted in small particles. The size of the particle of lubricant is inverse proportion to jet pressure. The higher the jet-pressure to lubricant is the smaller the particle size of lubricant is. The speed is also in a same situation to the size. Thus, high pressure of jet is required to make good wetting and cooling in machining. The following Figure 7 shows the effect of MQL amount on surface roughness. From the Figures for all three levels of feed rate, the surface roughness gets worse with the increase of MQL amount. This is because the amount of air in the MQL is getting less with the increasing the amount of MQL within one pressure level. The jet-pressure in the experiment is about three Bars, which may not be sufficient for MQL machining.

The surface roughness gets worse while the feed rate is increased. This is generally true for turning. The following Figure 8 shows the effect of feed rate on surface roughness. In general, surface roughness is best when feed rate gets smaller. However, the surface roughness is best when the feed rate is 0.15 mm/rev. with the 10 ml/h. MQL. When the feed rate is very small, it may cause buffing effect on the turned surface. Thus, it produces worse surface roughness.

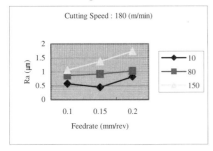
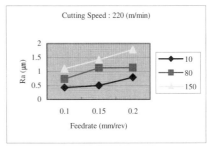
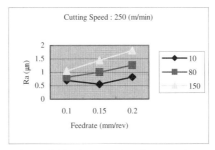

Figure 8. Feed rate and surface roughness

The cutting speed does not have great effect to the surface roughness. The following Figure 9 presents surface roughness versus to cutting speed. However, it is worth to note that cutting speed 220 M/min. gives good surface roughness. This is similar to wet cutting. Higher or lower than that speed produces worse surface roughness.

6. Concluding remarks

This research investigates effects on surface roughness while machining parameters vary in MQL machining. The work-pieces, SM45C, are turned on a CNC lathe. From the experiments above described, this paper draws the following conclusions:
- the amount of MQL, feed rate, and interaction between them have great effect on the surface roughness. The combination of the least amount of MQL and the slowest feed rate gives the best surface roughness at cutting speed 220 M/min.,
- turning with small amount of MQL maintains similar surface roughness in comparing with wet cutting. Thus, MQL machining is one of solutions to keep clean circumstance since it consumes the least amount of lubricant.

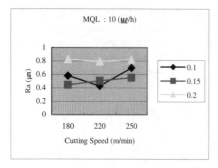
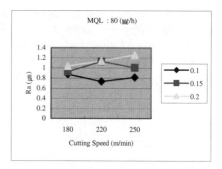
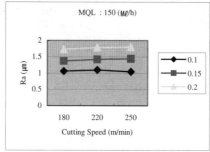

Figure 9. Cutting speed and surface roughness

This research does not consider as experimental factors the radius of nozzle tip and jetting direction of MQL from the nozzle. However these factors can greatly affect to surface quality. The future research should take count on these effects.

Acknowledgement
This work was supported by the Machine Tool Research Center at Changwon National University.

References

1. Baraga D.U., Diniz A.E., Miranda G.W.A., Coppini N.L.: *Using a minimum quantity of lubricant (MQL) and a diamond coated tool in the drilling of aluminum-silicon alloys*. Journal of Material Processing Technology, 2002; 122: 127-138.
2. Diniz A.E., Micaroni R.: *Cutting condition for finish turning process aiming: the use of dry cutting*. International Journal of Machine Tools & Manufacture, 2002; 42: 899-904.
3. Kelly J.F., Cotterell M.G.: *Minimal lubrication machining of aluminum alloys*. Journal of Materials Processing Technology, 2002; 120: 327-334.
4. Klocke F., Eisenblatter G.: *Dry cutting*. Annals of CIRP, 1997; 46 (2): 519-526.
5. Lee C.M., Hwang Y.K., Jung J.Y.: *The effect of cutting condition on surface roughness in MQL turning*. Proc. of 2004 Autumn Conference of the Korean Society of Machine Tool Engineers, 2004: 28-33 (Korea).
6. Shaw M.C.: *Metal cutting principles*. New York: Oxford Press, 2005.
7. Vieira J.M., Machado A.R., Ezugwa E.O.: *Performance of cutting fluids during face milling of steels*. Journal of Materials Processing Technology. 2001; 116: 244-251.

FUZZY MODELING AND APPROXIMATION OF SURFACE ROUGHNESS IN DRILLING

Sivarao P.S.

Abstract: The resurgence of interest in fuzzy logic over the past few decades has opened many new avenues in its applications. Fuzzy logic leads to greater generality and better rapport with reality. It is driven by the need for methods of analysis and design, which can come to grips with the pervasive imprecision of the real world and exploit the tolerance for imprecision to achieve tractability, robustness and low cost solution. This paper discusses a specific approach of modeling based on fuzzy logic, which are divided into two parts: fuzzy modeling and fuzzy approximation of internal surface roughness (Ra) in drilling. The paper concludes with model, rules and numerical outputs to show the effectiveness of the approach.

1. Introduction

Manufacturing has been the key to success among nations in the world economy in the 21st century. While responsiveness is very important to success of manufacturing firms, delivery of high quality products plays a vital role in manufacturing competitiveness. The internal surface roughness is the essential feature of drilling operation since the most applications of holes are assembly works, especially more focusing on the relative movement and tight size tolerance works. Therefore, high standard quality control is necessary to be introduced to assure quality [1, 2].

Many empirical methods have been developed to predict the internal surface roughness, but they are not efficient due to the numerous parameters of influential factor. Such factors are drill tool geometry, cutting condition, tool material, work piece material and vibration [3, 5, 13]. The author has carried out similar experiments for tool wear, surface roughness in drilling, involving paired and comparative analysis, whereby the statistical and inverse coefficient matrix methods were used to obtain the predicted values [18, 19, 20]. The measurement and stochastic modeling of torque and thrust force in twist drilling has been the main interest of many researchers as it was the best method of on-line tool wear (flank wear) sensing [14, 17].

Fuzzy modeling is based on the idea to find a set of local input-output relations describing a process. So, the method of fuzzy modeling can express a non-linear process better that any ordinary method [4, 6]. As more knowledge about the system is accumulated and the uncertainty diminishes the need for the fuzzy logic treatment and it can revert to a deterministic or statistical one. It means that, in practice there is a wealth of valid fuzzy logic applications and a freedom from the need to extend those based upon classical logic beyond their natural limits [8, 11, 12].

Fuzzy modeling and approximation for on-line detection to be studied with general approach and rules to accurately predict the internal surface roughness of drilled holes. Brief description of fuzzy approach, fuzzy model types, rules, membership functions, model and numerical output for specific drill job is discussed in this paper.

2. Fuzzy systems

Fuzzy modeling and approximation are the most interesting fields where fuzzy theory can be effectively applied. As far as modeling and approximation is concerned, one can say that the main interest is towards its applications. When we intend to apply fuzzy modeling and approximation to an industrial process, one of the key problems to be solved is to find fuzzy rules. Few ways to find fuzzy rules are [7]:
- the operator's experience,
- the production engineer's knowledge,
- fuzzy modeling and approximation of operator's action,
- fuzzy modeling of the process.

The rules fed to the fuzzy system for the purpose of modeling and approximations of this work are as shown in figure 1.

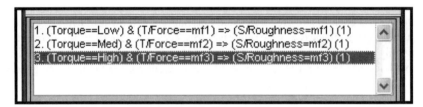

Figure 1. The rules used in the study

Through scientific practice, articulation of problems in fuzzy logic terms becomes more familiar and quite natural. It is then recognized as valuable tool in the expression and solution of cases where uncertainty, vagueness or ambiguity exists and where approximations are norm [9]. Basically, there are four types of different fuzzy systems. They are used at different conditions and rules as they have their own advantageous in the vast applications of fuzzy. They are:
1. Mamdani   2. Larsen
3. Tsukamoto   4. Takagi-Sugeno-Kang (TSK)

The one will be utilized in this paper is a special case of the Mamdani type in which the rules always have crisp consequence.

2.1. Algorithms in fuzzy

Fuzzy logic has a lot of applications in the real world. Basically the system will accept the input or some inputs and then pass the inputs to a process called fuzzification. In the fuzzification process, the input data (can be digital, precise/imprecise) will undergo some translation into linguistic quantity such as low, medium, high of physical properties. The translated data will be sent to an inference mechanism that will apply the predefined rules. The inference mechanism will generate the output in linguistic form. The linguistic output will go through defuzzification process to be in numerical form (the normal data form). Defuzzification is defined as the conversion of a fuzzy quantity represented by a membership function to precise or crisp quantity [10].

2.2. Fuzzy components

There are three basic components for a typical fuzzy system. They are input fuzzification, inference mechanism (rules application) and output defuzzification [13]. The entire fuzzy components of the system are as shown in figure 2.

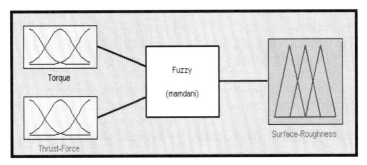

Figure 2. Typical components of fuzzy system

2.2.1. Input fuzzification

Input fuzzification 'translates' the system-input variables into universe of input memberships. Some refer to universe of input membership as degree of membership. The translating processes involve various sets of input membership. A series of fuzzy set is needed to be defined with analogue to the input range.

2.2.2. Fuzzy inference systems

Fuzzy inference is the process of formulating the mapping from a given input to an output using fuzzy logic. The mapping then provides a basis from which decisions can be made, or patterns discerned. The process of fuzzy inference involves all of the pieces, which called membership functions, fuzzy logic operators, and if-then rules. There are two types of fuzzy inference systems that can be implemented in the Fuzzy Logic; Mamdani-type and Sugeno-type. These two types of inference systems vary somewhat in the way outputs are determined [15].

Mamdani-type inference, as we have defined it for the Fuzzy Logic, expects the output membership functions to be fuzzy sets. After the aggregation process, there is a fuzzy set for each output variable that needs defuzzification. It is possible and in many cases, much more efficient to use a single spike as the output membership function which is known as a singleton output membership function. It enhances the efficiency of the defuzzification process because it greatly simplifies the computation required by the more general Mamdani method, which finds the centroid of a two-dimensional function.

Fuzzy sets and fuzzy operators are the subjects and verbs of fuzzy logic. These if-then rule statements are used to formulate the conditional statements that comprise fuzzy logic. A single fuzzy if-then rule assumes the form if X is A then Y is B where A and B are linguistic values defined by fuzzy sets on the ranges (universes of discourse) X and Y, respectively. The if-part of the rule "X is A" is called the antecedent or premise, while the then-part of the rule "Y is B" is called the consequent or conclusion [16].

2.2.3. Output defuzzification

Defuzz (X, mf, type) returns a defuzzified value out, of a membership function mf positioned at associated variable value X, using one of several defuzzification strategies, according to the argument, type. The variable type can be one of this:

1. Centroid: centroid of area method
2. Bisector: bisector of area method
3. Mom: mean of maximum method
4. Som: smallest of maximum method
5. Lom: largest of maximum method

In case, the type is not one of the above, it is assumed to be a user-defined function. The x and mf are passed to this function to generate the defuzzified output. In this work, weighted-centroid output method has been used to translate the linguistic-output into numerical form. Fuzzy membership function obtained is as shown in figure 3.

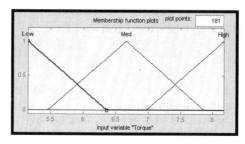

Figure 3. Membership function obtained by the fuzzy system

3. Experimental procedure

It is a metal drilling manufacturing operation. The machine is a single tool drill, repetitively carrying out a single identical task on parts as they arrive. The task involves the drilling of a single hole in each part. When a part is completed, another part is immediately available for drilling. As holes are drilled, the tool wears and is susceptible to breakage. It is assumed that there is a limit to the wear beyond which the tool is unacceptable as it directly affects the internal surface roughness of the drilled hole. Unmonitored tool wear may lead to tool breakage or bad surface roughness and may cause the almost finished product to be reworked or rejected. This will incur the manufacturing cost and machine down time. Evolution of tool wear necessitates the occasional replacement of the tool. Replacement of the tool involves some costs for both time and material. The spindle speed and the feed rate of the drill are fixed.

Eight identical experiments were carried out with 10 mm drill tool on a radial drilling machine in dry cutting condition to test the stability of the tool in producing required surface roughness and latter apply for on-line monitoring system to maintain the quality of drilled holes in unmanned drilling operation. For the purpose of experimentation, a mechanical type drill dynamometer with individual output of thrust force and torque were used. The work piece was clamped on the dynamometer with the help of chuck jaw. The values of torque and thrust force of an individual hole has been taken while drilling 30 holes onto the clamped work piece. The values of surface roughness are the average of three points along the drilled hole measured with the help of Mitutoyo surf-tester. The cutting conditions and tool specifications are as shown in table 1 and the worked samples are as shown in figure 4.

Table 1. Machining conditions and tool specification

Cutting Conditions		Tool Specification	
Cutting speed	12.69 m/min	Material / Dia.	HSS 10
Feed	0.285 mm/rev	Point angle	118°
Number of holes	30	Helix angle	32°
W/ piece material	Mild steel	Clearance angle	10°
W/piece thickness	25 mm	Hardness (3 pnt ave)	57 Rc
W/piece dimension	160X160X25 mm	Flute length	86 mm
Coolant	No	Overall length	136 mm

Figure 4. The worked samples

4. Results and discussion

The fuzzy modeling and approximation in this scientific research has been carried out successfully and the approximated values are very closely matching to the observed values of the experimental work carried out. The numerical output of the entire drill job for one of eight experiments carried out is as shown in table 2. The results confirm that the use of fuzzy based modeling and approximating is effective for tool condition monitoring in order to maintain the quality of the drilled holes. The deviation of observed and fuzzy output for all the holes is within strongly acceptable region.

Table 2. Fuzzy system numerical output as compared to the observed values

Hole Number	Thrust Force (N)	Torque (Nm)	Observed Ra (micron)	Fuzzy Estimated Ra (micron)
1	1366.15	5.15	4.20	4.60
2	1371.16	5.19	4.26	4.70
3	1379.46	5.26	4.38	4.71
4	1385.10	5.36	4.46	4.71
5	1396.29	5.44	4.56	4.72
6	1405.21	5.49	4.67	4.72
7	1416.17	5.56	4.91	4.98
8	1426.37	5.63	4.96	5.11
9	1431.20	5.66	5.10	5.32
10	1439.19	5.69	5.20	5.47
11	1443.10	5.74	5.28	5.54
12	1451.29	5.79	5.39	5.63
13	1459.28	5.88	5.48	5.72
14	1462.18	6.00	5.56	5.82
15	1466.27	6.10	5.69	5.90
16	1472.16	6.19	5.76	5.99
17	1476.29	6.28	5.89	6.05
18	1484.19	6.42	6.00	6.10
19	1491.24	6.64	6.15	6.10
20	1500.00	6.84	6.29	6.10
21	1519.88	6.92	6.36	6.10
22	1541.26	7.00	6.39	6.10
23	1566.16	7.19	6.42	6.10
24	1584.39	7.24	6.46	6.10
25	1619.21	7.32	7.49	6.25
26	1629.44	7.46	7.54	6.26
27	1648.10	7.69	7.62	6.67
28	1681.46	7.86	7.78	7.37
29	1705.26	8.00	7.92	7.47
30	1739.15	8.17	8.00	7.51

Figure 5 on the next page shows the rule viewer of the inference fuzzy system applied. It is clearly indicating how the input variables (yellow columns) of thrust force and torque is used in the rules to produce the output (Ra) of the internal surface roughness for particularly drilled hole. The red/thick vertical line in the box on the right bottom corner provides a defuzzified value and its plot shows how the output of each rule is combined to make an aggregate output and then defuzzified.

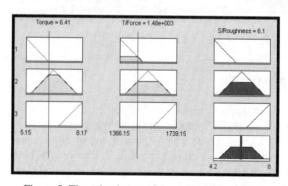

Figure 5. The rule viewer of the analysis carried out

The surface relationship of the input-output model is shown in figure 6, which clearly indicates the drastic stages of the surface roughness according to the inference applied to the fuzzy system. The surface matches exactly to the experimental result.

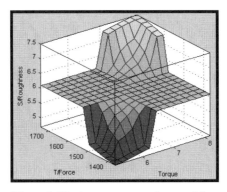

Figure 6. The input-output surface model

5. Conclusion

The conclusions of this scientific research can be summarized and drawn as follow:
1. A very general and direct approach of fuzzy system is much efficient and accurate as compared to statistical and mathematical methods used by the author in his previous research.
2. The basic three rules fed into the system were exactly fitting the fuzzy inference, which generates the excellent output of surface roughness studied within the range of experimental values.
3. The intuitions and experiences of a skilled machinist can be replaced by a set of fuzzy rules for drilling peripheral operation.
4. The relationship between the thrust force and torque with fixed speed and feed has shown an excellent output of fuzzy set as per the request by the project sponsorer.
5. The present work can be extended to obtain fuzzy model for tool wear and also on decision making of mainly contributing parameters which affect tool wear and surface roughness for the purpose of on-line tool condition monitoring.

Acknowledgement

The author would like to thank KP Engineering & Consultant for sponsoring the entire project. The author is also thankful to Mr. Ir. Vijayan.G and Professor Dr. Razali bin Mohamad the Dean of Manufacturing Faculty, KUTKM, for their sincere advise and guidance.

References

1. Armarego E.J.A., Wright J.D.: *Predictive models for drilling thrust and Torque: A comparison of three flank configurations.* Ann CIRP, 1984; 33: 5–10.
2. Chandrasekharan V., Kapoor S.G., DeVor R.E.: *A mechanistic approach to predicting the cutting forces in drilling.* Journal for Engineering Industry, 1995; 117: 559–70.

3. Galloway D.F.: *Some experiments on the influence of various factors on drill performance*. Trans. ASME 1957; 79: 191-237.
4. Huang.P.T, Chen J.C.: *Fuzzy Logic-Base Tool Breakage Detecting System in End Milling Operations*. Computers Ind. Engineering, 1998; 35: 37-40.
5. Hitomi.K, N. Nakamura, S. Inoue. Reliability analysis of cutting tools. ASME Journal for Engineering Industry, 1979; 101: 185-190.
6. Hung T., Nguyen, Nadipuram R., Prasad: *Fuzzy Modeling and Control*. Selected Works of M.Sageno. New York: CRC Press, 2000.
7. Hung T., Nguyen, Nadipuram R., Prasad: *Fuzzy and Neural Control*. New York: CRC Press, 2003.
8. Iwata K., Murotsu Y.: *A probabilistic approach to the determination of the optimum cutting conditions*. ASME Journal of Engineering for Industry, 1972; 94: 1099-1107.
9. Harris J.: *An Introduction to Fuzzy Logic Applications*. London, Kluwer Academic, 2000.
10. James J., Buckley E.E.: *An introduction to Fuzzy Logic and Fuzzy Sets*. New York: Physica-Verlag, 2002.
11. Kuo R. J.: *Intelligent tool wear estimation system through artificial neural networks and fuzzy modeling*. Artificial Intelligence in Engineering, 1998; 5: 229-242.
12. Liu T. I., Wu S. M.: *On-line detection of drill wear*. ASME Journal of Engineering for Industry 1990; 112: 299-302.
13. Merchant M.E.: *Basic mechanics of the metal-cutting process*. ASME Journal of Applied Mechnics, 1944; 15: 168–75.
14. Mauch C.A., Lauderbaugh L.K.: *Modeling the drilling process: An analytical model to predict thrust force and torque*. Comput Model Simul Manuf Process, ASME Prod Eng Div., 1990; 48: 59–65.
15. Russo M., Lakhmi C. J.: *Fuzzy Learning and Applications*. New York: CRC Press, 2001.
16. Orlovski S.A.: *Decision-making with a fuzzy preference relation*. Fuzzy Sets and Systems, 1978; 1: 155-167.
17. Pandit S. M., Kashou.S.: *A data dependent systems strategy of on-line tool wear sensing*. ASME, Journal of Engineering for Industry, 1982; 104: 217-223.
18. Sivarao T.S., Lee T.S., Chin C.W.: *Comparative analysis & modeling of surface roughness in drilling*. Proceedings of 7th Int. Conference on Deburring and Surface Finishing, 2004; 81-88.
19. Sivarao T.S., Lee T.S., Chin C.W.: *Comparative performance analysis and postulation modeling of tool wear in drilling*. The AiUB Journal of Science and Engineering 2004; 3: 75-82.
20. Sivarao T.S., Lee T.S., Chin C.W.: *Comparisons of statistical and mathematical method for tool wear in drilling*. Journal of Mechanical Engineering Strojnicky Casopis 2004; 55: 187-198.

PLANT-INPUT-MAPPING DIGITAL REDESIGN WITH COMPUTATIONAL DELAYS

Shimamura H., Hori N., Takahashi R.

Abstract: A method recently proposed as a digital redesign technique, which can take computational delays into account for the so-called the Plant-Input-Mapping (PIM) scheme, is tested using a benchmark problem by simulations for various design parameters not covered previously. The performance of the PIM method is compared with the widely used Tustin's method to show that, while the digital controller designed with Tustin method can easily become unstable for small delays and sampling intervals, the PIM method with delays taken into account remains stable for large delays and sampling intervals.

1. Introduction

The effect of computational delays is prevalent in applications where inexpensive microprocessors are used, such as in commercial products. A new mapping discretization method was proposed in (Shimamura and Hori, 2003) so that the delay can be compensated and that performance much superior to the popular Tustin's method can be achieved. This fact was experimentally confirmed for digital redesign of an analog driver for a stepping-motor. This method is, however, for the local digital redesign method where each controller block is discretized to obtain the corresponding digital controller block, and can be unstable as the sampling interval increases. Therefore, the global discretization method, such as the Plant-Input Mapping (PIM) method (Markazi and Hori, 1992), is preferred. The computational delay is taken into account in (Shimamura and Hori, 2005) for this PIM technique and the experiments are carried out for the stepping motor, which requires not only the computational delays but also disturbance attenuation. Since the effect of the delay compensation is obscured with the disturbance response improvement in that paper, the detailed simulation study is carried out in the present paper to show the effect of some design parameters.

2. PIM with computational delays

The analog control system shown in Figure 1, where the plant and analog controllers are given by:

$$\overline{G}(s)=\frac{\overline{n}_G(s)}{\overline{d}_G(s)}, \overline{A}(s)=\frac{\overline{n}_A(s)}{\overline{d}_A(s)}, \overline{B}(s)=\frac{\overline{n}_B(s)}{\overline{d}_B(s)}, \overline{C}(s)=\frac{\overline{n}_C(s)}{\overline{d}_C(s)}, \qquad (1)$$

is assumed to satisfy all the design specifications, including internal stability, good transient and steady-state responses, and robustness. This analog control system is implemented using a digital control system shown in Figure 2, where the delay due to computation is lamped together and cascaded just before the zero-order-hold. To relate discrete-time (DT) systems to continuous-time (CT) systems easily and to take advantage of superior numerical properties, the operator $\varepsilon = (z-1)/T$, where T is the sampling interval and z is the usual zee-operator, is used throughout this study (Middleton and Goodwin, 1990).

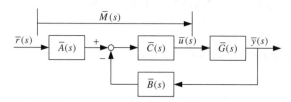

Figure 1. The analog control system

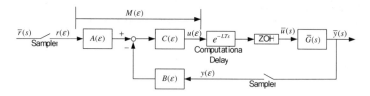

Figure 2. The digital control system with computational delay

Using a local discretization method, such as Tustin's, the controller blocks are discretized individually. The PIM method, on the other hand, discretizes the analog control system using the matched-pole-zero model (Markazi and Hori) of the transfer function from the reference input to the plant input, which is called the Plant-Input Transfer Function (PITF), denoted as $\bar{M}(s)$ in Figure 1. The PITF usually contains as its numerator the characteristic polynomial of the plant, while it contains as its denominator the closed-loop characteristic polynomial, which is known as controller pole and zero principles (Sain and Schrader, 1990). The DT PITF thus obtained becomes the target PITF and must be realized using the closed-loop configuration shown in Figure 2. The blocks are determined by solving a Diophantine equation, which usually results in three blocks (Markazi and Hori, 1992).

The design steps of this PIM method that takes computational delays into account (Shimamura and Hori, 2005) are summarized below:

Step 1. Calculation of the CT-PITF

The PITF of the analog control system shown in Figure 1 is gievn by:

$$\bar{M}(s) \triangleq \frac{\bar{u}(s)}{\bar{r}(s)} = \frac{\bar{A}(s)\bar{C}(s)}{1+\bar{B}(s)\bar{C}(s)\bar{G}(s)} = \frac{\bar{n}_M(s)\bar{d}_G(s)}{\bar{d}_M(s)} \quad (2)$$

where:
$$\bar{n}_M = \bar{n}_A \bar{d}_B \bar{n}_C$$
$$\bar{d}_M = \bar{d}_A(\bar{n}_B \bar{n}_C \bar{n}_G + \bar{d}_B \bar{d}_C \bar{d}_G). \quad (3)$$

Step 2. Discretization of the CT Plant Model

The computational delay is distributed inside the DT control system but lamped together and placed conceptually at the input of the plant, as shown in Figure 2. This leads to the step-invariant-model of the system with a delay (Middleton and Goodwin, 1990), and is denoted as $G_D(\varepsilon) = y(\varepsilon)/u(\varepsilon)$. This can be expressed, in the state space equations, as:

$$\begin{cases} \dot{\bar{x}} = \bar{A}_G \bar{x} + \bar{B}_G \bar{u} \\ \bar{y} = \bar{C}_G \bar{x} + \bar{D}_G \bar{u} \end{cases} \quad (4)$$

and the delay number be $L = N + \eta$, where N is the integer part and η is the fraction part with $0 \le \eta < 1$. This L is the delay number such that the delay time is $D=LT$. It should be noted that L does not have to be an integer. The transfer function of the plant with the delay can be expressed as:

$$G_D(\varepsilon) = (\bar{C}_G(\varepsilon I - \bar{A}_G)^{-1}(B_{G1} + (T\varepsilon + 1)B_{G0}) + \bar{D}_G)(T\varepsilon + 1)^{-(N+1)} \quad (5)$$

where

$$\bar{A}_G = \frac{e^{\bar{A}_G T} - I}{T}, B_{G0} = \frac{1}{T}\int_{\eta T}^{T} e^{\bar{A}_G(T-\tau)}\bar{B}_G d\tau, B_{G1} = \frac{1}{T}\int_{0}^{\eta T} e^{\bar{A}_G(T-\tau)}\bar{B}_G d\tau. \quad (6)$$

This can also be written as:

$$G_D(\varepsilon) = \frac{n_{G_D}(\varepsilon)}{d_{G_D}(\varepsilon)} = \frac{n_{G_D}(\varepsilon)}{(T\varepsilon + 1)^{N+1}d_G(\varepsilon)} \quad (7)$$

while the step-invariant-model without the delay is given by:

$$G(\varepsilon) = \frac{n_G(\varepsilon)}{d_G(\varepsilon)}. \quad (8)$$

Step 3. Discretization of the CT-PITF

The continuous-time PITF $\bar{M}(s)$ given by equation (2) is discretized using the Matched-Pole-Zero (MPZ) method as:

$$M(\varepsilon) = \frac{n_M(\varepsilon)d_G(\varepsilon)}{d_M(\varepsilon)} \quad (9)$$

Since the denominator of equation (7) includes the computational delay, the target PITF must be modified as:

$$M^*(\varepsilon) = \frac{n_M(\varepsilon)d_{G_D}(\varepsilon)}{d_M(\varepsilon)(T\varepsilon + 1)^{N+1}} \quad (10)$$

Step 4. Determination of the DT Controller Blocks

Digital controller blocks shown in Figure2 are to be determined such that the target PITF obtained in Step 3 could be achieved using $G_D(\varepsilon)$ given in equation (7). For the controller blocks written as:

$$A(\varepsilon) = \frac{n_A}{d_A}, B(\varepsilon) = \frac{n_B}{d_B}, C(\varepsilon) = \frac{n_C}{d_C}, \qquad (11)$$

the actual discrete-time PITF of the control system shown in Figure 5 is given by:

$$M(\varepsilon) = \frac{A(\varepsilon)C(\varepsilon)}{1 + B(\varepsilon)C(\varepsilon)G_D(\varepsilon)}$$

$$= \frac{n_A}{d_A} \frac{d_B n_C d_{G_D}}{n_B n_C n_{G_D} + d_B d_C d_{G_D}} \qquad (12)$$

which must match the target PITF. For this purpose, let:

$$d_A(\varepsilon) = d_B(\varepsilon) = n_C(\varepsilon) = \lambda(\varepsilon) \qquad (13)$$

$$\partial(\lambda) = \partial(d_M) - \partial(d_{G_D}), \qquad (14)$$

where λ is an arbitrary, stable design polynomial and ∂ denotes the degree of its argument. Equation (12) can, then, be written as:

$$M(\varepsilon) = \frac{n_A d_{G_D}}{n_B n_{G_D} + d_C d_{G_D}}. \qquad (15)$$

From equations (10) and (15), the following condition is obtained:

$$d_C(\varepsilon)d_{G_D}(\varepsilon) + n_B(\varepsilon)n_{G_D}(\varepsilon) = d_M(\varepsilon)(T\varepsilon + 1)^{N+1} \qquad (16)$$

$$n_A(\varepsilon) = n_M(\varepsilon). \qquad (17)$$

For a Diophantine equation (16) to be solvable, the degree condition:

$$\partial(d_M) \geq 2(\partial(d_{G_D}) - 1) \qquad (18)$$

must be satisfied. When this is not satisfied, a stable polynomial with an appropriate degree can be multiplied to n_M and d_M. The controller blocks are finally obtained as:

$$A(\varepsilon) = \frac{n_M}{\lambda}, B(\varepsilon) = \frac{n_B}{\lambda}, C(\varepsilon) = \frac{\lambda}{d_C} \qquad (19)$$

where d_C and n_B are the solutions of equation (16), which achieves the target PITF.

3. Benchmark simulation

The methods explained in the previous section are evaluated using the bench-mark simulation problem used in (Rattan, 1984), (Keller and Anderson, 1992), (Markazi and Hori, 1992), where the plant and the analog phase-lead controller are given by the following transfer functions:

$$G(s) = \frac{10}{s(s+1)}, A(s) = 1, B(s) = 1, C(s) = \frac{3s+7.2}{s+7.2}. \tag{20}$$

For the sampling period of $T = 0.1$ second and the computational delay number of $L = 0.9$, the transfer functions of the digital controllers are obtained as shown in Table 1. The design polynomial in the standard PIM method is $\lambda = \varepsilon + 1/T$, while that in the proposed PIM is $\lambda = (\varepsilon + 1/T)^2$. Figure 3 shows the response to the step reference input of analog and digital control systems when there is no computational delay. It can be seen that the PIM method gives the performance identical to the analog case, while Tustin's method results in a large overshoot. The proposed PIM method is identical to the standard PIM method for $L = 0$.

Table 1. Transfer functions of the digital controller blocks ($T = 0.1[s]$ and $L = 0.9$)

Tustin	$A(\varepsilon) = 1, B(\varepsilon) = 1, C(\varepsilon) = \dfrac{2.468\varepsilon + 5.291}{\varepsilon + 5.291}$
Standard PIM	$A(\varepsilon) = \dfrac{4.685\varepsilon + 10}{\varepsilon + 10}, B(\varepsilon) = \dfrac{4.993\varepsilon + 10}{\varepsilon + 10}, C(\varepsilon) = \dfrac{\varepsilon + 10}{1.999\varepsilon + 12.32}$
PIM with Delay	$A(\varepsilon) = \dfrac{4.685\varepsilon^2 + 56.85\varepsilon + 100}{(\varepsilon + 10)^2}, B(\varepsilon) = \dfrac{5.468\varepsilon^2 + 64.68\varepsilon + 100}{(\varepsilon + 10)^2},$ $C(\varepsilon) = \dfrac{(\varepsilon + 10)^2}{1.999\varepsilon^2 + 34.69\varepsilon + 165.7}$

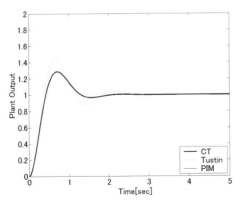

Figure 3. Step response to a unit-step reference input with for $T = 0.1[s]$ and $L = 0$

Figure 4 is the response when there is a computational delay of 0.9 samples. It shows that while Tustin design and the standard PIM design result in oscillatory responses, the proposed PIM method gives a response similar to the analog performance, with a slight delay at the beginning of the response, which is inevitable using a digital processor.

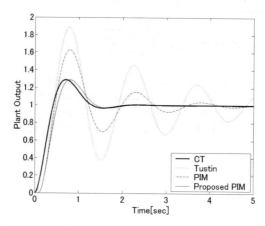

Figure 4. Step response for a unit-step reference input for $T = 0.1[s]$ and $L = 0.9$

The PIM method guarantees the stability for any computational delay time and non-pathological sampling interval. To demonstrate this fact, the step-reference response is shown in Figure 5 for $T = 2[s]$ and $L = 5$. Tustin's method is extremely oscillatory and even the standard PIM method does not work for this case. On the contrary, the PIM method with the delay taken into account gives a stable non-oscillatory response with no overshoot and offset.

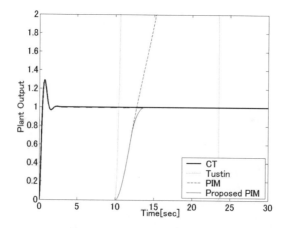

Figure 5. Step response for a unit-step reference input with $T = 2[s]$ and $L = 5$

4. Conclusions

It was shown in this paper that, through simulation studies using a benchmark problem, the computational delay, which is prevalent in industrial digital control systems where the cost-performance factor is important for commercial production, can easily and effectively be taken care of for the PIM based design.

References

1. Chen T., Francis B.: *Optimal Sampled-Data Control Systems*. Springer, 1995.
2. Franklin G.F., Powell J.D., Workman M.: *Digital Control of Dynamic Systems*. Addison Wesley, 1998.
3. Keller J.P., Anderson B.D.O.: *A new approach to the discretization of continuous-time controllers*. IEEE, Trans. Automatic Control, 1992; 37: 214-223.
4. Kuo B. C., Peterson D.W.: *Optimal discretization of continuous data control systems*. Automatica, 1973; 9: 125-129.
5. Markazi A. H. D., Hori N.: *A new method with guaranteed stability for discretization of continuous time control systems*. Proc. American Control Conference, 1992; 2: 1397-1402.
6. Middleton R.H., Goodwin G.C.: *Digital Control and Estimation: A Unified Approach*. Prentice-Hall, Englewood Cliffs, N.J., 1990.
7. Rattan K.S.: *Discretization of Existing Continuous Control Systems*. IEEE, Trans. Automatic Control, 1984; 29: 282-285.
8. Sain M.K., Schrader C.B.: *The role of zeros in the performance of multiinput, multioutput feedback systems*. IEEE Trans. Education, 1990; 33:244-257.
9. Shimamura H., Hori N.: *Digital redesign of a stepping-motor driver in the presence of computational delays and disturbances*. Trans. CSME, 2005; To appear.
10. Shimamura H., Hori N.: *Mapping discretization of a system involving delay*. IEEE Int. Conf. on Method and Models in Automation and Robotics, 2003; 1245-1249.

THE INTEGRATION OF HIGHER AND LOWER MANAGEMENT CONTROL TECHNIQUES IN MANUFACTURING COMPANIES

Perry D.J., Petty D.J., Smith S.A., Harrison D.K.

Abstract: Many manufacturing organisations use manufacturing planning and control (MPC) systems. Such MPC systems, however, have both high (strategic) and low- level (execution) functionality with specific regard to the planning, control and execution of manufacturing. Although there has been a substantial body of literature on particular MPC approaches, the relationship between different levels of control hat been the subject of relatively few studies. What is not well understood are the key determinants of the relationship between high and low level systems, as well as the manner in which such systems develop and evolve. This paper addresses two key research themes: first, that high-level and low-level systems are usually treated as distinct system types in relation to function, end users and response to change and therefore have a propensity to develop and evolve in isolation to one another. Secondly, that this lack of integration is detrimental to the overall efficiency and effectiveness of MPC systems.

1. Introduction

1.1. Research overview

This work was built upon a premise that was proposed by the industrial partner in this project: there is a need for research in the area of hierarchical control within manufacturing organisations. In particular, there is evidence, both from both observation and the literature that high and low-level systems in organisations are largely de-coupled. Notwithstanding the significant effort that has been applied to the development and implementation of distinct high and low level systems, many organisations have not performed in the expected, efficient manner. Both system types have been the subject of a substantial amount of research with specific regard to the individual system types. However, such research does not fully expound on the key determinants in the relationship of the two system types, as well as how best to develop and facilitate the evolution of the relationship and how this is reflected in organisational structure.

1.2. Background

High and low level systems exist in many organisations; however, in spite of this there is some commonality with regard to the overall function of such systems. The generic function of a system can be defined as a process which requires certain inputs which are processed in line with the system's characteristics and mechanisms with a view of achieving a specific set of outputs.

In an industrial context, high-level systems tend to be of a strategic nature with wide spans of functionality and a few specialist users. There systems aid the decision making process at a top level. Inevitably, these decisions have a significant impact on an organisation as a whole (see figure 1). For example, the output of a high level system can aid the Board of Directors in their deliberations on a rationalisation or investment programme. However, low-level systems tend to be tactical and specific in nature and can have a large number of relatively non-specialist users (see figure 1). Low-level systems can, for instance, monitor and feedback information about a specific manufacturing process with regard to cycle time or throughput.

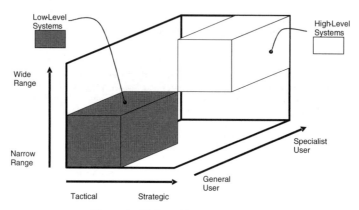

Figure 1. High and low level systems

The general characteristics of a low-level system will mean that it will occupy a position somewhere within the shaded cuboid in figure 1. Similarly high-level systems will occupy the space defined by the un-shaded cuboid. Although there is no need for the systems at the extremes of the high and low level continuum to be directly linked, there is a hierarchy of systems in place. It is through such a hierarchy that systems need to interface so that strategies and objectives may be deployed through policy and tasks/goals. Several issues can make such interfaces problematic; whether or not the system is computer, paper or human driven; the level of complexity of the system; ease of use; robustness and flexibility; reliance on large amounts of data etc.

Thus system interfaces and interactions are a significant issue to manufacturing industry. Indeed Davis (2003) writes, in relation to high level ERP systems and shop floor execution systems that *"integration is often considered to be industry's biggest challenge"*. This is further supported by the announcement (Deaves, 2003) that ABB, Accenture, Intel, and Microsoft have formed an alliance for the purpose of assisting manufacturers to close the gap between shop floor operations and enterprise information technology systems.

Manufacturing and Planning Control systems (MPC) incorporate several sub-systems including planning and control and production activity control. An examination of the use of MPC systems in practice highlights the problems of poor integration between high and low-level systems.

The basic function of an MPC system is to balance supply with demand in relation to the realisation of customer orders and the requisite raw materials and resource. Notwithstanding this very basic definition, MPC systems do have a variety of other functions, which support a myriad of business needs, e.g. interfacing with the management accounting systems. However, with the emergence of some fundamentally different MPC systems and the ever increasing emphasis that is being placed on Lean Manufacturing (LM) there is a propensity within an organisation to make use of the numerous and different MPC systems. Therefore many companies have employed low-level MPC systems such as Kanban in addition to MRP.

2. MRP (high level system)

2.1. MRP overview

The high-level system specifically examined in this paper is MRP. MRP has evolved into ERP via MRPII. For the purpose of this paper, however, MRP, will be used as a generic term for all three incarnations of MRP (MRPII and ERP will be used in specific circumstances). Such a generalisation is possible because the basic basic tenet for all 3 systems is the common task of managing production planning and control in order to meet customer demands. MRP is a methodology that facilitates the determination of future material and resource requirements through the use of Bills of Materials and product routings. In addition, MRP has numerous other functions, which support a variety of specific systems. As a high-level system, it interacts and is often closely integrated with accounting, engineering project management and maintenance systems.

2.2. MRP and organisational structure

Typically, an MRP system delivers certain desired results if a holistic approach is pursued in its implementation and use, such as reduction of inventory, purchase and transportation costs and overtime, whilst improving delivery schedule adherence and productivity (Wong and Kleiner, 2001). This, of course, is dependant on appropriate scheduling taking place, which MRP should assist. As aforementioned, this functionality feeds other systems from which strategic decisions are made, such as management accounting. This architecture can be used to drive into any organisation a high level of accountability, which through the correct management structure can yield a significant competitive advantage. As such, MRP demands (and indeed facilitates) an accurate and valid schedule to be devised, as well as enabling a relatively high degree of tractability. This provides a base line against which performance can be measured and therefore accountability can be established throughout the organisation. With such accountability being established, the realisation of the true strategic importance of the manufacturing becomes manifest. This should encourage all functional departments to become true supporting elements of the manufacturing operation.

However, herein lies a dichotomy. The traditional organisational structure and culture of many manufacturing companies tend to encourage a narrow view on the part of managers. Indeed many MRP packages utilise specific departmental boundaries within their own architecture. Therefore, the dominance of the manufacturing operation can be problematic to establish. Notwithstanding this, MRP fulfils a role of a high level system which can impose the requirement of a high degree of professionalism (Braglia and Petroni, 1999) and accountability about which a holistic management approach can be constructed. This is significant because when such a broad management front is established less adversarial relationships should be the result and a sense of organisational identity and unity should be promoted.

2.3. Limitations of MRP

Although there are many benefits in using MRP, like any other system there are inherent limitations. MRP suffers from deterministic lead-times, process times and product routings. Consequently any occurrences, which cause a deviation from the norm, can have significant detrimental effects not only on that particular order, but also many others. Similarly, MRP does not consider uncertainty such as supplier tardiness, poor quality of raw materials or components or indeed labour shortages and the subsequent effect on delivery schedule adherence (Koh et al, 2000). MRP also assumes infinite resources and even when capacity planning is available, MRP does not have the ability to resolve any capacity planning issues independently (i.e. without human intervention).

3. High and low level (lean manufacturing) systems

3.1. Requisites for low-level systems

Such limitations, as described above, on high-level systems cause many low order problems, which have significant effects. Yet even if real-time data is available, MRP lacks the structure for its utilisation and management. There are many compensation mechanisms which relate to low level (shop floor orientated) systems, e.g., the holding of large amounts of safety stock in the form of raw materials, semi-finished and finished goods, as well as the use of various weightings being applied to lead times and capacity. Sometimes even other planning and control systems are applied, in spite of such compensation tactics, contributing to the already numerous variables that influence manufacturing operations, thereby exacerbating the problem. Notwithstanding the prevalent use of such low level systems in this compensation role, significant uncertainty, as well as difficulty in exercising good coordinated planning and control of the manufacturing operation, still exists. As mentioned previously, there are many different low level systems. As most of them fit into the classic LM/MPC model, however, only LM techniques will be considered in this paper.

3.2. MRP/LM hybrid

Over the last 30 years MRP has become the dominant MPC system throughout the West. With the success of Japanese industry, which is attributed to their management methodologies and practices (Imai, 1986), LM has been readily adopted. Initially, the majority of practitioners and academics viewed LM and MRP as fundamentally opposed to one another and therefore mutually exclusive. Yet as the dynamics of the marketplace changed and interaction of the two systems took place in industrial practice, a greater understanding of LM and MRP developed. This has resulted in several authors proposing that LM and MRP are not incompatible but complementary. Table 1 details different attributes of LM and MRP (Lee, 1993; Hay, 1988; Slack and Correa, 1992; Spencer and Guide, 1995).

Table 1. Attributes of LM and MRP

LM	**MRP**
Ideal for repetitive and low variety manufacture	Can operate with ease not only in a high variety but also in repetitive manufacturing environments
Operates best with smooth production	Ability to deal with lumpy production
Incorporates changes in demand quickly and easily	Needs to be re-run to incorporate changes in demand
Promotes low levels of inventory	Allows moderate to high inventory levels
Promotes improvement to the Manufacturing process and quality	Neutral regarding improvement processes
Actively reduces manufacturing lead times	Neutral regarding length of lead times
Promotes single piece flow	Promotes batch manufacture
Good at scheduling in the short term	Good at long term planning

It can be seen that both MRP and LM have strengths and weaknesses and that the strengths of one may compensate for the weaknesses of the other. Nonetheless, some companies, who have implemented LM techniques into an existing MRP environment, have encountered significant problems. These differ depending on the extent to which LM techniques have been integrated and the nature of their specific goals. Some firms may be tempted to adopt LM techniques as an alternative to their existing MPC systems because it may suit their type of manufacture; LM is ideally suited to a repetitive, stable, low variety type of manufacture. Companies whose type of manufacture fits this profile can benefit greatly from the complete adoption of LM. On the other hand, MRP can more adequately deal with variable production with unstable demand and a high variety of products. Therefore, the emergence of hybrid LM/MRP systems is a logical step for companies whose manufacture is suited to both LM and MRP. Indeed Chung and Snyder (2002) suggest the use of a hybrid system leads to a better outcome than applying LM and MRP individually.

Irrespective of the extent of integration, changes to certain elements of the MRP system are necessary when LM techniques are successfully introduced to the shop floor. With increases in computer processing power, MRP systems are able to accommodate change relatively quickly, but the supporting infrastructure tends to inhibit such rapid changes being reflected on the shop floor. MRP is therefore a relatively flexible planning tool that has the drawback of rigid shop floor control. The integration of MRP and LM occurs for two reasons; as a transition from one MPC approach to the other or as a hybrid system. However, to accrue the full benefits of LM, a comprehensive set of techniques needs to be implemented, which can necessitate considerable organisational change. One of the main areas of concern is the transfer of customer orders and demand from the planning high level (i.e. MRP net requirements) to the shop floor execution low level, i.e. LM, as this process can be very complex. Accounting practices and performance measures are also of concern, as the changes in the Bill of Materials and product route, as well as LM techniques tend to blur the boundaries traditionally used.

3.3. Relationship of MRP and LM

When examining MRP/LM hybrid systems and their concomitant relationship, it is useful to consider the sequence of activities that a manufacturing organisation follows from customer order to dispatch and the manner in which they are be perceived by people associated with either high or low-level systems. Table 2 details the different perspectives of high-level and low-level system users with regard to specific activities.

The results detailed in Table 2 were obtained by the authors through informal discussions with the users of both high and low level systems in a collaborating company. The users consulted included both operatives and members of management.

This table is open to debate, as the classification of the activities will depend on the characteristics of the manufacturing organisation and the systems that are in place. With the exception of activities and roles that are clearly defined, some confusion is manifest due to the multi-variables that have influence on particular activities. For instance, the changing schedule priorities activity has the variables of lead times, due date, customer importance, inventory levels, available resource (raw materials, components and labour), set-ups, workload at other processes, manufacturing problems and insufficient capacity. Any combination or individual of these variables can have an effect on the priority of the schedule.

Table 2. Classification of Activities Based on Perception of High and Low Level Systems

Activity	High level systems (MRP) perspective		Low level System (LM) perspective	
	High level	Low level	High level	Low level
Determining Customer Demand	X		X	
Carry out Rough Capacity Scheduling	X		X	
Establish MPS	X		X	
Determine daily schedule	X		X	X
Determine sequencing of orders	X			X
Determine appropriate lots sizes	X		X	X
Determine due dates	X		X	X
Release work to the shop floor	X		X	
Start work on order	X	X	X	X
Change sequencing of orders	X	X	X	X
Change to schedule	X		X	X

Table 2 represents the perspectives of people within a single company and is therefore case sensitive. It does, however, demonstrates a particular point; communication and coordination in the manufacturing operation is difficult enough even when there is an understanding of the particular roles and responsibilities of the high and low level systems and their respective functionality. However, as the table indicates there is a certain amount of confusion and duplication with regard to the perception of the high and low level activities and the question arises where in fact they do (or rather should) sit. This situation is in part due to the multi-variables that affect the manufacturing operation, and also the narrow perspective of individual system users.

This problem can be further exacerbated when there is a combination of feedback systems (LM) and feed-forward systems (MRP is being classed as a feed-forward system, notwithstanding the internal closed loop between the MRP logic stage and that of determination of a realistic Master Production Schedule, which can exist in MRPII and ERP). Such systems are discussed at length in Fowler (1999). When LM is used to plan and control the manufacturing operation the system works through output sensitivity and the subsequent adjustment of the input. When this is applied to a number of processes, the input variable cascades upstream in the manufacturing process and therefore each input is mirrored by a measured output. Consequently the overall manufacturing operation operates very responsively. When MRP is being used at high level in combination with LM, there is little attempt to monitor the output of the manufacturing operation with a view of varying the MRP input, as the system characteristics are those of determining a static input in relation to the multi-variables that can impact on the process. Therefore such feed-forward systems are unable to react to changes in the circumstances at the execution level. Having different MPC forming such a feedback/feed-forward hybrid is likely to produce numerous conflicts, confusion and incompatibilities. This results in unclear and ill-defined links and relationship and is also exacerbated by the propensity for such hybrid systems to develop in an unstructured and non-deliberate manner (Fowler, 1999).

4. Organisation structure

All organisations are subject to change through both internal and external influences. Such change usually manifests itself as either slow unstructured change or radical structured change. Low-level systems have a tendency to develop in the former manner whilst high level systems in the latter. Indeed parallels can be drawn between high-level systems development and Business Process Re-engineering (BPR) and low level systems development and Kaizen. Therefore, the way in which an organisation is structured, as well as its characteristics and bias are key in the determination of the manner in which high and low level systems operate, interact and develop. It is, therefore, essential to consider the characteristics, structure and hierarchy of an organisation in relation to high and low level systems.

4.1. Relationship between control type and high and low level systems

A production system implies a type of control, which in turn implies certain organisational characteristics. However, when cumulative changes to the production system aggregate to a significance degree, then organisational change must be made to ensure effective operational management. Woodward (1970) devised two measures, which can be said to have a high and low level bias. He developed Unitary/Fragmented and Personal/Impersonal viewpoints and stated that all organisations lie somewhere between the two. Unit and small batch production are more suited to unitary and personal controls, for example, with one authority setting performance criteria and the same group or individual involved in all functions. This indicates a small structure, which has narrow spans of control. Large batch and mass production is better suited to fragmented and impersonal controls, with different authorities setting performance criteria. This implies a tall structure with relatively wide control spans.

As LM methodologies become readily adopted, there has been a transition from mass/large batch production to small batch production. Indeed in the UK Best Factory of the Year Awards 2002 (Matthew, 2002), 37% of all winners stated that they considered LM as the most significant success factor.

4.2. Organisation perspectives

Different organisations can have different perspectives of themselves. Some view themselves mechanistically, in which case emphasis is placed on the structure of the organisation. The rationale being that the correct structure will allow efficient functioning of the organisation. Other characteristics of this perpsective are strict divisions of labour, discrete functions and defined inter-relationships. The human element of the organisation therefore tends to be ignored, as people are expected to do as they are told. Such a view of the organisation is conducive with high-level feed-forward systems, as multi-variables that can affect the system are reduced. Due to this approach, such organisations tend to be relatively rigid and lack the ability to respond swiftly when novel situations arise. Such organisations high-level bias is in part due to the development process of high-level systems. That is to say, high level systems tend to be large encompassing monoliths with few people in executive control. Such systems often require a lengthy and highly structured development/implementation strategy. On the other hand, some organisations view themselves as a community. In this case, the emphasis is not on the structure of the organisation, but rather the human elements. The rationale being that when people are happy and have relative autonomy in their work they will perform efficiently and effectively. This viewpoint is supported by Moss (2003), who writes of a marine engine manufacturer Semco, where meetings are voluntary, employees set their own salaries and holiday entitlements, as well as choosing the managers for whom they will work. Moreover, the staff work in groups no larger than 12 people on the basis this corresponds to the size of an extended family. This approach requires a high degree of job satisfaction and a participative management style. Therefore, the whole organisation is aligned to the positive aspect of human behaviour. In such organisations, the propensity for human interaction and activity means that it is well placed to cope and operate effectively with systems that require a high level of human support. Such organisations have the ability to be responsive and flexible as they have implicit support of personnel. Such organisations tend to thrive with low-level feedback systems based in a real-time environment. Therefore the community perspective has a low-level system bias, which is in part due to the manner in which such low level systems develop, i.e. in immediate response to situations arising and in an ad hoc manner.

Burns and Stalker (1968) had a similar perspective and stated that all organisations were in a state between (in their terms) mechanistic and organismic states. They proposed that as an organisation was exposed to new and unfamiliar situations, the existing organisation structure that could not accommodate the required change would develop incompetent systems to compensate for the organisation's inadequacies. Such systems, due to the nature of the reaction, will tend to have a low level bias with the express purpose of circumnavigating elements of the organisation structure upon which effective high-level systems rely. Therefore, the very manner in which such systems develop is a result of these inadequacies (and potentially the prevalence of high-level systems) and can erode the effectiveness of the organisation as a whole.

4.3. Management style

A significant factor influencing organisational structure is the style of management that is adopted. This should not be confused with the style of individual members of management, but rather that of the overall organisation. Likert (1961) described two broad models. The Exploitative and Benevolent Authoritarian models are dominated by top management decision-making. Communication is downwards and performance is always under the miasma of penalty. Consequently team working and interpersonal relationships are poor. In such an organisation high-level systems will tend to operate efficiently and effectively. However, although low-level systems will manifest themselves this will only be after a stifling and difficult development process. Such development will be difficult because the requisite freedom, level of delegation and devolved authority does not exist. However, in the Consultative and Participative models, communication is both upwards and downwards with a consensus ethic. Performance is achieved through recognition and reward, and also team working and interpersonal relationships are good. Such an environment allows the relative conditions for the uninhibited development and maintenance of low-level systems.

4.4. Centralisation and system development

High-level systems require a certain amount of centralisation so as to minimise the number of multi-variables. As Lawrence and Lorsch (1967) pointed out, however, organisations have elements of de-centralisation as well. To satisfy the need to respond to change in an innovative, effective and appropriate manner, however, de-centralisation is needed. De-centralisation also has some detrimental aspects. Each functional department will view the world from a unique perspective. For example, when asking what would constitute a key supplier, a finance department may well apply a cost criteria, whereas, the procurement office may look at it from a lead-time perspective and production may consider it on usage/volume criteria. Therefore some integration is required, despite it being the potential source of conflict within the organisation. Consequently problems can arise if activity or system development is undertaken in an inappropriate functional department. An overly centralised organisation can have inhibiting qualities, as at department and section level strict adherence to rules and procedures can inhibit innovation and creativity and therefore impede the healthy development of low-level systems.

5. Conclusions

This paper has demonstrated that although there is a need for MPC systems, which consist of high level and low level systems; the understanding of the relationships between high level systems, low level systems and the traditional organisation structure is in its infancy and somewhat lacking in sophistication. High-level and low-level systems tend to exist as distinct and individual systems and interact in a piecemeal fashion, with the characteristics of the organisation structure tending to support high-level systems in a bureaucratic manner. For such a hybrid MPC to move forward, this relationship needs to be examined and better understood. In particularly, there are three issues that need to be addressed:

1. The relationship between high-level (MRP) and low-level (LM) systems needs to be considered with respect to organisation structures having a high-level system bias. This is due to many organisations introducing low-level systems into a bureaucratic environment. In addition, such low-level systems develop in an unstructured and implicit manner, which result in incompetent and detrimental systems being deployed.
2. As systems develop and are implemented there is an implied change in control. When such changes or cumulative changes become significant, it is necessary to make adjustments to the organisation structure. This is usually recognised when high-level systems are subject to change as it is a deliberate and structured process and the ramifications can be easily seen. However, due to the incremental nature of low-level systems development and change, the need for evaluating the organisation structure is often not recognised and overlooked.
3. There is a lack of clarity and definition regarding the roles of high and low level systems with specific regard to the activities carried out in the manufacturing function. This is due to the multi-variables that have significant influence.

Acknowledgements

Mr. Perry's and Mr Smiths' research is funded by EPSRC, Federal-Mogul and Strix Ltd. The authors would also like to thank the "I*PROMs Network of Excellence" for their support.

References

1. Braglia M., Petroni A. (1999): *Shortcomings and Benefits Associated With the Implementation of MRP Packages: a survey research*. Logistics Information Management. Vol. 2, Issue 6.
2. Burns T., Stalker G.M. (1968): *The Management of Innovation*. Tavistock.
3. Chung S.H., Snyder C.A. (2000): *ERP Adoption: Technological Evolution Approach*. International Journal of Agile Management Systems, Vol. 2 Issue 1.
4. Davis B. (2003): *Get Your Plant into a MES*. Professional Engineering, Vol. 26, No 6.
5. Deaves M. (Editor) (2003): *New Architecture for Process Control*. Manufacturing Engineer, April/May.
6. Fowler A. (1999): *Feedback and Feedforward as Systemic Frameworks for Operations Control*. International Journal of Operations & Production Management. Vol. 19, No. 2.

7. Hay E.J. (1988): *The Just-In-Time Breakthrough: Implementing the New Manufacturing Basics*. Chester: Wiley.
8. Imai M. (1986): *Kaizen (Ky'zen): The Key to Japan's Competitive Success*. Random House Business Division.
9. Koh S.C., Jones M.H., Saad S.M., Arunachalam S., Gunasekaran A. (2000): *Measuring Uncertainties in MRP Environments*. Logistics Information Management. Vol 13, Issue 3.
10. Lawrence P.R., Lorsch J.W. (1967): *Organisation and Environment*. Harvard.
11. Lee C.Y. (1993): *A Recent Development of the Integrated Manufacturing System: a Hybrid of MRP and JIT*. International Journal of Operations & Production Management. Vol. 13 No. 4.
12. Likert R. (1961): *New Patterns of Management*. McGraw-Hill.
13. Matthew G. (Editor) (2002): *Management Today*. Awards For Manufacturing in Association With Cranfield School of Management and the DTI. Management Today. November.
14. Moss S. (2003): *Idleness is Good*. The Guardian. Guardian Newspapers Ltd. April 17th.
15. Slack N., Correa H. (1992): *The Flexibilities of Push and Pull*. International Journal of Operations & Production Management. Vol. 12 No. 4.
16. Spencer M.S., Guide V.D. (1995): *An Exploration of The Components of JIT*. International Journal of Operations & Production Management. Vol.15 No. 5.
17. Wong C.M., Kleiner B. H. (2001): *Fundamentals of Material Requirements Planning*. Management Research News. Vol. 24, No. 3/4.
18. Woodward J. (1970): *Industrial Organisation: Behaviour and Control*. Oxford University Press.

OPTIMIZING THE VENDOR SELECTION PROCESS OF AN AUTOMOBILE INDUSTRY USING CRITICAL VALUE ANALYSIS – A CASE STUDY

Parthiban P., Arun J., Ganesh K., Narayanan S.

Abstract: Multi criteria evaluation of vendors is an already established milestone. Multi criteria evaluation of vendors is the foundation for the vendor selection method that has been proposed in this paper. In this paper we basically study the vendor selection method of an Automobile Industry, try to incorporate an iterative method called the Critical Value Analysis (CVA), in order to optimize and upgrade its vendor selection process. In optimizing the Vendor Selection method, we develop the Integrative Decision Making (IDM) software for Vendor Evaluation. Over the years, Vendor Selection has not considered hidden input performance measures, output performance measures and performance variability measures into the procedure. We introduce all these measures into the selection method and fill this void. The primary advantage of this technique is that it provides the buyer with effective alternate choices within a vendor group. The seller is given a part in the evaluation process for more accurate results.

1. Introduction

An Automobile industry is planning to upgrade its vendor selection process. The present method is a rating system. The Vendor Selection process (rating system) has its share of demerits. It does not select a vendor by equally weighing its pros and cons. On the contrary, Vendor Selection is based on lowest price quoted by the external supplier and a vendor rating system based on various generalized formulae devised by the purchase and quality control departments.

Vendor Selection in any organization is a very crucial task as it decides the profit of an organization, provides hitch free operation, quality assurance, helps lower the cost of production, enables cutting down additional costs, and delivers eco- friendly products. 'Green' vendor rating systems decide the supply of eco friendly products. The factors that govern these products' delivery are called Environmental Factors as proposed by Giuliano Noci, 1997 [1]. Optimized vendor selection also optimizes the supply chain performance to provide maximum possible return on investment.

Many researchers have studied vendor selection, vendor evaluation, vendor rating and other vendor assessment techniques. Their realm of study has been limited by various reasons and some of the criteria for the same have been neglected. In this study, the techniques and contributing factors that favor optimal vendor selection and those that have been neglected, but necessary will be discussed. Research materials have been studied to find the existing models for vendor selection. Those existing models provide us with information about realms that have not been investigated.

While traditional vendor performance measurement methods primarily considered financial measures in the decision making process, more recent emphasis is on manufacturing strategies such as Just-In-Time (JIT) has placed increasing importance on the incorporation of multiple vendor criteria into the performance measurement process (Chapman, 1989[2]; Chapman, Carter, 1990 [3]). There are a few short comings as to the scope of JIT. If the question rises as to determining lot sizes and delivery frequency in a buyer supplier relationship, the work of Israel David, Moshe Eben-Chaime, 2003 [4] has illuminated the fact that by forcing the vendor to 'JIT'- produce and deliver the buyer's order quantity, a double digit inflation occurs. It can be concluded from these studies that, vendor selection decisions must not be exclusively on least cost criterion and those other critical factors such as quality and delivery performance need to be incorporated into the selection process. Thus, researchers have sought the adoption of multi-criterion decision models for vendor selection purposes.

2. Need for optimization

2.1. Origin of proposed method

One of the vital areas of supply chain management is the Performance measurement (PM) of the supply chain entity (Banker and Khosla, 1995) [5]. There are numerous enterprises taking part in the supply chain from Supplier to end user. To manage a supply chain means to evaluate its performance quantitatively and qualitatively. To improve the overall performance of the supply chain one needs to assess the PM of every vendor entity from vendor one down to the final one of the chain. Although Multi criteria Vendor Selection has already been utilized already in traditional vendor selection methods, there has been little work in incorporating performance variability measures into the selection process.

Talluri, Narasimhan, 2003 [6] defined performance variability as the vendor's multi factor variation when evaluated against targets set by the buyer. The method was a max- min productivity method and considered weaknesses of vendor to evaluate them. We try to incorporate a method that houses these important issues into it-the Critical Value Analysis (CVA).

2.2. Definition of the problem

The scenario of the company presents before us a single product-multi supplier-single stage supply chain. The vendor selection process is a very tedious process and needs to be simplified in order to make the selection process relatively easier. The performance of the supply chain entity must be evaluated against best target measures set by the buyer. We create vendor selection models that will take care of the hidden input and output performance measures and include customer expectation and performance variability measures. The buyer uses ideal targets for selection of vendors and if these targets are not achieved exactly by the vendors, the values are used as benchmark values

3. Restructuring the industry's vendor selection method

3.1. Weighted method

We consider the inbound supply chain of the automobile industry. This includes the external supplier of the organization and the departments of the assembly unit. The organization devices vendor selection methods based on field inspections and analytical calculation.

As we investigate the literature of various vendor selection methods we find a very important work, that satisfies the objectives that are proposed in the paper. In the Analytically based method proposed by Layel Abdel-Malek and Nathapol Areeratchakul, 2004 [7] the authors propose a selection process that considers salient performance attributes necessary for a reliable supply chain. More important, they compare a pool of suppliers, assign weights to each attribute depending on its importance and rank them accordingly. This method plays an important role in the optimization process that is undertaken; only difference being the fact that the optimization method followed in the paper is an iterative method and not an analytical method. Various norms are maintained for rating the vendor and deciding the existence of a vendor as a permanent supplier. Replacement of vendors is also done using this method.

3.2. Outline of the activities that are to be carried out

The different activities that are carried out from the start to the end are as follows:
- discussions were carried out with the Stores personnel, Quality personnel, and Materials Management experts, Plant Engineering Personnel in order to investigate the customer expectations from both the internal and external customers,
- study the existing method being followed in the organization for Vendor Selection, Vendor Evaluation, Vendor Rating and finally Vendor Management,
- incorporate the CVA and finalize the forms required for updating the decision support system.

4. Vendor management of the organization

In studying the vendor management process of the industry, we migrate from their expectations to their rating method to evaluate and mange their vendors.

4.1. Expectations of Materials Management Department

Any organization looks for three major criterions from their internal and external Customers. They are: quality, delivery, pricing. This is the broad categorization of priorities in their industry. This is further classified into various segments. In the organization, the Vendor Management is carried out as follows. The constituents can be termed as: vendor development, vendor selection, vendor evaluation, vendor rating.

Seven tools of quality control department: histogram, graphs, scatter diagram, Pareto diagram, cause and effect-diagram, control charts, process capability.

4.2. Vendor quality rating

In the organization, the procedure followed is termed as Vendor Quality Rating. The Vendor Approval is represented pictorially in the flow chart that has been drawn in fig. 1.

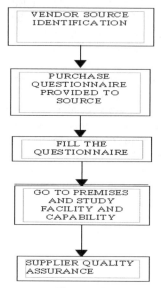

Figure 1. Vendor approval plan

The process consists of identifying the vendor and subjecting him to varying degrees of evaluation process and finally rating him. Rating only gives a chance to the vendor for improving his efficiency, and not setting any benchmark values.

4.3. Varying degrees of approval

There are various degrees of approval for vendor rating. They can be termed as follows: approval, provisional approval, restricted approval, to be re-assessed, rejected.

4.4. Input performance measures, output performance measures and performance indicators

There are technical terms in every organization, which are unique to those organizations only. The Specific terms and methods used for vendor performance measurement, vendor rating, vendor evaluation needs to be studied. Those terms are: goods receipt note, vendor quality rating, feedback on improvement, rejection reduction, line stoppage, supplier quality assurance

4.5. Vendor quality rating (VQR)

VQR is expressed as a percentage. It is calculated using the formula:

$(Q_1 \times w_1 + Q_2 \times W_2) \times 100/Q$

where:
- Q1   Normally accepted quantity
- Q2   Concessionally Accepted Quantity
- W1   Weightage for normally accepted quantity
- W2   weightage for quantity accepted with concessions.

weightage values: W1- 1; W2- 0.6.

4.6. Type of components – purchased

In the stores of Quality department, these are the type of components purchased: panels, fasteners, fabricated items, machine items, electrical and pipe items, castings, forgings, plastic, rubber, trim items, glass

4.7. Process plan for data collection

The main objective is prioritization of models. Prioritization of evendors. Prioritization process sequence.

4.8. Data collection

Data from: purchase department, various assembly shops. Overall rating: LCL and MRP rating, JIT rating. Rating is done on a monthly basis. They are classified under:
- Vendor Quality Rating,
- Vendor Delivery Rating,
- Vendor Commercial Rating.

In the end, evaluated values of all the three ratings and weighted values of materials according to the LCL, MRP and JIT delivery are incorporated to arrive at the final Vendor Rating. The formula for final rating is:

Final Rating = [0.9[VQR1 X W X VQRP] + 0.1 X VQRR]

where:
- VQRP   Vendor Quality Rating Production,
- VQRR   Vendor Quality Rating Reliability,
- VQRI   Vendor Quality Rating Inward.

4.9. Quality objectives

Maintaining Quality is linked directly with Total Customer Satisfaction. The main priorities in maintaining Quality is the following: conform to drawing and specification, fit for use in intended application, less or no failures, the parts are repairable or replaceable in an easy manner, very well packaged, eco-friendly, priced right and competitively, Just in Time delivery, life cycle cost is economical, queries and complaints handled in a proper manner.

4.10. Delivery objectives

Delivering the materials at the required time is also of vital importance to an organization. The priorities are tabulated in the below table. The ratings for LCL, JIT and MRP products are different from each other, Table 1, Table 2.

Table 1. Delivery Ratings

+1 days	100%
+2	50%
+3	0%
-2	75%
-3	50%
-more than three	0

Table 2. LCL Ratings

LCL	
+3	100
4-6	50
More than 6	0
-More than 4-8	50
8-10	25
10	0

MRP, Delivery Rating = 100-((quantity supplied-schedule)/quantity scheduled) X 100.

4.11. Pricing objectives

The pricing strategy of a supplier must be such that it serves as an optimum balance between cost effectiveness and low magnitude. Consider two vendors A&B, Table 3.

Table 3. Price Rating

A&B same price	100%
B>A at 5-10%	90%
B>A at more than 10%	75%

5. Problems in the present method

The problems in the present method followed are that, all the terms and conditions are inclined towards one side of the supply chain. There is a need for a method that provides equal opportunity for both the participants to play equally important roles in the supply chain management process. Sandy D. Jap, 2001[8] states that there can be competitive advantages in maintaining good relations between the buyer and the seller. The method of rating the vendors has another important problem. The problem is that they do not consider hidden performance indicators, contribution factors and performance variability measures. Talluri, Narasimhan, 2001 [6] have contributed much to the above cause. They propose a max-min approach to maximize and minimize the performance of a vendor against best target measures set by the buyer: Tables no 4 and no 5.

Table 4. Ranking with respect to profitability

Product name	No. of units	Cost per unit	Total cost	Profitability ratio	Ranked products

Table 5. Ranking with respect to customer expectation

Ranked vendors	Profitability ratio	Customer expectation variants	New ranking
		Weighted values	

6. Steps followed in the critical value analysis

The critical value analysis is as follows:
1. It is an iterative method.
2. It ranks vendors with respect to profitability
3. It ranks vendors with respect to customer expectation.
4. It ranks vendors with respect to customer rating
5. It trades off these vital values to reach an optimum vendor ranking taking all the values into consideration.

7. Development of decision support system

Visual Basic 6.0 is used to develop user interfaces with MS-ACCESS as the back end.

7.1. Plan for software tool

Input unit values and prices, for profitability ratio. Enter customer expectation values for prioritizing vendor. Customer weight age values are given to each vendor. These three are traded-off for the final optimized value.

8. Conclusion

There is a need to generalize this software for all single-product-multi supplier-single stage supply chain problems. Test data has not been acquired from the organization to validate the DSS. This also needs to be done for further development of the software tool for varied applications. As such the DSS using the critical value analysis is a process derived from the past literature. The selection procedure is just an improvement over the present method. Further up gradation is possible by further research and industrial survey, which will be done in the near future.

References

1. Noci G.: *Designing 'green' vendor rating systems for the assessment of a supplier's environmental performance.* European Journal of Purchasing & Supply Management, 1997, vol.3, No.2, pp. 103-114.
2. Chapman S.N.:. *Just-in-Time supplier inventory: An empirical implementation model.* International Journal of Production Research, 27 (12), 1989.
3. Chapman S.N., Carter P.L.: *Supplier/Customer inventory relationships under just-in-time.* Decision Sciences, 21 (1), 35-51, 1990.
4. Israel D., Moshe E.C.: *How far should JIT vendor- buyer relationships go?* Int. J. Production Economics, 81-82, 361-368 2003.
5. Banker R.D., Khosla I.S.: *Economics of operations management: A Research perspective.* Journal of Operations Management,12, 423-425, 1995.
6. Talluri S., Narasimhan R.: *Vendor evaluation with performance variability: A max-min approach.* European Journal of Operations Research, 146, 543-552, 2003.
7. Layel A.M., Areeratchakul N.: *An analytical approach for evaluating and selecting vendors with interdependent performance in a supply chain.* Int. J. Integrated Supply Management, Vol. 1, No 1, 2004.
8. Sandy D.J.: *Perspectives on joint competitive advantages in buyer-supplier relationship management system.* Expert Systems with Applications, xx 2002.

Appendix

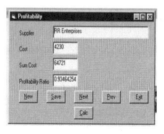
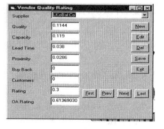

Figure 2. Invoking the program from Windows

Figure 3. form for profitability ratio (data provided is imaginary)

Figure 4. Form for weighted customer expectations

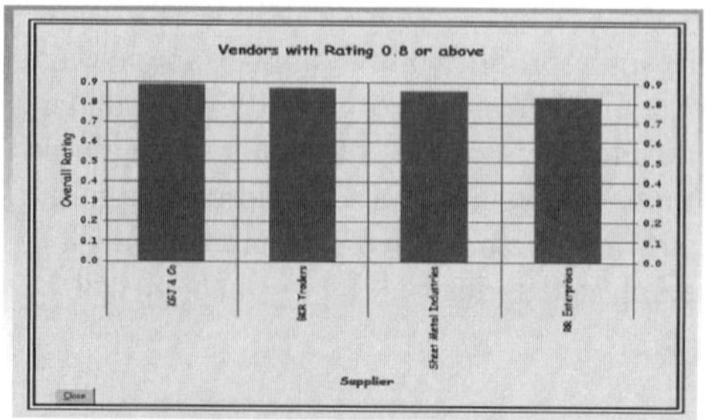

Figure 5. Bar diagram depicting the vendors with rating above 0.8 (data is imaginary)

Figure 6. Spread sheet showing calculated values (data is imaginary)

DECISION SUPPORT SYSTEM FOR THE EVALUATION OF ENTERPRISE RESOURCE PLANNING

Parthiban P., Narayanan S., Dhanalakshmi R.

Abstract: Decision-making is the process that leads to a choice between a set of alternatives.. Justification of multi-criteria decision-making problem becomes a complex, when it is involved with a number of alternatives and in turn, which involves a number of criteria at different levels for different alternatives. Multi criteria decision-making problems are commonly categorized as continuous and discrete. In case of continuous problem, solution space is continuous and it is defined by constraints. In the case of discrete problem, there will be a choice between numbers of discrete alternatives. To aid the choice between discrete alternatives, many approaches are useful in the process of narrowing down a long list of alternatives. An approach is required which will aid the decision maker in analysis and synthesis of detailed information in a way, which is consistent with their value judgments about the relative importance of the decision maker objectives. One such approach has been considered called Analytic Hierarchy Process (AHP), for evaluation of complex integrated system like Enterprise Resource Planning (ERP) ,. The justification of ERP is much more complicated, because the entire system is integrated and the benefits of one department/entity will depend upon another. ERP attempts to integrate all departments and functions across a company onto a single computer system that can serve all those different departments' particular needs.. The objective of this research is the development of an analytical methodology to incorporate both tangible and intangible criteria to evaluate ERP in accordance with the customer service preferences. The application of AHP in aspect of Customer Relationship Management (CRM) for the evaluation of ERP is the new era of research.

1. Introduction

The strategic level decision making process is very complex in nature because it consists of numerous variables and most of which cannot be quantified. The management, based on the available information, takes decisions. When the available information, which is both qualitative and quantitative, is not considered for the decision making process it results are inaccurate. There are very few models available in the literature, which will consider both tangible criteria and intangible criteria of decision problem. Among them, two models are considered called AHP developed by Thomas L. Saaty and Multi Attribute Utility Theory (MAUT) considered to be developed by Keene and Raiffa and emphasis is given to the AHP for its extensive applications and ease in use. This model is applied upon a decision-making problem as mentioned below. As an example where an organization is in a situation of choosing the alternative information system, which is mainly supportive in nature to the entire organization is considered. For the contrast the decision maker has to compare the alternative with the existing system and should be able to give a justification for his decision. The alternative of ERP consists of various elements both tangibles and intangibles and the decision may be wrong if one only considers tangibles benefits. Generally it is difficult to quantify this intangible benefits, because they are often complicated and subjective. Without a systematic approach, the decision maker may address the problem intuitively, and thereby suffer from inaccuracy and inconsistency. So there is a need of methodology to evaluate both intangible and tangible benefits simultaneously.

2. Solving multi criteria decision making problem through AHP

An extensive survey on multi criteria decision problem solution results as follows. The complication of a multi attribute decision problem, which demands each alternative must be judged on a multidimensional scale spanning risk, performance, time, and cost, can be resolved by the application of two approaches. The first is linear weighting scheme developed by Thomas L.Saaty known as the AHP. It uses pair-wise comparisons and a ration scale to arrive at a cardinal ranking of the alternatives. The second one is based on the work of Keeney and Raiffa and is known as MAUT. It provides an ordinal ranking of the alternatives through the use of either a linear or multiplicative model.

Procedure of AHP:
1) define the problem and determine the objectives,
2) identify the factor, which influences the decision,
3) the factors are groups based on their interdependence as criteria sub-criteria, sub-criteria criteria etc.,
4) formulate the hierarchical structure i.e., the criteria on the top level and sub criteria and sub-sub criteria are arranged in the intermediate and lower levels,
5) this can be done by, writing all factors in row wise and column wise. Compare the first factor in the first row with all the factors in the column using the standard Thomas L.Saaty's qualitative scales, which range from (1-9). During comparing, if factor 1 in the row dominate over the factor 2 in the column then the whole integer (ranging for 4 to 9 based on the preference of one over the other is assigned in the cell (1,2) and the reciprocal is entered in cell (2,1). If elements being compared are equal 1 is assigned to both the positions. Otherwise integer 2 or 3 is assigned in the cell 1,2 the reciprocal is assigned in cell 2,1,
6) there are n (n-1) judgments are required to develop the set of matrices in step 5,
7) having made all the pair-wise comparisons of the data the consistency ratio (CR) is determined. The consistency ratio (C.R) is an approximate mathematical indicator, or guide, of the consistency of pair-wise comparisons. It is a function of what is called "maximum eigen value" and size of the matrix, "n" (called a consistency index) which is then compared against similar values if the pair-wise comparisons had been merely random (called a "Random Index" (RI) formulated by Thomas L.Saaty) . If the ratio of the Consistency Index to the Random Index is no greater than 0.1 (with in 10%), Thomas L. Saaty suggests the consistency is generally quite acceptable for pragmatic purposes. The Consistency index (C.I) can be calculated using the equation C.I. = $(\lambda max-1) / n - 1$. Here "λmax" s the maximum eigen value of the pair-wise comparison matrix and "n" is the size of the pair-wise comparison matrix,
8) steps 5 to 7 are performed for all levels and clusters in the hierarchy,
9) hierarchical composition is now used to weight the eigen vectors by the weight the criteria and the sum is taken overall weighted eigen vectors entries corresponding to the next lower levels of hierarchy.

3. Problems in justification of ERP

In studying the vendor management process of the industry, we migrate from their expectations to their rating method to evaluate and mange their vendors. An enterprise is a group of people with common goal, which has certain resources at its disposal to achieve the goal. The group has some key functions to perform in order to achieve its goal. Resources included are money, man-power, materials, and all the other things that are required to the enterprise. Planning is done to assure that nothing goes wrong. Planning is putting necessary functions in place and more importantly, putting them together. Therefore ERP is a method of effective planning of all the resources in an organization.

Let the ERP procured by a Public sector company, is SAP R/3. The SAP R/3 system is a group of sophisticated business modules that span a range of highly specific business processes. The business and data processes implemented in these modules are exposed to application developers through the Business Application Programming Interface (BAPI), a group of objects that provide an object-oriented view of the SAP R/3 system. The package is developed by SAP (System Applications and Programming), Germany. There are so many of consultancies, which will implement these ERP packages in the companies according to their need.The Company formulated a task force and has done detailed vendor evaluation for selecting the best vendor. And they come up with System Application and Programming, version R/3 among the 16 vendors tendered for the implementation. The implementation work in the primary department of the Company is a pilot project. In the implementation, molding the package according to the needs of the company is very important. The cost involved in the implementation is also very high. The Company has to pay Annual Maintenance Cost of about 18 to 20% of the total cost. The product range is 8 and among them about 1400 variations are there. Depending upon the material used, sealing arrangement, flow direction and controlling aid used etc the variations are existing. After successful implementation, ERP will be implemented in the company.

4. Deriving criteria for evaluation of ERP through AHP

For the application of AHP to evaluate the ERP and Conventional System the following criteria are considered. The criteria, which can be considered, should represent the alternatives, which we are considering for the evaluation. For the evaluation of ERP Vs Conventional System, the criteria which are, excavated by thorough elicitation of the various responsible employees from the company at different levels of hierarchy in the ERP cell are as follows. For the elicitation process the expertise are considered from various departments. The criteria as analyzed by the experts are clustered in to two sets. They are Tangible Criteria and Intangible Criteria, and each cluster is further sub divided as follows. Tangible Criteria are Annual Inventory savings and Manpower Reduction. Intangible Criteria are Cycle Time Reduction, Billing cycle Improvement, Response time to customer queries and Timeliness (Decision Making Speed)

5. Pair wise comparison matrix for first level

The first level in the hierarchy is the criteria which is common for the both alternatives. The pair-wise comparison matrix gives the preferences of the criteria one over the other. In the opposite cell of each cell the inverse value will be occupied, which means that if one criterion is preferred than other with some factor, inverse of the factor will occupy in the opposite cell saying that the other criterion is preferred inverse times of the factor.

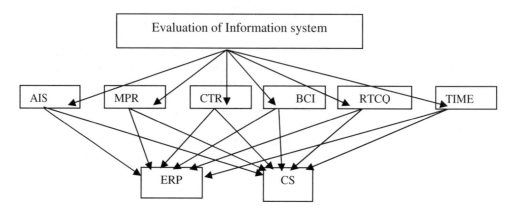

Figure 1. Structuring the problem in to hierarchy, where: AIS - Annual Inventory Savings, MPR - Manpower Reduction, CTR - Cycle Time Reduction, BCI - Billing cycle improvement, RTCQ - Response Time to Customer Queries, TIME - Timely ness in decision making, CS: Conventional system

Table 1. Pair-wise comparison matrix for the level 2

Criteria	AIS	MPR	CTR	BCI	RTCQ	TIME	EV
AIS	1	3	2	3	2	1	0.29
MPR		1	0.2	0.3	1	0.8	.07
CTR			1	1	2	3	0.23
BCI				1	2	2	0.19
RTCQ					1	1	0.10
TIME						1	0.12

CTR	ERP	CS	EV
ERP	1	7	0.87
CS	1/7	1	0.13

The eigen values of the pair-wise comparison matrix are found by multiplying the pair-wise comparison matrix with the unit column vector repeatedly until we get the maximum eigen value same to the previous iteration. For example in the level 2 of the hierarchy procedure for finding eigen values is explained as follows - Table 2.

Table 2. The eigen values of the pair-wise comparison matrix

BCI	ERP	CS
ERP	1	5
CS	0.25	1

The eigen values for the remaining criteria in the second level of the hierarchy are as shown in the table 3.

Table 3. The eigen values for the remaining criteria in the second level

CRITERIA	AIS	MPR	CTR	BCI	RTCQ	TIME	TOTAL WT'S
WEIGHTS	0.29	0.07	0.23	0.19	0.10	0.12	
ERP	0.83	0.80	0.87	0.83	0.83	0.86	0.841
CS	0.17	0.20	0.13	0.17	0.17	0.14	0.159

Finding the Consistency Ration and Global Weightages:
- Consistency Ratio (CR) = Consistency Index / Random Index,
- Consistency Index (CI) = (Max eigen value – 1) / (n – 1).

For (2 X 2) matrix the CR becomes "1" because of the random index.

In the hierarchic structure one eigen vectors of bottom level are multiplied with the corresponding eigen vectors of the next level as following. The total weightings are as follows:

ERP = (0.29*0.83) + (0.07*0.80) + (0.23*0.87) + (0.19*0.83) + (0.10*0.83) + (0.12*0.86) = 0.841

CS = (0.29*0.17) + (0.07*0.20) + (0.23*0.130) + (0.19*0.17) + (0.10*0.14) + (0.12*0.14) = 0.159.

As shown above the total weightage for the Enterprise Resource planning is consistently high. It results as 0.841.

6. Conclusion

A software package can be developed for Analytical Hierarchical Process and is used for the evaluation of an information system called ERP. The software can readily be applied for any type of multi criteria decision making problem, having any number of levels and criteria in the hierarchic structure. The credibility of the results of the evaluation process using the Analytical Hierarchical Process depends upon the input, which should be a broad spectrum of opinion and expertise. Here the expertise is of the team of decision makers or experts who are involved in the evaluation process. Further the results will also depend upon the quality of the information extracted and the relevance of the criteria extracted out, from the alternatives This research developed a methodology that qualitatively and quantitatively defines and measures the level of Enterprise integration on an intra-company and intercompany basis. This is achieved through a structured search for alternatives.

References

1. AHP Series by Thomas L. Saaty T.L., Volume II, 315 pp, 1995/ 1996.
2. Fawcett S.E., Cooper M.B.: *Logistics performance measurement and customer success*. Industrial Marketing Management 27 (4), 341–357, 1998.
3. Gunasekaran A., Patel C., Tirtiroglu E.: *Performance measures and metrics in a supply chain environment*. International Journal of Operations and Production Management, 21 (1&2), 71–87, 2001.
4. New S.J.: *A framework for analyzing supply chain improvement*. International Journal of Operations and Production Management, 16 (4), 19–34, 1996.
5. Sistach F., Pastor J.A., Fernández L.F.: *Towards themethodological acquisition of ERP solutions for SMEs*. Proc. of the First Int. Workshop on Enterprise Management and Resource Planning: Methods, Tools and Architectures (EMRPS'99), Venice, 1999.
6. Dyer J. S.: *Remarks on the Analytic Hierarchy Process*. Management Science, 36, no. 3, 249-258, March 990.
7. Saaty T. L.: *An Exposition of the AHP in Reply to the Paper Remarks on theAnalytic Hierarchy Process*. Management Science, 36, no 3, 259-268, March 1990.
8. Bailey W., Norina L.: *Supply Chain Management for Single DeskSellers*. 11th Annual Food and Agribusiness Symposium, Sydney, June 2001.

The International Journal of **INGENIUM** 2005 (3)

ENGINEERING ACHIEVEMENTS ACROSS THE GLOBAL VILLAGE

edited by

Janusz SZPYTKO

Engineering Ergonomics, Education and Training

Cracow - Glasgow - Radom, 2005

TABLE OF CONTENTS

page

4. Engineering Ergonomics, Education and Training **285**

4.1. The distribution of computer based training systems,
Ellis R.L.A., Persad P. 287

4.2. A better solution for DRG-national application,
Stanescu L., Burdescu D.D. 297

4.3. Active noise cancellation strategies based on digital signal processing - an overview,
Ádám T., Kane A., Varga A., Vásárhelyi J. 305

4.4. Advantages of using engineering ergonomics in the design of office/ computer workstations to control back problems,
Lewis W.G., Ameerali A.O. 313

4.5. Integrated methods for manual wheelchair design,
Ariff H., Samsudin A.S. 321

All rights reserved. No part of this book may be reproduced, stored in a retrieval system, or transmitted, in any form or by any means, without prior written permission from the Publisher.

The International Journal of INGENIUM
Chief Editor: Professor David K. Harrison, Glasgow Caledonian University, UK

© GCU Glasgow

ISSN 1363-514x

A CIP catalogue record for this publication is available from British Library

Publishing cooperation: Instytut Technologii Eksploatacji – PIB w Radomiu

THE DISTRIBUTION OF COMPUTER BASED TRAINING SYSTEMS

Ellis R.L.A., Persad P.

Abstract: After the development and testing of a Computer Based Training (CBT) system, comes the time to distribute it. The mode of development is dependent upon the needs of the client, for whom the work was developed. The best CBT incorporates learning theories and various types of media to be able to meet the needs of learners of various learning styles. This necessitates the use of audio, video, animations and software external to the CBT for tracking and monitoring the progress of the learners. The options, which are available to organizations that are seeking to leverage CBT in order to ensure that their staff members are effectively trained, are: the use of Compact Disks (CDs), distribution via an Intranet and distribution via the Internet. This paper discusses all three options and the techniques and tools that are needed with each option in order to avoid software conflicts and ensure that the CBT is distributed in an efficient and effective manner.

1. Introduction

The Personal Computer and it's associated hardware and software technologies are changing the way in which training is undertaken in the public sectors, private organizations and for personal development. More and more are the managers of education based institutes and departments of training seeking to find effective ways of meeting the demands of their clients for flexibility, relevance and autonomy in the curricular in which they are engaged (Gery 1993).

One way of accomplishing this is in the employment of CBT. This is because CBT can be used:
- when and where the learner wants to use it (Flexibility),
- to elicit information or knowledge that is immediately applicable (Relevance),
- at a pace that does not affect others, nor affect the other commitments that may be competing for the learner's time (Autonomy).

Thus it becomes imperative that, in the design, development and final distribution of a CBT system, the issues associated with *the learner, the curriculum* and *the technologies* be carefully addressed.

2. The learner

Today, the learner is viewed as being highly discriminating, and consisting of individualized mechanisms that contribute to the manner and rate that he/she learns. There are many available models that can be used to determine individual learning styles, among these (Ellis and Persad, 2004):

- Socio-Cultural Models: which describe how the interaction of the learner with the teacher and other learners, influences the learning process. These models examine the differences between male and female and younger and older learners,
- Personality and Psychological Models: These models describe the levels at which the deepest personality traits shape the orientations that the learner has towards knowledge and the world. They describe the how the persistent potential abilities, which vary from individual to individual, influence the process of learning,
- Cognitive Models: these describe how the individual acquires knowledge and how an individual processes information. They are related to the mental behaviors, habitually applied by an individual to problem solving, and generally the way that information is obtained, sorted and utilized.

The influence of these models is made manifest in the types of media and opportunities for interaction that are included in the CBT, for example, the use of audio, text, graphics, animations, movies, simulations and chat facilities. The inclusion of elements of each, influence the size and complexity of the final CBT, and thus the mode of distribution.

3. The curriculum

Since the 1930s psychologists such as J.B. Watson and B.F. Skinner have tried to understand the manner in which human beings learn. As Gary DeMar (2004) reports, that behaviorism originated with the work of John B. Watson, who claimed that psychology was not concerned with the mind or with human consciousness. Instead, psychology would be concerned only with behavior. His work was based on the experiments of Ivan Pavlov, who had studied animals' responses to conditioning. B.F. Skinner made his reputation by testing Watson's theories in the laboratory. His studies led him to reject Watson's almost exclusive emphasis on reflexes and conditioning. Skinner concluded that people respond to their environment, and they also operate on the environment to produce certain consequences. Like Watson, however, Skinner denied that the mind or feelings play any part in determining behavior. Instead, our experience of reinforcements determines our behavior.

In the 1970s behaviorism began to wane as the dominant psychological theorem and cognitive psychology began to take over. Cognitive psychology takes its name from the word cognition, which means the process of knowing. It places emphasis on unobservable constructs, such as the mind, memory, attitudes, motivation, thinking, reflection and other presumed internal processes (Allessi and Trollip, 2001). Cognitivism itself, originated in the early part of the twentieth century when the Gestalt psychologist of Germany, Edward Chase Tolman of the United States, and Jean Piaget of Switzerland imposed their influence on psychology, and caused a shift away from behaviorism. There are many schools of thought that exist in cognitivism, the major ones are:

- Cognitive Information Processing Theory. In the cognitive information processing theory of learning, the environment is regarded as having a great role in the learning outcome. Information processing theories sought to explain how information in the world enters through our senses, become stored in memory, is retained or forgotten and is used,

- semantic network theory attempts to parallel how biologists view the connections of the human brain. According to the semantic network theory, our knowledge consists of nodes that are connected in countless ways. The activities of remembering, thinking acting, problem solving and all other cognitive activities consist of nodes being activated via relationships or connections to other nodes, that in turn activate other information. In semantic networks, such as the brain, adding or removing links between nodes or by creating or changing nodes may represent learning,
- schemas are considered to be organized collections of information and their relationships in a manner, which is similar to that if the semantic network. The schema theory began with Sir Frederick Bartlett and postulates that out existing knowledge comprises of collections of schemas, each schema being specialized for different aspects of our daily living. Learning is therefore thought to occur when the schemas are modified to accept new knowledge either through assimilation or accommodation, or a combination of both.

In 1966 J. Bruner developed the constructivist theory of learning. In it he stated that a theory of instruction should address four major aspects: (1) predisposition towards learning, (2) the ways in which a body of knowledge can be structured so that it can be most readily grasped by the learner, (3) the most effective sequences in which to present material, and (4) the nature and pacing of rewards and punishments. Good methods for structuring knowledge should result in simplifying, generating new propositions, and increasing the manipulation of information. Bruner's constructivist theory is a general framework for instruction based upon the study of cognition (Bruner, 1960, 1973).

In the development of a CBT system, all three theories are relevant, and such any CBT should include elements of all. This is termed as an eclectic approach to curriculum design. Behaviorism allows for observable changes in the disposition of the learner through the use of knowledge reinforcement through repetition and questioning at different taxonomical levels as predicated by Bloom in 1956 and Romiszowski in 1981 (Carter, 1985, Ellis and Persad, 1999). Cognitivism allows for the testing of the learner to determine what it is that he/she knows so that action can be taken to provide opportunities for learning and understanding the unknown areas in the target knowledge domain. Constructivism provides a guide for developing the learning environment, which will give the learner the autonomy over what he/she learns.

4. The technologies

The choice of technology that is used to develop the CBT system should be able to facilitate distribution of the CBT via CDs, the Intranet or the Internet. The technologies must also be able to facilitate the inclusion of multiple media, which is necessary to cater for learners of different learning styles. In addition to these requirements, there may also be a need to monitor the progress of the learners, thus requiring the inclusion of the recording of relevant feedback such as quiz scores, current status of the learner, in terms of the area of the curriculum that has been completed and areas of difficulty that may need remedial work. In order to accomplish this it is necessary to include in the CBT some form of database capability.

Many of the available technologies are specific to the Internet these technologies include Blackboard and WebCT. One can use object-oriented languages such as Java and C++ to develop CDs Interactive software or material for the Intranet or Internet. The disadvantages of using object-oriented programming languages, is that once the programming has been done, the mode of distribution is fixed and would require a lot of time to revise the CBT in order to distribute it via any other mode.

Authoring software such as Macromedia's Authorware, allows one to publish the CBT for CD, Intranet or Internet distribution, with little modifications to the original CBT (Schifman, Van As, et. al., 1999). This is quite advantageous as it allows for the CBT developer to concentrate on the development of the CBT with the relevant or expected levels of interaction and multimedia, while knowing that the development environment will facilitate the distribution of the CBT via any chosen mode.

5. The ECOTech experience – an example

ECOTech – The Emission Control Technology Tutor, which has been developed at the Department of Mechanical and Manufacturing Engineering at the University of the West Indies, includes multimedia and databasing capabilities. It is an eclectically designed CBT, facilitating different forms of interaction with the learner. The aim of the current research is to develop a model for the types of media and interaction that are to be included in a CBT, which will optimize the learning time and performance of individuals of varying learning styles and pre-knowledge (Ellis and Persad, 2000, 2004).

Figure 1 shows the model for the ECOTech Tutor observable is the interface between Macromedia Authorware and Microsoft Access via the ODBC (Open Database Connectivity). ODBC is a programming interface that enables programs to access data in database management systems that use Structured Query Language (SQL) as a data access standard. This interface comes as part of the Microsoft Windows Operating System, and can interface with many types of databases.

5.1. Connecting to Databases

Utilizing ODBC could be a complex process for the uninitiated, which includes copying relevant database to the user's computer and setting up the System (DSN) Data Source Name in the ODBC interface (see Figure 2 and Figure 3). This is oft times difficult for the learner whose objectives are to utilize the CBT to obtain knowledge that is relevant to the task at hand.

Fortunately for the designers of CBT there exist an XTRA for Macromedia Authorware that allows the CBT to automatically register the database on the learners' personal computer.

When distributing a CBT, it is necessary to determine the files that are to be distributed in order to ensure that the piece runs seamlessly via any mode of distribution. Macromedia Authorware includes XTRAS (see Figure 4), which are used when distributing images, sounds, movies and animations (Macromedia, 2001). Authorware uses a plug-in called tMsDNS (see Figure 5) to automatically register a database with the Operating System. It is imperative that both the plug-ins and the XTRAs be included in any distribution of the CBT.

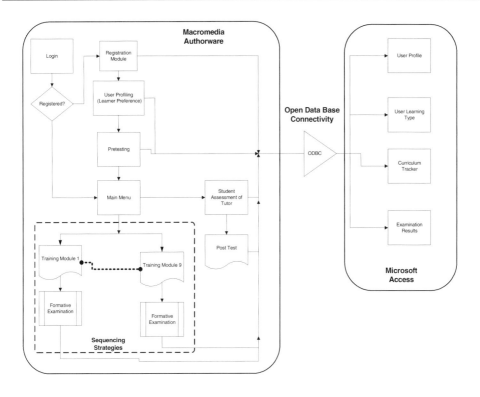

Figure 1. Model of ECOTech

Figure 2. ODBC Administration

Figure 3. ODBC Microsoft Access Setup

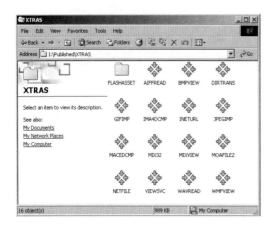

Figure 4. XTRAS for Authorware

Figure 5. Lists of Plug-ins

5.2. Distributing on CDs

In order to distribute a piece via CDs, it is best to package the CBT and include all the external files in the packaged piece (Figure 6). Figure 7 shows the elements of the CBT that are needed to register the Database with the OS. Of importance is the line:

$$dbList:=dbList^{\wedge}"DBQ="^{\wedge}FileLocation^{\wedge}"Ecot\_Base.mdb;"$$

The *FileLocation* variable, tells the computer where to look for the relevant database during execution.

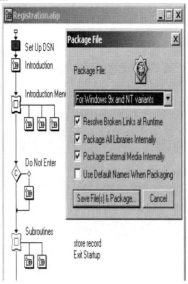

Figure 6. Packaging an Authorware CBT for CD Distribution

Figure 7: Automatic Registration of CBT using tMsDSN

5.3. Internet distribution

In order to distribute a CBT system via the Internet, consideration must be given to the available infrastructure, specifically the bandwidth that are available to the learners. Where the learners do not have DSL connections and are reliant upon 56 kbs dial-up modems, CBTs containing audio and movies would take a lot of time to be downloaded before the learner can start utilizing it (Schifman, Van As, et. al., 1999. This can lead to frustration and hinder the progress of the learner. In order to eliminate this, Macromedia has come up with a technique whereby the CBT system is "shocked", thereby reducing it into smaller segments so that they can be rapidly transmitted over the Internet.

Again, the process starts by packaging the piece, similarly as in Figure 6 but omitting the runtime environment selection as shown in Figure 8. The file that is created is then 'shocked' using the web-packager or 'afterburner.' During the packaging process an Authorware Map File is created in which all the segments that the CBT was broken into are listed, along with instructions for the learners' browser to effectively and efficiently put the CBT together.

Figure 9 shows the segmentation process. The designer has the option of varying the sizes of the segments in order to improve the efficiency of the delivery of the CBT via the Internet.

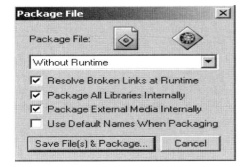
Figure 8. Packaging without Runtime

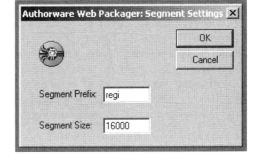
Figure 9. Web Packaging Settings

Figure 10 shows the details of the Map File, listing all the segments, XTRAS, Plug-Ins and instructions for the control of the streaming of the CBT.

In order that it is possible for the leaner to view the CBT as it was created, Macromedia has created a set of plug-ins, called Web Players for different web browsers that the learner will have to install. The web-players are available free from Macromedia - at http://www.macromedia.com/shockwave/authorware/form/download

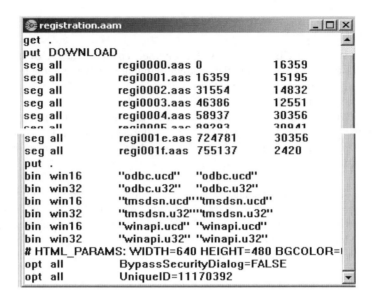

Figure 10. Map File of CBT

In distributing the CBT via the Internet, the major concern is in being able to monitor the performance of the learner by capturing the data in a database. If one is using Microsoft Access as the backend database, then a couple of problems will arise. Firstly there will be a need to utilize either Active Server Pages (ASP) or some other Common Gateway Interfacing (CGI) technique (Ellis and Persad, 1999) to interface the CBT with the database over the Internet. In order to avoid using any of the CGI techniques, one can obtain 'Hot Sockets,' a piece of software that interfaces between Authorware and Microsoft Access in the web environment. The disadvantage of 'Hot Sockets' is its price.

Alternatively, one can utilize the Integration New Media V12 Database Engine for free, and employ scripts written in LINGO in order to do the interfacing. The V12 Database Engine gives a seamless interface with Authorware and the Web and increases the flexibility from a designers' perspective (see http://www.integrationnewmedia.com/)

5.4. Intranet distribution

Most education institutions and corporate bodies possess their in-house client-server network or intranet. Thus for the developer, this option of distributing the CBT is a viable as well as a feasible one. Intranets allow the CBT to be hosted on a server and be accessed by those who are authorized to do so.

When distributing a piece via an Intranet, there is no need to set-up a web server, nor is there need for the learners to install any unwanted plug-ins on there computers. This is because the executable, with the runtime environment, as described in Figure 6, is used.

Due to the oft times small amount of network traffic on an Intranet, the CBT can be distributed without being 'shocked', and with little or no loss in speed or quality.

Again, the major concern is with monitoring the learners' progress with the use a database. Given that over the Intranet, the piece does not run on the server, but rather on the learners' computer, the *FileLocation* Variable, as shown in Figure 7, will attempt to force the CBT to look on the client computer for the database. If there are a large amount of learners using the CBT, it will be uneconomical to allow the database to be copied unto every machine. Also, it would be very inefficient for resources to be employed to configure the ODBC interface on every client machine.

What we have discovered is that by using the tMsDSN plug-in, and changing the *FileLocation* Variable, in the DSN setup (See Figure 7), to the *NetLocation* Variable, then Authorware automatically registers the database with the ODBC and allows the client to connect to the database each time the CBT is used.

For security reasons, one can embed the executable of the CBT in a web page, thus eliminating the need for the learner to have to browse the network. In this case, all that will be necessary is to set up a web server, with one page that points to the CBT.

6. Conclusion

As the need for individual, corporate and institutional training increases, there will be greater pressure to quickly distribute CBTs. This is because the CBT can cater for the variances in the learning styles among individuals, can be designed based on strong theoretical education philosophies and principles, and can employ modern technologies, which will individualize the learning experiences.

The key factor in distributing such a CBT, is in the recording and monitoring of the progress of the learners. We have shown that by using a common database management system such as Microsoft Access, it is possible to record and monitor the learners' performances via CD, Internet or Intranet Distribution. The choice of which mode of distribution to use is therefore dependant upon the need to centralize the monitoring of learner performance, the availability of a client-server environment, and the geographical location of the learner relative to the provider of the CBT system.

Some of the pitfalls and technologies that are associated with each form of distribution were discussed and solutions to some of the more common issues have been presented. These will go a long way in assisting the CBT practitioner in deploying his or her work and having organizations reap the vast benefits of having training programs readily accessible to their employees and educational institutes having remedial courses available to those who need them.

References

1. Alessi S.M., Trollip S.R. (2001): *Learning Principles and Approaches*. Multimedia for Learning – Methods and Development, pp.16-47.
2. Bruner J. (1960): *The Process of Education*. Cambridge, MA, Harvard University Press.
3. Bruner J. (1973): *Going Beyond the Information Given*. New York, Norton.
4. Carter R.: *A Taxonomy of Objectives for Higher Education*. Studies in Human Education, Vol.10, No.2, pp. 135-149.
5. DeMar, Gary; Behaviorism, http://www.forerunner.com/forerunner/X0497_DeMar_-_Behaviorism.html ; Accessed December 2004.
6. Ellis R., Persad P. (2004): *Incorporating Individual Learning Styles in CBT Designs*. (CARS&FOF2004), pp. 304-310.
7. Ellis R., Persad P. (1999): *Evaluation of a Web-Based Tutoring System*. (CARS&FOF'99)
8. Ellis R., Persad P. (2000): *Empirical Evaluation of a Web-based Tutoring System*. (CARS&FOF2000)
9. Gery G. (1993): *Making CBT Happen*. Cambridge: Ziff Communications Company.
10. Macromedia (2001): *Packaging an Authorware Piece*. Macromedia AUTORWARE 6- Using Authorware, pp. 175-182/
11. Schifman R.S, van As S., Ganci J., Kernman P., McGuire J., Well W. (1999): *Authorware in a Network*. The Ultimate Authorware Attain Tutorial, Springer, pp. 111-178.

A BETTER SOLUTION FOR DRG-NATIONAL APPLICATION

Stanescu L., Burdescu D.D.

Abstract: This paper presents a modern and more efficient solution for the DRG-National application, used by the Romanian Government to finance the hospitals. The proposal implies an on-line application based on JSP technology and a MySQL database in order to replace the old application realized in MS Access 2000. With such an application, the DRG National Bureau can see, at any time, real-time data, and even send it to the international organizations. This solution keeps the structure and many menus from the old application, so the users can easily get accustom to it. The accent is on the data security because the users have limited rights, which are set by an administrator.

1. Introduction

The Romanian Government uses nowadays the DRG classification system as a base to finance 185 hospitals, and it plans to extend this number in the following years. The DRG is a system which permits to classify the patients based on the diagnosis, the procedures and other information (the complexity of each case) and to link this type of patients that each hospital treats to the expenses needed (DRG, 2005).

The development of the DRG started at the end of the 60's at the Yale University in the USA. The initial reason for the development of DRG was to create an efficient way to monitor what is happening in a hospital. Consequently, first of all, the DRGs must be considered more like a classification system and not like a financial one. If the DRGs are used as a start for the finance system, then many other concepts must be implemented. Ever since, over 30 countries use the DRGs for measuring the clinic activity, the improvement of the clinical management, and to management of financing the hospitals. The necessary data for the patient classification on the basis of the diagnosis and the procedures in DRG categories are: age, sex, hospitalization period, principal diagnosis, secondary diagnosis, procedures, health condition when leaving hospital, the birth weight (in the new-born child case). These data define the DRG classification system.

Through the system of the diagnosis group (DRG) the characteristics of each patient who left the hospital are analyzed, and in accordance with these, the patients are classified in a different category. This way, the DRG system makes an ,,image" to the hospital results, trying to standardize the results of this activity. The diagnosis groups have two essential characteristics:
1) the clinical homogeneity, meaning that in a certain DRG the cases (the patients) are similar from the clinical point of view, but not identical,
2) the costs homogeneity, meaning that each DRG has cases that need similar usage of resources.

The diagnosis groups are medical and surgical ones, varying according to the presence or the absence of a surgical intervention and they are conceived to guarantee the pathology associated with the acute patients that request the hospitalization. To be able to classify each patient that left the hospital in a diagnosis group there have to be run four phases:
1) the disponibility of the clinical data for the patients that left the hospital,
2) the codification of the necessary data for the diagnosis and the procedures in order to have a standardized language for these variables and to be able to use them easily,
3) gathering these data in an electronic manner,
4) the automatic classification of each patient in a diagnosis group.

Nowadays, to gather the information for the patients it is used the DRG-National v4.0 application that is delivered through the district agencies (DRG, 2005). This application must be installed on every computer used for gathering data about the patients. The electronically registration for a patient, one for the whole period the patient stays in hospital, is in concordance with the new clinical observation form introduced by the Romanian Health and Family Ministry. Once collected, the data are added to a database which has to be sent monthly to the DRG department from the National Health Institute for Research and Development, Bucharest.

Nowadays, the application DRG-National v4.0 is delivered on a CD as a Runtime application implemented in Microsoft Access 2000, that collects information about the patients at the department level, encrypt them and after that send them to the DRG National Bureau using the e-mail. The data centralization from all departments on a single computer is also possible, where the application also has to be installed, and send them in this centralized manner to the DRG National Bureau. The application has minimal requirements at the hardware level and at the level of the Internet connection.

2. The description of the new solution for the application

2.1. The general presentation

The paper presents a new solution to implement the DRG-National application, which eliminates the difficulties of the one used at present. First of all, MS Access 2000 is an administration system for desktop databases and it administrates with efficiency a slightly reduced number of recordings: around 20000-30000 records, generally used in activity management in small companies. That is why, instead of using MS Access 2000 it is proposed a much stronger system of database administration, named MySQL (Welling, 2003).

Secondly, an on-line application that offers more advantages is proposed:
❑ it cancels the encrypt phase of data and the transmission through the e-mail because it can bring a lot of problems; the data received by e-mail at the DRG National Bureau must be saved in a certain format, eventually a database, in order to be consulted either with visualization, or with the elaboration of some reports, statistics or graphics, this being the most efficient way; when adding in the same database information resulted from different sources there might be problems; for example: the primary key values

might be identical in several different files. In this case certain processing must be made before being saved in database; all these require a lot of time,
❑ the data that DRG National Bureau sees are always up to date; it is not necessary to wait for the end of a certain period of time to receive those data; anytime the up-to-day data can be seen as reports, statistics or graphics, which is a very important thing for the medical domain; these up-to-day data could be placed even at the disposal of some international medical institutes, an usual and absolutely necessary task,
❑ the application, database and the database server must be installed on a single computer (server); it will be accessible for any user in the range of his rights from his own computer, using a browser; in this manner it is not needed to install the application on each client's computer; of course, the server's performances must be very high.

The application described below is a client/server application and represents the ideal solution to implement a competitive database with minimum costs (software costs are zero). It is based on a MySQL database motor, installed on a Linux server, which ensures high performance and high speed as well as outstanding security. The system also encompasses the problem of soft licensing. It is realized with JavaServer Pages technology (Matthews, 2003, Goodwill, 2000, Forta, 2001).

2.2. The database structure

Applying the normalization process (Burdescu, 2004, Elmasri, 1994), the database proposed for this on-line application will be described further on. First of all there is a series of set-up tables which contain encoding. Each of these tables contains an id field which identifies uniquely the records and a name:
1) table *tblStatus* allows encoding of the patient status by means of two fields status_id (primary key) and status name,
2) for encoding the diagnoses there are three tables *tblClass*, *tblSubclass* and *tblDiagnoses* connected by a 1:m relationship because in one class there are several subclasses, and in one subclass there are several diagnoses,
3) for encoding the XR investigations and the functional explorations there is table *tblXr*, respectively *tblInvestFunct*,
4) there is also a table which allows the encoding of the surgical procedures, *tblSurgical*
5) the table *tblInt* allows encoding of the different types of hospital registration.

Some other tables in the database are:
1) in the table *tblHospital* there are data about the hospital from the DRG network: hospital_id (primary key), hospital name, county, city, number of beds,
2) in the table *tblDepart* there are data about the departments from each hospital: department_id (primary key), department name, hospital_id (foreign key) used to implement the relationship 1:m between the tables tblHospital and tblDepart,
3) in every department works a number of doctors about whom there are data in the table *tblDoctor*: doctor_id (primary key), name, specialty, department_id (foreign key) used to implement the relationship 1:m between the tables *tblDepart* and *tblDoctor*,
4) for a certain department and a certain doctor are registered the patients who have a new sheet at every hospital record. For their data a new table *tblPatient* was created having the following structure: sheet_id (primary key), personal_id, first name, last name, sex, birth weight in the new-born child case, data of registration, data of release from the

hospital, doctor_id (foreign key) in order to know the doctor who attended the patient, the code of the first principal diagnosis, the code of the second principal diagnosis, the code of the secondary diagnosis, the code of the hospitalization type, the code of the patient status.

The database also contains a series of tables as the result of the implementation of some m:m relationships between the above tables:
1) the table *tblPatient_XR* stores for every observation sheet identified by sheet_id the codes of the effectuated XR investigations,
2) the table *tblPatient_FuncExpl* stores for every observation sheet identified by sheet_id the codes of the effectuated functional explorations,
3) the table *tblPatient_Surgical* stores for every observation sheet identified by sheet_id the codes of the surgical procedures made during the hospitalization time.

2.3. User Interface

The main menu of this application is: Patient File, Search, Reports and Exit. Each of these menus has few submenus. The first option groups the web pages managing the sheets of paper for the hospitalized, the second allows searching patient details or searching data through other tables (for example the diagnosis table), and the third allows to see the reports, statistical data, graphics. After the user connects to the database (he enters the username and password) he can insert records for new patients, or modify existing ones.

When adding a new record, the user must write the name and surname of the patient, date and time of hospitalization. When the user presses Submit button, the new record is inserted in tblPatient table. After that, a new submenu with the following options is generated:
- general information,
- diagnoses,
- surgical interventions,
- functional explorations,
- XR investigations.

When General Information option is called, a new form appears allowing to enter general information about the patient: name, surname, birthday, sex, status (insured, uninsured, foreigner), personal ID, address, type of hospitalization, name of the doctor.

The Diagnoses option allows entering two main diagnoses, the diagnosis when leaving the hospital and other complications, if any. The other three options represent auxiliary information regarding surgery, radiography, analyses (if any). As it can be seen, data about patients are added at different moments of time, so the update function is very used and therefore very important, as searching function matching one of several criteria. The proposal in order to find information about patients is to search for record number, or patient details: personal ID, name, surname. Once the patient identified, the necessary information can be updated using the option above.

To ease the operator's work filling the data, certain data can be selected from drop-down controls. For example, selecting a diagnosis or a type of analysis can be done in such a manner. The data added in forms are validated. For example, it is checked if a field is date type or, if some fields aren't null (name, surname). When the patient leave the hospital all the data needed now are added (diagnosis, health condition), and the flag which indicates this is set to true.

Regarding the reports, they can be obtained from one department, hospital or even DRG National Bureau. In the first case the data reflect information only from that department, in the second case data from all departments in the hospital, and in the last case from all data in the database.

The reports may consider the patients from a certain month in order to match the financial part of the application, or all the patients who have ever been (or still are) hospitalized in that department. The application wants to make a few statistics regarding the number of patients with a certain diagnosis in every district, over a period of time, or in accordance with health condition when leaving the hospital. It has to be said that when a user logs on, and tries to update the database, he will see only those records that the next superior level administrator permits him to modify. A user from a department will see only the records with patients in his department.

In figure 1 there is the application window opened for a user identified by name and password who can proceed to the insertion of a new observation sheet being entitled to this. In figure 2 there is the window which allows the update of the data concerning the principal and secondary diagnoses of the patient.

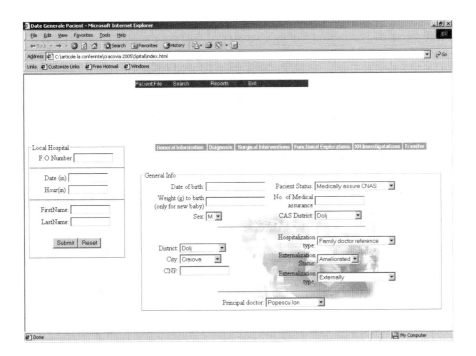

Figure 1. The application window allowing the insertion of a new observation sheet

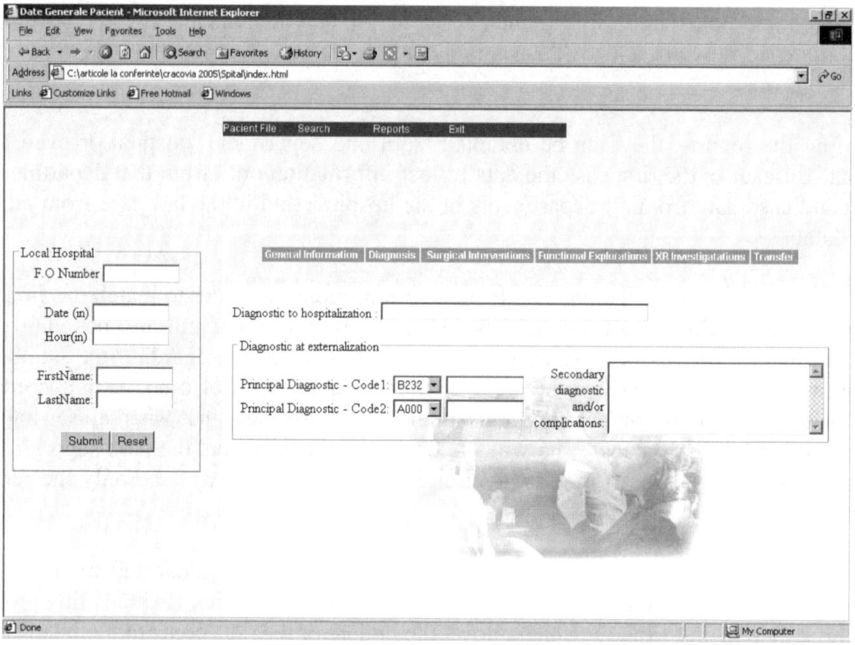

Figure 2. The application window allowing the updating the principal and secondary diagnoses

2.4. Administrators, users and rights, security

In order to manage the application there have been defined several levels that shall describe further on. The highest level is level 0, where the database administrator who has absolute rights is. For identification he uses a username and a password.

One of his tasks is to introduce in the tblHospital table a new record for each hospital that is added in the DRG network. For each hospital he will create a new administrator who has absolute rights regarding the information from his hospital. This is what is called the layer 1 of management. The administrator from layer 0 is the one who has as a task the management of the data shared by all users (for example the diagnoses table, status table)

The hospital administrator (from level 1) has the right to add a new record for each department from his hospital in the tblDepart table. He also has to create for each department one or more users who can update the information regarding their department. Each of the new users must be identified by username and password. This is level 2 of management. These 3 levels are showed in the following figure 3.

The emphasis is on the fact that every person who interacts with the database has limited access; he has certain rights over a small part of it. Level 0 administrator has rights over the whole database, level 1 administrator has rights over the information regarding his hospital, and level 2 users have rights only for information regarding their department. In this way, a user from Department 11 has no rights over the data from Department 12.

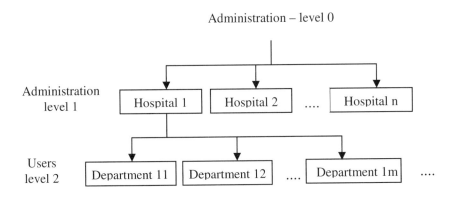

Figure 3. The security levels

Besides the fact that the persons who interact with the database are restricted to see only certain records (in accordance with the level), they can be restricted to execute certain operations. The database operations that are usually executed are: inserting new records, update records, delete records, and find records. For example, some users may not have the right to delete records because this is an operation that must be executed carefully as it may result in losing important data. Other users may not have the right to modify data, but only to see it.

The administrator for each level is the one who grants those rights to the next level users. For example there are users at level 2, from the DRG National Bureau, who have rights only for viewing data. All the aspects presented above, guarantee a good security for the database which has a national importance. This is the reason why it is such an important matter.

3. Conclusion

The paper presents a new solution to implement the DRG-National application, used by the medical system in some hospitals in Romania as the basis for the establishment of the degree of finances they need. The application is used successfully in many countries and in Romania it is intended to extend the number of hospitals that are involved in this system. At present a MS Access 2000 application is used. This application must be installed on each computer where the data are inserted and updated. At the end of the month, these data must be encrypted and sent by e-mail to the DRG National Bureau which makes o series of synthetic reports. This paper has presented not only the drawbacks and the inconvenients of the used application, but also a new solution. This one supposes a client-server application working on-line, uses a MySQL database, a Tomcat Web server and JavaServer Pages Technology.

The new solution is not expensive and offers a series of advantages:
- the MySQL database has performances superior to those offered by MS Access,
- the data viewed by the DRG National Bureau are every moment the up-to date ones, without having to wait for the end of the month,
- the application and the database are installed on a single computer, from where is accessed by everyone who has the rights,
- the administration and the security of the database have been realized by means of some security levels and of some views on the database defined for each type of user.

In the end, it has to be mentioned that the application is in the developing stage.

References

1. Burdescu D., Ionescu A., Stanescu L.: *Databases*. Universitaria, Craiova, 2004.
2. Elmasri R., Navathe S.B.: *Fundamentals of Database Systems*. Addison - Wesley Publishing Company, 1994.
3. Forta D., Smith E., Stirling S., Kim L., Kerr R., Aden D., Lei A.: *Dezvoltarea aplicatiilor*. JavaServer Pages, Teora, 2001.
4. Goodwill J., Pure J.S.P.: *Java Server Pages: A Code-Intensive Premium Reference*. Smas, 2000.
5. Matthews M., Cole J., Gradecki J.D., Gradecki J.: *MySQL and Java Developer's Guide*. Wiley, 2003.
6. DRG Romanian Oficial Page http://www.drg.ro, 2005.
7. Welling L., Thomson L.: *MySQL Tutorial*. MySQL, Press, 2003.

ACTIVE NOISE CANCELLATION STRATEGIES BASED ON DIGITAL SIGNAL PROCESSING - AN OVERVIEW

Ádám T., Kane A., Varga A., Vásárhelyi J.

Abstract: Basics of Active noise cancellation using digital signal processing technology are introduced. The main cancellation strategies such as single channel broadband and narrowband feedforward, and feedback ANC systems are presented. Several algorithms for ANC systems are also shown. The problems of acoustic duct model planning and different kind of models are presented. A realized single-channel broadband model is shown.

1. Introduction

Acoustic noise in the environment can be classified into two groups. The first one is the turbulent noise, where the energy is distributed across the total frequency bands. It is referred to as broadband noise. Examples are the low-frequency sounds of jet planes and the impulse noise of an explosion. The second type of noise is called narrowband noise that concentrates most of its energy at specific frequencies. This type of noise is related to rotating or repetitive machines, so it is periodic or nearly periodic.

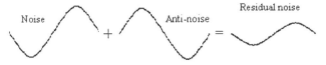

Figure 1. Theory of ANC

There are two ways of decreasing acoustic noise: passive and active. The traditional approach to acoustic noise attenuating uses passive techniques such as enclosures, barriers, and silencers.

These passive silencers attenuate the noise over a broad frequency range. However, they are relatively large, costly, and ineffective at low frequencies. The new approach is discussed in this paper is active noise control. The active noise control system contains an electro acoustic device that cancels the noise by generating an antinoise of equal amplitude and opposite phase. The acoustic combination of the original sound and the antinoise signal results the cancellation of both sounds, as it is shown in the Figure 1.

2. The main application fields of active noise control

From a geometric point of view, active noise control applications can be classified in the following four categories:

1. Duct noise can be characterized as one-dimensional application: ventilation ducts, exhaust ducts, etc.
2. Interior noise, where noise within an enclosed space has to be reduced.
3. Personal hearing protection that is a compacted case of interior noise.
4. Free space noise: noise radiated into open space.

Typical applications of active noise control are:
❑ automotive (car, van, military vehicle), electronic muffler for exhaust system, noise attenuation inside passenger compartment and heavy-equipment operator cabin, active engine mount, air conditioner, refrigerator, washing machine, furnace, air duct, transformer, compressor, pump, chain saw, noisy plant, airplane, ship, diesel locomotive, etc.,
❑ the algorithms developed for active noise control can also be applied to active vibration control. As the performance and reliability increase and the initial cost decreases, active systems may become the preferred solution to a variety of vibration-control problems, too,
❑ because the characteristics of an acoustic noise and the environment are not constant, the active noise control system must be adaptive; therefore active noise control applications were influenced by the development of powerful DSPs and the adaptive signal processing algorithms. The active noise control systems are based on digital adaptive signal processing technology, but adequate consideration of the acoustical elements is also very important. If the acoustical design of the system is not optimized, the digital controller may not be able to attenuate the noise adequately.

3. Types of ANC systems

Broadband ANC. The knowledge of the noise source is necessary for broadband noise cancellation requires in order generating the antinoise signal. The primary noise is used as a reference input to the noise canceller. Primary noise that correlates with the reference input signal is canceled downstream of the noise generator (a loudspeaker) when phase and magnitude are correctly modeled in the digital controller Narrowband ANC. For narrowband noise cancellation (periodic noise of rotational machinery), active techniques do not rely on causality (having prior knowledge of the noise).

Instead of using an input microphone, a tachometer signal provides information about the primary frequency of the noise generator. Because all of the repetitive noise occurs at harmonics of the machine's rotational frequency, the control system can model the known noise frequencies and generate the antinoise signal. There are two kinds of active noise control systems. In feedforward control a reference noise input is sensed before it propagates past the canceling speaker.

In feedback control the active noise controller attempts to cancel the noise without the reference input. Systems for feedforward ANC are further classified into two categories:
❑ adaptive broadband feedforward control with an acoustic input sensor,
❑ adaptive narrowband feedforward control with a nonacoustic input sensor.

3.1. The Broadband Feedforward System (BBFFANC)

Noise is often produced in ducts such as exhaust pipes and ventilation systems. A feedforward control system for a long, narrow duct is shown in Figure 2. A reference signal x(n) is sensed by the input microphone. The noise canceler uses the reference input signal to generate a signal y(n) of equal amplitude but $180°$ out of phase. This antinoise signal is used to drive the loudspeaker to produce a canceling sound that attenuates the primary acoustic noise in the duct. In the case of the BBFFANC the propagation time delay between the input microphone and the active control source offers the opportunity to electrically reintroduce the noise at a position in the field where it will cause cancellation. The spacing between the microphone and the loudspeaker is a key parameter: it must satisfy the principles of causality and high coherence, meaning that the reference must be measured early enough so that the antinoise signal can be generated by the time the noise signal reaches the speaker.

Also, the noise signal at the speaker must be very similar to the measured noise at the input microphone; otherwise the acoustic channel cannot significantly change the noise. The error microphone measures the error e(n), which is used to adapt the filter coefficients to minimize this error. The use of a downstream error signal to adjust the adaptive filter coefficients does not constitute feedback, because the error signal is not compared to the reference input. Actual implementations require many additional considerations to handle acoustic effects in the duct.

3.2. The Narrowband Feedforward System (NBFFANC)

In special cases, where the primary noise is periodic and is produced by rotating machines, a nonacoustic sensor such as a tachometer, or an optical sensor can replace the input microphone.

This replacement eliminates the problem of acoustic feedback. The nonacoustic sensor signal is synchronous with the noise source and is used to simulate an input signal that contains the fundamental frequency and all the harmonics of the primary noise. An error microphone is still required to measure the residual acoustic noise. This error signal is then used to adjust the coefficients of the adaptive filter.

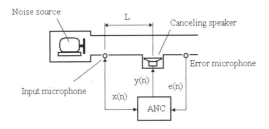

Figure 2. Single-channel broadband feedforward ANC system in a duct

In the NBFFANC systems the nonacoustic sensors are insensitive to the canceling sound, resulting robust control systems. Moreover, environmental and aging problems of the input microphone are automatically eliminated. This is especially important, because sometimes it is difficult to sense the reference noise in high temperatures or turbulent gas ducts like an engine exhaust system. The periodicity of the noise enables the causality constraint to be removed, since the noise waveform frequency content is constant, and only adjustments for phase and magnitude are required. Selective cancellation is possible, that is, each harmonic can independently be controlled. It is necessary to model only the part of the acoustic plant transfer function relating to the harmonic tones. A lower-order FIR filter can be used, making the active periodic noise control system more computationally efficient.

3.3. The feedback ANC system (FBANC)

In FBANC system (See Figure 3) a microphone is used as an error sensor to detect the undesired noise. The error sensor signal is returned through an amplifier (electronic filter) with magnitude and phase response designed to produce cancellation at the sensor via a loudspeaker located near the microphone. This configuration provides only limited attenuation over a restricted frequency range for periodic or band-limited noise. It also suffers from instability, because of the possibility of positive feedback at high frequencies.

3.4. The multiple-channel ANC system

Many applications can display complex modal behavior, for example active noise control in large enclosures, or active noise control in passenger compartments of aircraft or automobiles.

When the geometry of the sound field is complicated, it is no longer sufficient to adjust a single secondary source to cancel the primary noise using a single error microphone. The control of complicated acoustic fields requires both the exploration and development of optimum strategies and the construction of an adequate multiple-channel controller. These tasks require the use of a multiple-input multiple-output adaptive algorithm. The general multiple-channel ANC system involves an array of sensors and actuators.

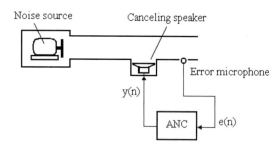

Figure 3. Feedback ANC system

3.5. Algorithms for BBFFANC systems [1, 4]

The following section discusses the algorithms used in BBFFANC systems. Broadband active noise control can be considered as a system identification procedure, as it is shown in Figure 5. The system uses an adaptive filter W(z) to estimate the response of an unknown primary acoustic path P(z) between the reference input sensor and the error sensor.

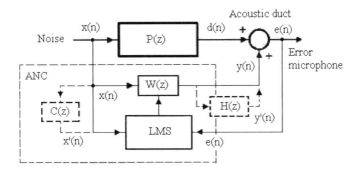

Figure 4. FLMS algorithms

The error signal e(n) can be expressed as:

$$E(z) = D(z) - Y(z) = X(z)[P(z) + W(z)] \quad (1)$$

where: *X(z)* is the input signal, and *Y(z)* is the adaptive filter output.

If the adaptive filter *W(z)* has converged, *E(z) =0*. Equation becomes: W(z) = – P(z) and y(n) = – d(n)

Therefore, the adaptive filter output *y(n)* has the same amplitude but is 180° out of phase with the primary noise *d(n)*. When *d(n)* and *y(n)* are acoustically combined, the residual error becomes zero, resulting in cancellation of both sounds.

3.6. Secondary-Path Effects

The antinoise signal can be modified by the secondary-path function *H(z) (dotted lines, Fig.4.)* in the acoustic channel from *y(n)* to *e(n)*, just as the primary noise is modified by the primary path *P(z)* from the noise source to the error sensor. Therefore, it is necessary to compensate for *H(z)*. The error signal *e(n)* is:

$$E(z) = X(z)P(z) + X(z)W(z)H(z) \quad (2)$$

Assuming that *W(z)* has sufficient order, after the convergence of the adaptive filter, the residual error is zero (that is, *E(z) =0*). This result requires *W(z)* to be:

$$W(z) = -\frac{P(z)}{H(z)} \quad (3)$$

The adaptive filter *W(z)* has to model the primary path *P(z)* and inversely model the secondary path *H(z)*. It can be seen that to invert the delay caused by H(z) is possible, if the primary path P(z) contain a delay of at least equal length. This is the overall limiting causality constraint in broadband feedforward control systems. Furthermore, the control system is unstable if there is a frequency, where $H(\omega) = 0$. Also, the system is ineffective if there is a frequency, where $P(\omega) = 0$ (unobservable control frequency). Therefore, the secondary path H(z) have significant effects on the performance of an ANC system.

3.7. Filtered-X Least-Mean-Square (FXLMS) algorithm

The effects of the secondary-path transfer function H(z) can be compensated by the Filtered-X Least-Mean-Square (FXLMS) Algorithm. The input to the error correlator is filtered by a secondary-path estimate C(z). This results in the filtered-X LMS (FXLMS) algorithm to compensate for the effects of the secondary path in ANC applications. The output y(n) is computed as:

$$y(n) = \underline{W}^T(n)\underline{X}(n) = \sum_{i=0}^{N-1} w_i(n)x(n-i) \quad (4)$$

where: $\underline{W}^T(n) = [w_0(n) \quad w_1(n)... \quad w_{N-1}(n)]^T$ is the coefficient vector of W(z) at time n and $\underline{x}(n) = [x(n) \quad x(n-1)... \quad x(n-N-1)]^T$ is the reference signal vector at time n.

The filter is implemented on a DSP in the form: $y(n) = \sum_{i=0}^{N-1} w_i(n)x(n-i)$

The FXLMS algorithm is: $\underline{w}(n+1) = \underline{w}(n) - \mu e(n)\underline{x}(n)h(n)$

where: μ is the step size of the algorithm that determines the stability and convergence of the algorithm and h(n) is the impulse response of H(z).

Therefore, the input vector x(n) is filtered by H(z) before updating the weight vector. However, in practical applications, H(z) is unknown and must be estimated by the filter, C(z). Therefore:

$$w_i(n+1) = w_i(n) - \mu e(n)\underline{x}'(n-i) \quad \text{and} \quad \underline{w}(n+1) = \underline{w}(n) - \mu e(n)\underline{x}'(n-i) \quad (5)$$

$x'(n)$ that is computed as: $x'(n) = [x'(n-1)...x'(n-N+1)]^T$ and $\underline{c} = [c_0 c_1 ... c_{M-1}]$ is the coefficient vector of the secondary-path estimate, C(z).

The transfer function H(z) is unknown and is time varying due to the aging of the loudspeaker, changes in temperature, and airflow in the secondary path. Therefore, different on-line modeling techniques are developed. If the H(z) is time-invariant, off-line modeling technique can be used to estimate H(z) during a training stage.

3.8. Acoustic feedback effects and solutions (FBFXLMS algorithm)

The antinoise output to the loudspeaker not only cancels acoustic noise downstream, but unfortunately, it also radiates upstream to the input microphone, resulting in a contaminated reference input x(n). This acoustic feedback introduces a feedback loop or poles in the response of the model and results in potential instability in the control system. To compensate the acoustic feedback, different solutions are proposed. For example usage of fixed compensating signals to cancel the effects of the acoustic feedback, or usage of an adaptive IIR filter can be the solution. The filter is an estimate of the feedback path from the adaptive filter output to the output of the reference input microphone. The filter removes the acoustic feedback from the reference sensor input; the filter (C(z) in Fig. 4) that compensates the secondary-path transfer function H(z) in the FXLMS algorithm is also necessary. Removal of the acoustic feedback from the reference input adds a considerable margin of stability to the system if the feedback channel model D(z) is accurate. The models C(z) and D(z) can be estimated simultaneously by an off-line modeling technique using an internally generated white noise. The expressions for the antinoise y(n), filtered-X signal x'(n), and the adaptation equation for the FBFXLMS algorithm are the same as that for the FXLMS ANC system, except that x(n) in FBFXLMS algorithm is a feedback-free signal that can be expressed as:

$$x(n) = u(n) - \sum_{i=0}^{L} d_i y(n-i) \qquad (6)$$

where: u(n) is the signal from input microphone, d_i is the i-th coefficient of D(z), and L is the order of D(z).

3.9. FXGAL algorithm

An alternative to FxLMS is the FxGAL algorithm [3], which can be seen as a version of the gradient adaptive lattice (GAL) algorithm modified to be used in the context of active noise control. The aim of FxGAL algorithm is to obtain faster and much less signal dependent convergence than FxLMS, at the expense of an increase in computational cost. In the GAL algorithm, also known as the Griffiths' algorithm, the tapped delay line (TDL) of an FIR filter is substituted by an adaptive lattice predictor (ALP).

Thus, approximate self orthogonalization of the input data is performed in the time domain, since the correlated sequence of delayed samples of the input signal:

$\{x[n], x[n-1], ...x[n-M+1]\}$ is transformed by the ALP in the uncorrelated sequence of backward prediction errors: $\{b_0[n], b_1[n], ...b_{M-1}[n]\}$.

Without loss of information. This uncorrelated sequence is the input to an adaptive linear combiner, which finally provides the output of the system.

The linear combiner is adapted following the LMS algorithm. Due to the orthogonality property of the backward prediction errors, convergence modes are uncoupled. Thus, using the same normalized step size for all of the modes; it is possible to make them converge at the same rate, speeding up convergence of the whole system with respect to LMS. So, in the GAL algorithm faster convergence can be achieved at the expense of an increase in the computational complexity. The ALP structure is modular, in such a way that the predictor of order M consists of $M-1$ identical cascaded stages. The inputs to the stage m are the forward and backward prediction errors of the m-th order predictor. The ALP system itself is an adaptive filter, where a stochastic gradient method is used to adjust the filter coefficients, also named PARCOR, independently at each stage, so as to minimize the mean square of the sum of forward and backward predictor errors at the output of that stage.

3.10. Genetic algorithms in ANC systems

Instead of classical adaptive control (FBFXLMS) GA can also be used in ANC systems [2]. The control strategy to combine GA and GP has shown high potential for self-structuring control models in real time applications specially for autonomous systems by exploring artificial intelligence and the parallel processing capabilities.

The combined methodology is termed GENETIC CONTROL (GC) for real time applications. The realization needs high performance DSP systems.

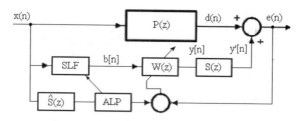

Figure 5. FXGAL algorithm

3.11. Experimental system

In Figure 6. the laboratory system is shown, that is used for test and examines simple BBFFANC system performances. The ANC is realized by high performance floating-point DSP. The steps of the program executions are:
1. Input the reference signal x(n) and the error signal e(n) from the input ports.
2. Compute the antinoise y(n).
3. Writing y(n) signal to the output to drive the speaker.
4. Compute the filtered-X version of x'(n).
5. Update the coefficients of adaptive filter W(z) using the FXLMS algorithm.
6. Repeat the procedure for the next iteration.

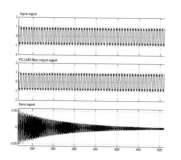

Figure 6. FX-LMS filter simulation   Figure 7. Experimental ANC system

References

1. Kuo S.M.,. Panahi I., Chung K.M., Mark T.H., Chyan N.J.: *Design of Active Noise Control Systems With the TMS320 Family*. Texas Instruments, SPRA042, June 1996.
2. Suchar R., Schalling B., Ciocoiu I., Brezulianu A.: *Active noise control with artificial neural experts*. Seoul 2000 FISITA World Automotive Congress, June 12-15, 2000, Seoul, Korea.
3. Vicente L., Masgrau E., Sebastián J.M.: *Active Noise Control Experimental Results with FxGAL Algorithm*, [N485]. The 32nd Int. Congress and Exposition on Noise Control Eng., Seogwipo, Korea, Aug. 25-28, 2003.
4. Välimäki V., Antila M., Rantala S., Linjama J.: *Adaptive Noise Cancellation in a Ventilation Duct Using a Digital Signal Processor*. DSP Scandinavia '97 Digital Signal Processing Conference, Vesa Välimäki, 1997.

ADVANTAGES OF USING ENGINEERING ERGONOMICS IN THE DESIGN OF OFFICE / COMPUTER WORKSTATIONS TO CONTROL BACK PROBLEMS

Lewis W.G., Ameerali A.O.

Abstract: Some facilities that people use and the methods of work that are followed have a major impact on posture. In this regard, the most important consequence of improper posture is with respect to cumulative trauma disorders related to the back. Incorrect posture also causes fatigue and inefficiency in the workplace. *In light of the problem, the authors conducted an investigation into the effective selection of office chairs for various office tasks using ergonomic principles in order to circumvent these problems. It was also necessary to encourage users to understand and make correct use of these items. Using the principles of ergonomics in the study, the authors were able to generate a comprehensive list of the major factors that contribute to back pain, faced by computer intensive users, and make recommendations to alleviate these problems with the proper selection of office chairs to carry out the tasks. The study resulted in a system to identify ergonomic chairs, which improve lumbar support for relief of back pain for the specific task.*

1. The science of seating

More than half of the working population in the major industrial nations spends most of their time sitting at a desk [Bendix 1987]. This is confirmed by the latest work analysis published by the federal institute for occupational safety (Table 1). For the three listed areas of activity, the following percentages emerged:

Table 1. Time spent sitting, standing & moving for various tasks

AREA OF ACTIVITY	SITTING	STANDING	MOVING
Sales processing	65%	22%	13%
Purchasing processing	84%	6%	10%
Organization programming	96%	3%	1%

With office equipment, a proper selection of the chair is necessary in order to maintain a proper posture while working as improper designs of chairs and seats have been found to affect the work performance of people and also contribute to backaches and back problems. The back problems triggered by improper posture arise principally from the pressures exerted by the vertebrae on the discs between them. According to the National Institute for Occupational Safety and Health (NIOSH), in 1998, 3.6 million people were treated with occupational injuries in U.S. Hospital Emergency Departments, with 2.5 million being male and 1.1 million being female. Of these numbers 348,500 (approximately 10% of the total annual estimate) were specifically related to lower back problems and in 1999 the number of persons with lower back problems increased to 401,100. In Trinidad, there have been no major studies conducted on the actual number of persons suffering from lower back problems, but the situation is somewhat similar to that of the United States.

It is a requirement of good seating that the person sitting should be able to maintain a good posture, which will not cause straining of any particular group of muscles. Continual use of one particular group of muscles can cause fatigue. Thus, a well-designed seat should enable the user to change his/her posture at intervals so that different muscle groups may be called into play. Sitting posture is a non – standardisable form of movement and therefore differs with different users.

2. Functional and anatomical criteria for a healthy seated posture

Good posture, as defined by the Posture Committee of the American Orthopedic Association [Rajendra 1995], *"is that state of muscular and skeletal balance which protects supporting structures of the body against injury and progressive deformity irrespective of the attitude in which these structures are working or resting."*

To gain a comprehensive understanding of what ergonomic seating posture entails, we first have to acquaint ourselves thoroughly with the functional and anatomical criteria for a healthy seated posture. Generally speaking, the spine has three main biomechanical functions:
- the spine bears and transfers the weight,
- the spine allows for a variety of important movements,
- the spine protects the spinal cord.

The spine is literally the supporting structure of the body. In profile, the spine is curved in a faint S – shape which increases flexibility, enhances the shock absorbing capacity of the spine and ensure that there is adequate stiffness and stability at each spinal joint helping it cope with the different kinds of load. It is this natural form that should be supported during sedentary work. Any modern office chair can do this but there are natural limits to it. We are not designed to sit still, even in the correct position, for long periods of time, as this will create problems. Backaches will remind you when enough is enough.

There are two important aspects of the vertebral discs that we must understand if we are to appreciate the importance of correct posture. In the first place, the invertebral discs do not have a vascular system of their own. They are supplied with nutrients only via diffusion, i.e. by the inflow and outflow of fluids from neighbouring tissue. This is accomplished by the smallest movement between the individual movement systems. Invertebral disc nutrition depends mainly on alternating hydrostatic pressures, above and below a critical hydrostatic value [Grieco 1986, 1989].

The second important factor is the mechanical load bearing capacity of the invertebral discs. Even though the fibrous ring (inside of the disc) is capable of withstanding a great deal of pressure, it does not posses great tensile strength. This means that it is susceptible to rupture if subjected to tension for long periods of time. This happens especially when the surface of the two vertebral bodies are not parallel to each other e.g. when a person sits with a curved back.

The implications of seated posture are clear: a rigid, unchanging posture exerts a detrimental affect on the metabolism of the invertebral discs. This is the reason why static seated posture is inevitably unhealthy. Orthopaedics as well as ergonomics recommends frequent or at least occasional changes of position from leaning forward to leaning back and vice – versa. This calls for a dynamic chair, which allows easy changes of sitting posture. The elements of good seating will depend on the length, width and shape of the seat; to a limited extent on the material of which the seat is made; on the shape and height of the back rest and height of the seat above the floor.

2.1. Seat height

If pressure is to be avoided on the under side of the thigh, the seat height must be related to the length of the lower leg from the under side of the knee to the heel and to the curvature of the thigh. Due to variations in the thickness of the thighs of different individuals there may be a difference of as much as 7.5 cm between the lower leg length and the height of the heels and it would seem that increased plumpness of the thighs in the female may be to some extent offset by the higher heels which are normally worn. Most of the recommendations, which are made on seat height, are based on the assumption that the sitter will have the feet flat on the floor and the lower leg vertical.

On the whole, therefore, the seat height should be at or near the mean of the population for which it is intended. Height should be such that it adjusts to fit the height of the user and/or the work surface.

2.2. The size and shape of the seat

The depth of a seat should be sufficient to allow the buttocks to move to permit changes of posture but should not be so great that the seat cuts into the back of the knee. The depth of the seat should adjust to provide support for the legs. There should be a 2.5 to 5 cm space between the front edge of the seat and the back of the knees, and the user should be able to slide the seat backward and/or forward to set the proper depth position.

Some padding on seats is desired, but for seats, which are intended for use at work it should not be too soft, otherwise the buttocks will sink in too far and the thighs will be elevated. The seat of the chair should have a forward and backward tilt adjustment, allowing the thighs to slope slightly downward with the spine being straight.

2.3. The back rest

The backrests in most industrial seats are for support in the lumbar region. The backrest should have up and down adjustments to fit the curves of your spine to provide support and comfort. Adjustable lumbar support is often necessary to accommodate flat, average or deep spinal curves.

The back of the chair should make full contact with the back when in a normal working position. If the individual reclines, the backrest must be able to extend to support the upper back and the neck. The backrest should be such that it can move backward and forward. The chair should have a built-in lower back support. The back should adjust independently of the seat tilt to provide back support for a variety of work positions.

2.4. Adjustments

All the adjustments to the chair should be made from a seated position. It is recommended that the chair be pneumatic. The adjustment of the seat height should be made with the feet planted firmly on the floor.

With this in mind, a few guidelines are given below for a healthy posture on a chair:
1) the feet should rest flat on the floor and not on the base,
2) always keep in contact with the backrest. the spine should be upright,
3) as soon as you notice that you are sliding in the seat, correct your posture,
4) avoid leaning to far to the sides,
5) do not lean to the side to pick up things from off the floor. turn the chair towards the object and lean forward, over the legs to pick up,
6) do not cross the legs when sitting upright as the pelvis will then tip backwards and the lumbar vertebrae will be in the wrong position. you can, however cross the legs when reclining, as the angle between the body and the pelvis is then wide enough and the pelvis already tilted backwards,
7) practice dynamic posture as often as possible,
8) do exercises at your desk whenever you have time.

3. Assessment of office tasks and workstations

An assessment of office tasks and workstations was carried out as part of this investigation into factors that impact on good-seated posture. The following groups were developed for a typical company / engineering plant and related to the level of task component required at the workstations (Table 2)

The time spent at workstations by groups with finance, information technology and clerical workers are very high (90 – 95% of the time spent at work in most cases). Thus the chair must be carefully selected to meet ergonomic specifications. Table 3 shows the recommended adjustments that should be on chairs for the category of tasks identified. It is assumed that all the chairs must have the following minimum requirements:
- five (5) legs on castors,
- height adjustments,
- back tilt and spring tension.

4. Manufacturing considerations for ergonomic seating

Factors such as price range, activity type, amount of time spent at the work-station and rank within the organization will influence furniture selection thus, it can be concluded that no one chair design will satisfy all the requirements of management and worker. Thus a number of chairs with different features must be manufactured to satisfy the wide and diverse range of needs of potential customers.

Table 2. Typical Office/ Computer Task Components Requirements

Task Category	Writing/ Editing	Reading	Keyboard Operations	Telephone Use	Face to Face discussion	Time Spent at workstation
Management	**	**	*	***	***	*
Finance	**	***	***	**	*	***
Information Technology	**	***	***	**	*	***
Engineering /Drafting	*	*	*	***	***	*
Logistics/ Laboratory, Workshop	*	*	*	***	***	*
HR, HESQ, Manufacturing	**	**	**	**	***	**
Control Room Operators	**	*	***	***	**	***
Secretarial	*	**	**	***	**	**
Clerical	**	***	***	**	*	***

Key: *** High Level   ** Medium Level   * Low Level

Table 3. Recommended adjustments on chairs for the categories of tasks identified

Task Category	Seat Depth	Seat Angle	Backrest Angle	Lumbar Support	Armrest	Padded	Posh	High Back
Management	***	***	***	***	**	***	***	***
Finance	***	***	***	***	***	***	**	**
Information Technology	***	***	***	***	***	***	**	**
Engineering/ Drafting	*	*	*	**	**	**	**	**
Logistics/ Laboratory, Workshop	*	*	*	**	**	**	**	**
HR, HESQ, Manufacturing	*	*	*	**	**	**	**	**
Control Room Operators	***	***	***	***	**	***	**	***
Secretarial	**	*	**	***	*	**	*	*
Clerical	**	*	**	***	*	**	*	*

4.1. Ergonomic Seat with no Armrest

The first category is the ergonomic seat with a low back and no armrest (Figure 1).

Figure 1. An ergonomic seat with no armrest   Figure 2. An ergonomic chair with armrests

This seat provides the following features:

- lumbar support,
- backrest support,
- tilt tension,
- seat depth adjustment,
- pneumatic height adjustment.

4.2. Ergonomic seat with armrest

The second category is the ergonomics seat with a low backrest and armrest. (Figure 2). This chair has all the features of the ergonomic seat design before with an adjustable armrest included. From the point of view of manufacturing one additional attachment is required and the rest is standardized.

4.3. Ergonomic Seat with High Back and Armrest

The third category is the ergonomic seat with a high backrest and an armrest (Figure 3). This chair has all the features of the second design mentioned with the only difference being the type of backrest utilized. Again from the standpoint of manufacturing, the seat and base of the chair are identical in construction to the first two.

Thus the three categories mentioned above can be considered to be one seat with options for armrests, or high or low backrests. An obvious alternative not mentioned is the high back without the armrest, which again gives some flexibility in choice and expands the alternatives from three to four.

It should be emphasized that once the employee spends more than 50% of his/her time at the workstation, or more than four (4) hours per day, and the furniture does not meet ergonomic requirements, there will be a stress build – up which will ultimately result in injury to the employee.

Figure 3. An ergonomic chair with high backrest

5. Discussion

The primary functions of the ergonomic chairs presented in this project are to support a balanced work posture and to minimize fatigue. A balanced work posture means the torso and hips are supported so that minimal force is exerted in holding the torso stable. Torso stability is crucial because the torso's sole purpose is to be a firm anchor for working limbs. If the torso is a firm anchor for the arms, hands and legs, then the limbs don't have to move as far, resulting in less fatigue and less risk of cumulative trauma disorders.

The designs presented here have pneumatic height adjustments so that a variety of workers can use one chair with minimal adjustments to be made. The mechanism to operate the pneumatic height adjustment is also very easy to reach and operate. With all the chairs, mobility was aided by allowing the chair to swivel and placing the base on castors. The base had five legs since this arrangement had better stability than the four-legged frame.

The backrest was also cushioned for lumbar support hence reducing back pain. The backrests were also able to move backward and forward. This is because reclined sitting puts more of the person's weight onto the chair's backrest.
If the chair's backrest holds up more weight, then the discs in the lower back hold up less weight thus reducing stress in the lower back.

6. Conclusions

All sitters should move around. In addition to helping the muscles relax and recover, this alternately squeezes and unsqueezes the vertebral discs, which results in better filtration of fluids into and out of the cores of the discs. Discs stay plumper and in the long run, healthier. However, short breaks are not always possible for sitters (e.g. long board meetings, lectures etc.) and so proper seat design had to be implemented in the chair. Seat design affects comfort, safety and health of the occupant. Posture of the back, however, was very complicated but current scientific and medical opinion believes that an erect posture is healthier than a slumped one. Thus the chairs were designed with this point particularly in mind.

The most important chair adjustments are:

- seat height from the floor – the feet must be able to rest flat on the floor. However this does not mean that the feet always have to stay on the floor. The designed allowed the legs to be free to stay in different positions,
- depth from the front of the seat to the backrest – this ensures that sitters are able to use the backrest without any pressure behind the knees,
- lumbar support height – this was important based on the fact that every person is shaped differently.

It should be noted that the proper chair adjustments and chair posture are greatly influenced by the rest of the work area. In particular, the eyes can affect posture, especially if the work material is too far, low or high. Hand positions (especially working far from the body) can also affect body position particularly the posture of the upper back. With these ergonomically designed chairs it is believed that most of these stresses can be relieved since the chairs can conform to almost anyone in any type of seated work environment. This will result in healthier workers and possibly less absenteeism in the workplace due to cumulative trauma disorders.

References

1. Bendix T.: *Adjustment of the Seated Workplace – With special reference to heights, inclinations of seat and table.* Laboratory for Back Research, Department of Rheumatology, Rigshospitalet, University of Copenhagen, 1987.
2. Rajendra P.: *Effects of office layout and sit stand adjustable furniture: a field study.*, USA Technologies and Concepts Group, 1995.
3. Grieco A.: *The Ergonomics Society - Sitting posture: An old problem and a new one.*, Ergonomics, 29(3):345-362, 1986.
4. Grieco A., Occhipinti E., Colombini D., Menoni O., Bulgheroni M., Frigo C., Boccardi S.: *Muscular effort and musculo-skeletal disorders in piano students: Electromyographic, clinical, and preventive aspects.* Ergonomics, 32(7):697-716, 1989
5. Lewis W.G., Imbert C.A.C.: *Safety in the workplace through the use of Ergonomics.* Regional Conference on Protecting the Industrial and Natural Environment, UWI, St. Augustine, 1993.
6. Lewis W. G., Narayan C.V.: *Design and Sizing of Ergonomic Handles for Hand Tools.*, Applied Ergonomics, 24(5): 353 – 365, 1993.
7. National Institute for Occupational Safety and Health, Work – RISQS Injury Results, "www2.cdc.gov/risqs".

INTEGRATED METHODS FOR MANUAL WHEELCHAIR DESIGN

Ariff H., Samsudin A.S.

Abstract: Manual Wheelchair is a complex system therefore needs proper design for better performance. This paper describes the steps in the design process which use integrated methods such as identify customer requirement, product design specification, conceptual design and development, detail designs including stability of wheelchair etc. This paper describes preliminary research of transfer problems among individuals who use manual wheelchairs. Elevating seat was proposed with pneumatic concept as mechanism to raise or lower a seat in order to offer wheelchair users more independence in daily activities and improve them physically.

1. Introduction

Wheelchair is a tool that enhances quality of life. Wheelchair users are not always ill, not condemned to a life of no meaning or pleasure. Even for people with life-threatening illness, the right wheels facilitate their ability to be out in the world, continuing their life, partaking of the many realms of human experiences. There are many reasons why people may require the use of a wheelchair. Paralysis due to spinal injury (such as paraplegia), amputation of both legs, acute Multiple Sclerosis (MS), Cerebral Palsy, and chronic arthritic conditions being just a few of the more common reasons. Whatever the reason, "full time" wheelchair users share the problem of being unable to walk. There has been a little change in the design of the manual wheelchairs few years back, particularly in allowing users transferring easily.

The investigation of the manual wheelchair in term of transferring has become increasingly important because the population of individuals using wheelchairs is growing and requires efficient transferring and mobility to maintain a quality of life equivalent to the general population. Thus, this paper focuses on the transferring problems and come out with specific feature to improve the existing manual wheelchair design.

2. Methodology of research work

Figure 1 shows the design methodology (integrated designs) that was followed in carrying out this work. Detail each stage of the work are given below.

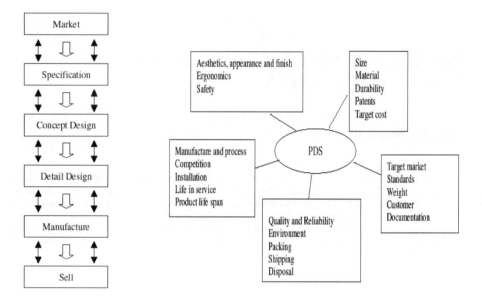

Figure 1. Integrated design methodology   Figure 2. Elements of the PDS

3. Problems identifications

Getting out of a chair, getting in and out of bed, or moving from one surface to another (i.e. moving from wheelchair to toilet) are tasks called "transfers". Some of wheelchair users become aware of a problem when they had experienced arthritis aches and pains or have difficulty with transfers as a result of an injury or chronic condition. The wheelchair questionnaire is done where targeted area includes students and staffs of Loughborough University, Queen Medical Centre (QMC) and Loughborough Town Centre. From survey of wheelchair questionnaire, more than 33% of wheelchair users faced transferring problems.

3.1. Product design specification (PDS)

To design a wheelchair, which is easily, and safe to transfer from and into a wheelchair to another surfaces such bed, car seat, toilet seat, chair, etc. There are 23 elements (figure-2) that are considered in the product design specification.

3.2. Conceptual design

The product takes a common sense look at the issues surrounding the creation and generates ideas in order to achieve the needs of the wheelchair users. For this purpose, brainstorming is used to generate ideas to meet the product design specification. Several factors must be considered when designing a manual wheelchair such as what are the intended uses, what are the abilities of the users, what are the resources available and what are the existing products available. These factors determine if and how the manual wheelchair will be designed and built.

3.3. Wheelchair measurements

The important in the design of wheelchair is measurement of the human body. To support the sitting body and to extend its functions, a wheelchair should be comfortable. The first requirement for comfort is fit and the first condition of fit is correct size. Approximate measurements are found in table-1. In addition to standard size, wheelchairs will be custom-built in any desired dimension. Basic measurement of manual wheelchair is illustrated in figure-3.

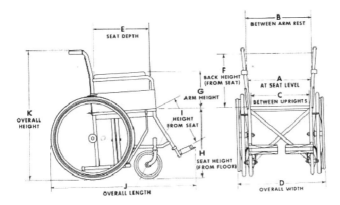

Figure 3. Wheelchair measurements

Table 1. Dimensions of standard wheelchair (*Principle dimensions of standard wheelchairs in inches, Approximate average measurements compiled from several catalogs*)

No.	Small Child	Large child	Junior	Adult	Oversize
Seat Height (H)	18	20	19	20	20-22
Seat Width (C)	12	14	16	18	18-22
Seat Depth (E)	11	11	13-15	16	16
Leg Length (I)	3-9	7-11	13-18	15-20	15-20
Arm Height (G)	6	7	9	10	10
Back Height (F)	17	15	16	16	16
Overall Height (K)	35	35	35	36	36-38
Overall Length (J)	30	33	39	40	41-42
Overall Width (D)	19	21	22	24	26-30
Width Folded	10	10	10	10	11-12
Weight in Pounds	35	41	41	43	46-48

3.4. Concept evaluation

To effectively evaluate concepts, an agree set of criteria is needed. The criteria are deduced from the product design specification (PDS) elements. Table-2 below shows the evaluation chart for the 7 comparable concepts. The chosen datum (concept 7) is the existing manual wheelchair in the market, which has been developed over many years and widely used in the world.

Table 2. Concept evaluation

Criteria \ Concept	1	2	3	4	5	6	7
Easy to use	S	S	S	S	S	S	
Maneuverability	S	S	S	S	S	S	
Good stability	-	-	S	S	S	-	
Easy to transfer from and into wheelchair	-	+	+	+	+	+	
Easy to fit accessories	-	-	S	S	-	S	
Safety	S	S	-	-	-	S	
Low manufacturing costs	S	S	S	S	S	S	
Durability	S	-	S	S	S	S	
Easy of repair and availability of spares	S	S	S	S	S	S	
Easy of transportation and storage	-	-	-	+	-	-	Datum
Weight	S	S	S	S	S	S	
Size	S	S	S	S	S	S	
Life in service	S	-	S	S	-	S	
Easy cleaning	-	S	-	+	S	S	
Good appearance	S	-	S	+	-	S	
Environment	S	-	+	+	-	S	
Adjustable seat	-	-	+	+	-	-	
Detachable or swing away components	S	S	-	S	S	S	
Σ +	0	1	3	6	1	1	
Σ -	6	8	4	1	7	3	
Σ s	12	9	11	11	10	14	

3.5. Concept development

The concept evaluation revealed that concept design-4, which necessitated further developments and analysis (figure 4). Several factors must be considered when designing a manual wheelchair such as what are the intended uses, what are the abilities of the users, what are the resources available and what are the existing products available. These factors determine if and how the manual wheelchair will be designed and built.

3.6. Detail design

This wheelchair (figure-5) is designed to make user easier to transfer from or into a wheelchair to another surfaces. The elevating seat allows the user to raise the seat to gain access to different height for a wider range (maximum 120mm) of activities at home, work or school (figure-6). It also allows the user to have eye contact with people standing up. The elevating seat on wheelchair is pneumatic operated by pushing of a pneumatic handle/lift lever and require some power of one lower limb but even if this is done by another person, it is most often still easier than helping the user to transfer. This wheelchair will enable user to function independently in virtually any environment, and will give user the freedom to do what he or she wants to do, without the need for assistance from other people.

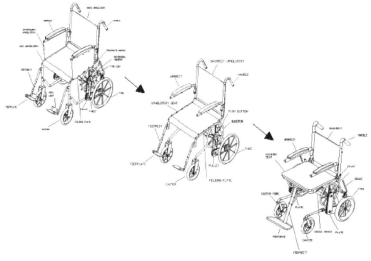

Figure 4. Wheelchair development

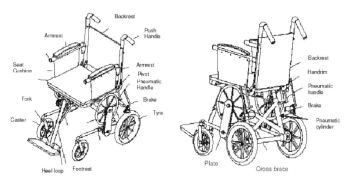

Figure 5. Final design

For the new design, frame can be folded into more convenient shape (figure-7) and size for lifting and transporting. The ergonomically handle also allows a caregiver to push a wheelchair without awkward, better positions the caregiver's arm to reduce wrist and arm fatigue and provides increased control. The stability of wheelchair is important to make the users seat with comfortable and stable Generally the stability of a wheelchair will be increased by moving the front and rear wheel as far as possible and can also be improved by ensuring that the centre of mass of the user and wheelchair is as low to the ground as possible.

Materials have been the basis of major evolution in wheelchair products. Materials impact characteristics such as durability, strength, cost, appearance, design and manufacturing flexibility, and weight. Most of the users who have benefited from material impact characteristic have been in manual wheelchairs. CES 4 software and PLASCAM software are used to determine material for the frame, seat and wheel/tyre. We also can analysed the weight and material cost for a new design and their calculation.

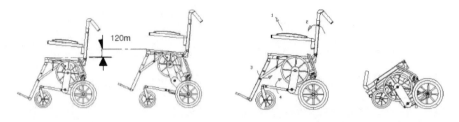

Figure 6. Elevating seats   Figure 7. Frame can beFolded

4. Stability of wheelchair

The world is not flat. Hills, ramps, curb cuts and sidewalks with side slopes are just a few of the environments that will test how stable the wheelchair will be. Figure-8 shows the basic wheelchair stability (existing manual wheelchair) when a wheelchair user in a wheelchair on a gentle slop with the brake applied. In this situation the centres of gravity of the wheelchair and user both fall in front of the point of contact of the rear wheel with the ground. The weight of the user (U) and wheelchair (W) apply turning moments in a clockwise direction that are counteracted by the turning moment generated by the reaction force (R) through the front castor of the wheelchair and hence the user and wheelchair are stable.

The same wheelchair user is on a steeper gradient in figure-8. This time the centre of gravity of the user falls behind the point of contact between the rear wheel and the ground. Although the weight of the wheelchair (W) is still applying clockwise about the rear wheels, the weight of the user (U) is now applying an anti-clockwise turning moment. At the limit of stability the anti-clockwise turning moment due to the weight of the user equals the clockwise turning moment due to the weight of the wheelchair. The front caster will barely be in contact with the ground and so there will be no weight taken through the front caster and all the weight of the user and wheelchair will be taken through the rear wheels.

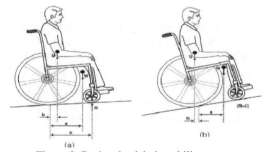

Figure 8. Basic wheelchair stability

Figure 8a: Turning moments about the point where rear wheel is in contact with the ground resulting in stability (U x b) clockwise + (W x a) clockwise= (R x c) anticlockwise.
Figure 8b.: Turning moments about the point where rear wheel is in contact with the ground at limit of stability: (U x b) anti-clockwise= (W x a) clockwise.

4.1. The stability of a new design

In normal position, the angle of a2 is greater than a1 but b2 is less than b1. However, the stability of a wheelchair will be increased if the distance between front and rear wheel is set as far as possible This means the stability of a new design is greater than the existing standard manual wheelchair.

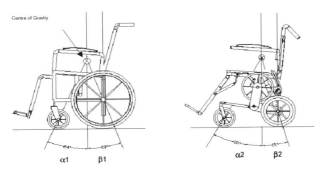

Figure 9a. Existing wheelchair   Figure 9b. New design

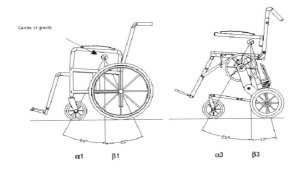

Figure 9c. Existing wheelchair   Figure 9d. New design

When the seat is raised to maximum height, the angle of b3 (13°) is less than b1(22°) and a1 is equal to a3. This means the stability of a new design is less than the existing standard manual wheelchair when the seat is elevated to the maximum height. Generally the stability of a wheelchair will be increased by moving the front and rear wheel as far as possible and can also be improved by ensuring that the centre of mass of the user and wheelchair is as low to the ground as possible. The stability of the wheelchair standing still not only indicates how stable the wheelchair is when at rest but also indicates how stable the wheelchair will be when it is moving. So, wheelchair should be tested in some environments to be certified before used.

This wheelchair design also needs further investigation and development due to more adjustability without compromising their performance. Further work can look into the mechanism of the pneumatic system for the elevated seat in more detail. Front casters also required further study. What are the desired rolling and suspension properties of front casters and how do they depend upon the position of the centre of mass of the wheelchair and rider? Stability of wheelchair is also important and need further investigation. How wheelchairs become more stable? Material also needs further considerations to make the wheelchair lighter and stiffer.

5. Conclusion

Improvements in wheelchair design have provided increased ease and range of mobility especially when transfer from wheelchair to another seat. This design allows the user to elevate the seat easily to the maximum distance (120mm). The elevating mechanism used to change the seat level is operated by pneumatic concept by pushing pneumatic adjustable handle. Analysis on material and manufacturing process were used to determine the weight and material cost of a wheelchair. This design also has more adjustability, easy storage for transportation and comfortable thus will increase the user ability and at the same time will also improve user physically due to the improvement of their quality of life.

Acknowledgements

The authors would like to thank to Mr. Andy Taylor, Ergonomic and Safety Research Institute & Disabilities and Additional Need service (DANS), Loughborough University for their upport and information.

References

1. Pugh S.: *Total design.* Pearson, 1991. p. 46.
2. Kamenetz H.L.: The wheelchair book. Charles Thomas, 1969, p. 132-134.
3. Rosalind H., Patsy A., Porter D.: *Wheelchair users and postural seating.* Churchill, CES 4 selector software, Material universe: /Metal/Non-Ferrous Alloy.
4. PLASCAM software, PLASCAMS-elimination search.
5. Roslind H., Patsy A.: *Wheel chair users and postural seating.* 1998, p. 4-13.
6. Cooper R.A.: *Rehabilitation engineering applied to mobility and manipulation.* 1998, p. 256-267.
7. Kamenetz H.L.: *The wheelchair book.* 1969, p. 74-98.
8. Mott R.L.: *Machine elements in mechanical design.* Prentice Hall, 2004, p. 12.

The International Journal of **INGENIUM** 2005 (3)

ENGINEERING ACHIEVEMENTS ACROSS THE GLOBAL VILLAGE

edited by

Janusz SZPYTKO

Product Testing

Cracow - Glasgow - Radom, 2005

TABLE OF CONTENTS   page

5. Product Testing 329

5.1. Overview of product testing within the constructional fixings industry,
Bisland G.G., Cadden S., Harrison D.K., Wood B.M., Temple B.K. 331

5.2. On-line measurement of fatigue crack growth rate in servo hydraulic testing machine using LabView software,
Dasharathi M.S., Annamalai K., Adithan M., Ranganath V.R., Manjunatha C.M., Rao Patange P.S.S. 339

5.3. Velocity and static analysis of parallel manipulators,
Bouanane K., Khaldi A. 347

5.4. Development of softness equipment and softness measuring method,
Nagao M. M., Sakai Y., Yokota O. 359

5.5. Design, construction, control and tests of an experimental diesel fuel high pressure burner cannon type,
Tibaquirá G.J.E., Burbano J.J.C., Holguín L.G.A. 367

5.6. Reducing the sampling-rate of digital drivers for a stepping-motor by frequency-response reshaping,
Fukuzawa M., Hori N., Smo I. 375

5.7. Monitoring of laser alignment based on induced ultrasonic waves in laser-assisted jet electrochemical machining,
Dąbrowski J., Pająk T.P., Harrison D.K., Szpytko J. 383

5.8. The integration of vision based measurement system and modal analysis for detection and localization of damage,
Kohut P., Kurowski P. 391

5.9. Testing motional accuracy of a manufacturing machine - a task imposed on modern maintenance,
Blacharski W. 399

All rights reserved. No part of this book may be reproduced, stored in a retrieval system, or transmitted, in any form or by any means, without prior written permission from the Publisher.

The International Journal of INGENIUM
Chief Editor: Professor David K. Harrison, Glasgow Caledonian University, UK

© GCU Glasgow

ISSN 1363-514x

A CIP catalogue record for this publication is available from British Library

Publishing cooperation: Instytut Technologii Eksploatacji – PIB w Radomiu

OVERVIEW OF PRODUCT TESTING WITHIN THE CONSTRUCTIONAL FIXINGS INDUSTRY

Bisland G.G., Cadden S., Harrison D.K., Wood B.M., Temple B.K.

Abstract: This paper aims to outline the critical role that product testing takes within the constructional fixings industry. To achieve this it examines Artex-Rawlplug ltd. and more specifically the testing process undertaken for the R-HPT Throughbolt to achieve ETAG approval. The paper proposes areas which the company may pursue to improve upon the current procedures and facilities.

1. Introduction

1.1. General

Within the constructional fixings industry the role of product testing is paramount primarily due to the inability to use theoretical testing to sufficiently analyse a product (EOTA, 1997). This paper provides an overview of the role product testing takes within the constructional fixings industry. To this end focus is placed upon Artex-Rawlplug a fixings specialist with worldwide sales, through review of testing within this company a broader idea of the industries issues can be identified. This paper also aims to present potential areas of advancement within this field by utilising a case study of a product that has been through this process.

1.2. Company

Artex-Rawlplug was formed in 2001 merging both Artex-Bluehawk (a leading company in textured finishes) and Rawlplug under the BPB (British Plaster Board) banner. It is the Rawlplug side of this business which shall be investigated through the course of this paper. Rawlplug was founded in 1919 with only one fixing product however now Rawlplug manufacture and sell numerous products suitable for various applications and sold to both the construction industry and the DIY Market. This makes Rawlplug a prime example of a business within the constructional fixings industry. This paper places emphasis on the test centre within the company.

1.3. Product Testing

The following are the primary sources of workload received by the test centre at Artex-Rawlplug, these inputs are ranked in order of scale of work required:
1. Approvals - Gaining European technical approval on products.
2. NPD (New Product Development) – Working on testing for New Product Developments.
3. Marketing – Conducting testing for publications etc…, closely linked to NPD.

4. Technical Support/Assistance – Conducting specific tests for applications for sales etc... Linked to procurement.
5. Procurement – Supplier testing etc...
6. Customer Complaints – Helping in the resolution of customer issues through product testing, although low priority there are targets which must be made in resolving these issues to maintain quality standards.

This suggests a great deal of diversity and demand on the work handled by the test centre. Due to the irregular frequency of these tasks it is difficult to develop a firm schedule for work this is particularly true for external demands (i.e. not approvals or NPD). For this reason this paper shall focus on only internal demands and as approvals are the main priority of the test centre they shall be the main focus of this paper.

2. Product overview

2.1. Introduction

Artex-Rawlplug products can be very generally categorized into two group's -light/medium weight fixings and heavy fixings. Light and medium weight fixings (mainly plastic and chemical fixings) are aimed primarily towards the DIY sector with the latter being used for constructional work where the fixing is subjected to heavier loading. Due to the broad range of products available within the heavy fixings range only one shall be discussed here. The product chosen for this purpose is the R-HPT Throughbolt as it has undergone the most rigorous product testing to gain a European Approval and therefore can also be used as the subject of a brief case study.

2.2. R-HPT Throughbolt

The R-HPT Throughbolt is an example of a torque-controlled expansion anchor developed roughly seven years ago for high performance applications. Figure 1 shows a rough representation of the anchor in application and Figure 2 shows the actual anchor itself.

As can be seen in Figure 1 the anchor is located through the fixing and into a clean hole drilled to fit the anchor diameter. The anchor nut is then tightened down the thread to a recommended torque value at which point the expansion sleeve will have traveled down the cone expanding into the substrate creating a clamping force which can withstand high tensile and shear loads from the fixing.

To further prove the R-HPT's performance it gained an option 1 European technical approval meaning that it is suitable for use in cracked concrete applications. Option 1 approval is the hardest approval to receive.

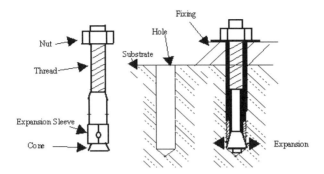

Figure 1. R-HPT Throughbolt (EOTA, 1997)

Figure 2. R-HPT Throughbolt (Rawl fixings, 2004)

3. Resources

The test centre at Artex-Rawlplug has various apparatus suitable for use on a wide range of tasks that the test centre may be required to perform. The equipment available allows the test centre to conduct the following:
- metrology: shadow graph, slip gauges, etc…,
- installation tests,
- tests on various substrates: Concrete grades, cracked concrete, brick, etc…,
- corrosion tests: Salt spray cabinet,
- material tensile tests.

The test centre also has various drills, grinders and saws that can be used when and as needed in testing. Within this report focus is placed upon the apparatus used to perform the necessary tensile testing on metal anchors.

The apparatus used to perform tensile tests on the R-HPT Throughbolt is shown in Figure 3 and explained in the simplified diagram in Figure 4.

Figure 4 shows how a regulated and monitored air supply provides the necessary pressure to a piston that in turn applies a tensile load to the anchor. Located atop the anchor is a displacement transducer that monitors the distance that the anchor traverses. Both the applied load and the anchor displacement data is automatically recorded by virtual instrument software on a computer that charts a load/displacement curve of the anchors performance. Figure 5 shows a sample of potential curve shapes.

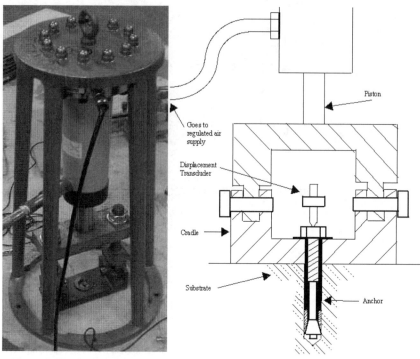

Figure 3. Tensile Test Rig

Figure 4. Simplified Diagram of Tensile Test Rig

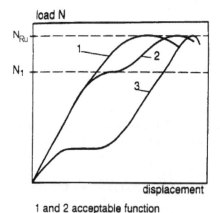

1 and 2 acceptable function
3 non-acceptable function

Figure 5. Requirements for the load/displacement curve (EOTA, 1997)

The load displacement curve should show a steady increase until failure of either the substrate or of the anchor itself. This can be seen in curve 1 of Figure 5. If the curve shows a short plateau then this is indicative of an uncontrolled slip of the anchor. The European Technical Approval Guidelines (ETAG) deems that if this slip occurs above a percentage (N_1) of the maximum load (N_{Ru}) then that curve can pass, as is shown in curve 2 of Figure 5. If however the slip occurs below this percentage then the anchor will have failed, shown in curve 3 of Figure 5. This also demonstrates how ETAG's influence the testing process.

4. Testing process

4.1. Introduction

This section of the paper briefly examines the test procedures used by the test centre at Artex-Rawlplug to gain European Technical Approval (ETA). As this procedure is broadly based around ETAG's then for the most part this section will discuss ETAG's and how they dictate the test procedures.

4.2. European Technical Approval Guidelines

The ETAG's act as instructions to being awarded ETA's, which in turn primarily aims to provide:

"An independent system for harmonising technical assessment and approval of construction products throughout Europe."

As stated on the Artex-Rawlplug website. This hints towards one of the main reasons why companies drive towards gaining ETA's. Companies wish to remain competitive and by having products awarded these approvals a company can gain a competitive edge or at least compete. Another factor is that the majority of governments within Europe are expected to in the future remove national legislations and force companies to seek ETA.

4.3. Testing Planning

As a part of the detail design of a product Waizeneker et al (2000) propose including a stage of verification and validation from which a strategy is developed and implemented to ensure that:
- the completed product is tested comprehensively,
- the components are individually tested,
- the tests are coordinated with any other necessary requirements (inspection, public verification for CE Marking standards, etc...).

Although this is intended for NPD it can be equally applied to any product testing and in this regard much of what Waizeneker et al discuss can be assisted greatly by the ETAG document that can be used as comprehensive means of planning the tests required to achieve approval.

The necessary testing to achieve an approval can be grouped into three categories:
1. Suitability testing – Establishes if an anchor is capable when accounting for adverse conditions to function safely and effectively.
2. Evaluating admissible service conditions.
3. Durability testing - To ensure that factors which cause degradation (corrosion, etc...) of the product do not overly hinder performance.

The test of admissible service conditions allows the company to determine how extensive a testing program they wish to conduct. Influencing factors which can determine which path a company will take can be cost, competition and anticipated product performance. There are 12 "options" stated within the ETAG from which the manufacturer can choose, included within the ETAG for each option is a basic test program.

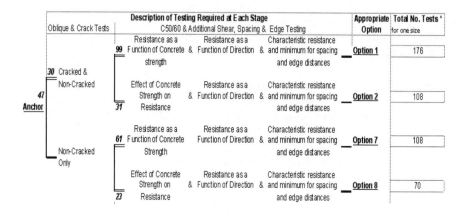

Figure 6. ETAG Option Tree

Figure 6 shows the 4 most popular options (1,2,7 and 8). Figure 6 also states the number of tests required for each option however it only states the minimum ideal number of tests that each option would require. In reality these test figures would be significantly greater. These numbers do demonstrate the additional testing required in achieving an option such as 1 or 7 which allows a broad array of service conditions.

5. Case study

The R-HPT is a prime example of a product which has undergone this testing process. For this product Artex-Rawlplug achieved an option 1 part 2 ETA. One of the major drivers behind this was competition. The company needed an approval for a product to function in cracked and uncracked concrete if it was to compete with its competition who had gained such an approval.

Therefore as stated on the company website over one thousand tests were conducted on the M8, M10, M12, M16 and M20 R-HPT. These tests determined how the product would function when its installation deviated from the manufacturer's intentions. To do this the product was tested in various concrete grades and torque strengths, hole diameter tolerances and under sustained and repeated loading to name some of the tests conducted. Furthermore testing had to be done on the characteristics of the product, i.e. spacing and edge distances as well as tensile, shear and combined load strengths.

To assist in conducting such a comprehensive testing scheme and to provide impartiality in the testing process some of the testing was outsourced and the results then collated, reviewed and submitted for approval.

In addition to this testing Artex-Rawlplug had to satisfy an inspection from an independent control body to be awarded a certificate of attestation. This is a further requirement to gaining an ETA. Its aim is to ensure that once the ETA has been awarded the manufacturer has the capabilities to control the manufacturing processes for the product so that its performance will not fall. It is worth note that this inspection process is far more complex than the standard BS EN ISO9001.

Although the end results of the testing for this product were favorable due to the necessary speed to get the product to market there was some issues which could have been improved. The most notable of these is the actual product cost, ideally the company would have liked to have spent some more time on reducing the material and manufacture cost needed to achieve product performance. This is a potential source of future product development.

6. Discussion

Through interviewing the test centre manager at Artex-Rawlplug some points were raised regarding areas of refinement within the test centre. Although out with the scope of this paper it was felt that with the release of an individual ETAG for plastic anchors and the growing use of plastic anchors that plastic anchor testing was becoming a more important issue. It was felt that with this in mind the facilities within the test centre to conduct plastic anchor testing required upgrading.

With a growth in the company on NPD spending it was discussed that the use of Computer Aided Design (CAD) software as an analytical tool may prove beneficial. Although the theoretical testing of anchors using CAD would not be comprehensive enough to accommodate the numerous variables affecting anchor performance and thus reduce the test programs. CAD may prove more beneficial as a means of reducing tooling cost (primarily in metal anchors) by performing preliminary testing on designs to determine if they merit the investment in tooling. It is possible then that design optimisation is an area of research that the company may find beneficial.

Also in terms of management of the test centre it was felt that a clearer process was required particularly with the upcoming additional demands on the test centres resources. Areas of particular improvement could be in prioratising strategy for external department work to minimise scheduling issues. In addition by implementing CAD into the test centre it allows the possibility to integrate the test centre procedures and the design process more fluidly.

7. Conclusion

The testing process used at Artex-Rawlplug demonstrates sound hands on fundamental engineering and in this regard is indicative of the majority of the constructional fixings industry. That being said however there is an opportunity to refine this process and the potential for changes such as greater software utilisation as an evaluation tool to reduce cost. With a continued push in this sector of NPD the subsequent demand and workload growth to the test centres will require that facilities be both amply equipped and organized in terms of structures and management to meet these new demands head on.

References

1. EOTA (1997), ETAG No.001: Guideline for European Technical Approval of Metal Anchors For Use In Concrete Part one: Anchors In General", UK Spokesbody in EOTA.
2. Rawl Fixings(2004), Product Specification & Design Guide.
3. Waizeneker, John, De Newton, John, Simpson, Roger, Woolston, Helen (2000), Croner CCH: Product Design and Development, George Over Ltd.
4. EOTA (1999) What is an ETA? [Online] available from: http://www.EOTA.be/.

ON-LINE MEASUREMENT OF FATIGUE CRACK GROWTH RATE IN SERVO HYDRAULIC TESTING MACHINE USING LABVIEW SOFTWARE

Dasharathi M.S., Annamalai K., Adithan M., Ranganath V.R., Manjunatha C.M., Patange P.S.S.R.

Abstract: This paper demonstrates virtual instrument software that enables the capability of on-line measurement of fatigue crack growth for CT, MT and TPB specimens. The fatigue crack growth is being monitored with the help of compliance based measurement. For measuring the compliance, crack opening displacement (C.O.D) gauge is being used. In addition to the online measurement, the software can perform DAQ, being done at intervals of change in crack-length. The software was developed in LabVIEW 7.1 with implementation using NI's 6014 card. The validation of the software is being done by comparisons with visual measurement. The author's intention is to apply the software for laboratories and for the educational demonstration of FCGR testing.

1. Introduction

Cracks compromise the integrity of engineering materials and structures. Under applied stress, a crack exceeding a critical size will suddenly advance breaking the cracked member into two or more pieces. This failure mode is called fracture. Even sub-critical cracks may propagate to a critical size if crack growth occurs during cyclic (or fatigue) loading. Crack growth resulting from cyclic loading is called fatigue crack growth (FCG). Because all engineering materials contain micro structural defects and the fabrication process might also induce some crack like defects (unavoidable stress risers, weld heterogeneity etc.) that may produce fatigue cracks, a damage tolerant design philosophy was developed to prevent fatigue failure in crack sensitive structures. Damage tolerant design acknowledges the presence of cracks in engineering materials and is used when cracks are expected. Both sudden fracture, and fracture after FCG, must be considered as failure modes. Because initial critical defects are rare in well designed engineering structures.

Fatigue crack growth prediction under constant loading requires the knowledge of various parameters such as material's FCGR properties, applied stresses, crack closure behaviour, etc. The major share of the fatigue life of the component may be taken up in the propagation of crack. By applying fracture mechanics principles it is possible to predict the number of cycles spent in growing a crack to some specified length or to final failure. The role of computers in fatigue and measurement is increasing day by day. (Lincoln, 1986).

In a fatigue crack growth experiment, the progress of a crack growth under a cyclic load is measured, and the results are plotted as a fatigue crack growth rate curve, "a versus N". Fatigue crack growth rate relates crack tip velocity to applied cyclic loading. This relationship in many circumstances can be used to predict the life of a structure susceptible to crack growth under cyclic loading. Standard test methods for measuring fatigue crack growth rates are well established for homogeneous, isotropic materials such as metals and plastics (ASTM- E 647, Volume03, 01, Ewalds, 1985). These methods take advantage of analytical fracture mechanics stress analysis solutions in front of the growing crack tip, making it possible to relate cyclic stress intensity to crack growth. One popular relationship often employed is the Paris-Law equation (Paris, 1972). That expresses a linear relationship between crack growth rate and stress intensity factor on a log-log scale: $da/dN = c (\Delta K)^m$, where c and m are power-law fitting constants. The other methods of fatigue crack measurement is being discussed in (Tarbiat EIS'86, Lincoln.A.P, 1986).

2. Need for fcgr test and its automation

FCGR test is useful for materials that would undergo high cyclic loading stresses such as an airplane wing or a helicopter rotor. From the structural safety viewpoint, there is clearly a need to quantify the rate of fatigue crack growth, after development of crack but prior to the point where separation becomes a possibility. In this way the remaining safe life of the structure can be assessed. The major share of the fatigue life of the component may be taken up in the propagation of crack.

By applying fracture mechanics principles it is possible to predict the number of cycles spent in growing a crack to some specified length or to final failure. Thus, by knowing the material growth rate characteristics and with regular inspections, a cracked component may be kept in service for an extended useful life. There is clearly a need for automation of such a software package to provide on-line information of the way in which the crack propagates, which is important in assessing the behavior of the material under suitable loading.

3. Experimental set-up

A 50 kN Instron make computer-controlled servo hydraulic test system was used to perform automated FCGR tests. The fatigue crack growth cycle fatigue tests is performed using sinusoidal waveform and is controlled with constant load range. The fatigue crack growth is being monitored with the help of compliance-based measurement. For measuring the compliance, crack opening displacement (COD) gauge is being used. COD gauge consists of two arms on which two strain gauges have been wired to each of the arms so as to make a wheat-stone bridge. On one arm, a known excitation voltage is being supplied and because of the strain/stress (i.e. because of change in displacement) in the other arm, there will be an imbalance created in the bridge. Because of this imbalance, an output in terms of change in voltage (Delta V) is obtained, which is calibrated in terms of change in displacement.

4. Test software and setup characteristics

4.1. General

The general setup needed for FCGR testing consists of a precise signal generator, a COD gauge, a controller, UTM and a PC. This setup must be integrated with convenient software, which must control the test, collect and process the recorded data, as well as present and archive the results. The general purpose FCGR software (FCS control systems) does not contain on-line monitoring of fatigue crack growth, and they enable only data processing. The on-line fatigue crack growth monitoring software (MTL – FCP software, 2002) however enables data acquisition at intervals of number of cycles, and not at desired crack length achievements. This is particularly important since the crack grows rapidly towards the end, leading to absence of critical data. This led to the decision of developing new special virtual instrumentation software. The developed software has capability of doing FCGR tests on CT, MT and TPB specimens. The developed software has capability of loading the specimens as per the entered dimensions of the specimen, stress ratio and initial crack length. The flowchart of the developed software is illustrated in Figure 1.

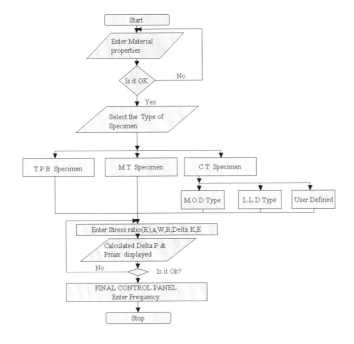

Figure 1. Flowchart of developed software

4.2. Initial Panel

The Initial Panel of developed software prompts user for inputs such as Material Name, Specimen ID, Poisson's ratio, Ultimate Tensile Strength, Young's Modulus and Yield Stress. This panel also displays the entered values for confirmation and on confirmation the details will be stored. This will lead the user to the main panel.

4.3. Main Panel

The main panel prompts the user to select among CT, MT, and TPB specimens. In case of CT specimen, user will be prompted to select mouth opening displacement (MOD), load line displacement (LLD) or user defined type. In case of user defined type, user has to modify the governing equations for calculation of fatigue crack, whereas it has been defined in other cases according to ASTM standards (ASTM- E 647, Volume03, 01). On selection of any particular specimen, the software prompts for thickness (B), width (W), initial crack length (a), ΔK (stress intensity factor range), R (stress ratio). The software also includes a check that prompts user to maintain **a/W** ratio according to ASTM standards (MTL-Windows 7.0, 2002) for particular specimens as illustrated in Figure 2. On breaking the check, an error message is being displayed. On maintaining correct a/W ratio, the software calculates and displays ΔP and maximum load. If the displayed value is found to be satisfactory, then user is guided to the final control panel with calculated values being transferred as the load amplitude.

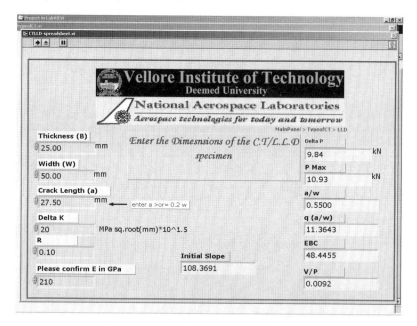

Figure 2. Front panel for CT-LLD specimen

4.4. Final Control Panel

The final control panel as shown in Figure 3 enables:
- loading the specimens with appropriate amplitude calculated from the previous panel,
- retrieving and displaying the COD waveform from the controller,
- Data Acquisition (DAQ) with raw data and processed data in two separate files, with the location of files being pre-decided by the user,
- automated DAQ, i.e. without the need for operator's interference,
- manual DAQ i.e. when user wishes he can log data by clicking DAQ button available in front Panel,

- auto saving of the logged data, i.e., on loss of power, the logged data can be retrieved back,
- on-line plot of "Crack Length Vs No of Cycles".

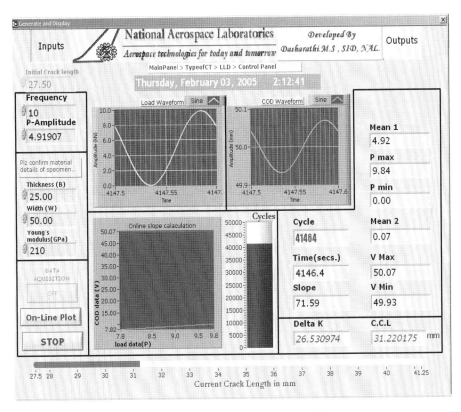

Figure 3. User front panel of the developed VI software

5. Example of achieved test results

5.1. Test conditions

The tests were conducted, at room temperature in constant amplitude load control. The stress ratio, R for sinusoidal load was 0.1.

5.2. Test Set-Up/Apparatus

Tests were conducted on an Instron make servo-hydraulic universal test machine with a 100kN load range. Crack opening displacement was measured with a knife-edge mounted COD gage having a 2.5-12.5 mm range. Control was achieved by employing a BiSS servo-hydraulic controller. Function generation and data acquisition were accomplished by employing a National Instruments PCI-6014 A/D-D/A board in conjunction with in-house developed LabVIEW software. Visual crack measurements were made using magnifying lenses in conjunction with Meta vis 2D, an Image analyzer software.

5.3. Test procedure

Fatigue tests were initiated with an initial compliance check to establish a growth from initial crack length. After observing crack growth, visual measurements of crack length were taken periodically for comparison purposes. After the initial stages of the test, specimen compliance was checked and was recorded if there was a change from previous measurements. During each compliance check the complete load-COD history of the cycle was recorded for post-test analysis. The results of comparison is found to be agreeable as shown in Figure 4, which validates the capability of the developed software.

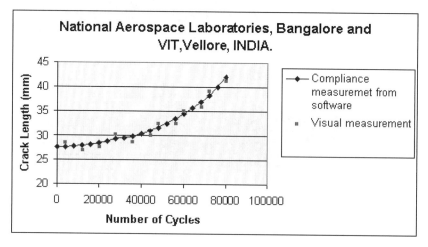

Figure 4. Comparison of software based compliance measurement and visual measurement

Table 1. Sample of acquired raw data. (To be read as 1 – Sample number, 2 - Time in seconds, 3 - Load-amplitude and 4 - COD- amplitude respectively)

1	2	3	4
200	1.3	31.75571	50.73243
201	1.301	30.92789	50.68085
202	1.302	30.07246	50.62755
203	1.303	29.196	50.57627
204	1.304	28.28751	50.51364
205	1.305	27.36249	50.45871
206	1.306	26.41887	50.39992
207	1.307	25.45904	50.34012
208	1.308	24.48542	50.27946

Table 2. Sample of acquired processed data. (To be read as 1 – Time in seconds, 2 - Cycle number, 3 – Max load in KN, 4 - Slope and 5 – Crack length in mm respectively)

1	2	3	4	5
3999.6	39996	9.838136	72.64488	31.09874
3999.7	39997	9.838136	72.64415	31.09883
3999.8	39998	9.838136	72.64343	31.09891
3999.9	39999	9.838136	72.6427	31.09899
4000	40000	9.838136	72.64197	31.09907
4000.1	40001	9.838136	72.64125	31.09915
4000.2	40002	9.838136	72.64052	31.09923
4000.3	40003	9.838136	72.6398	31.09931
4000.4	40004	9.838136	72.63907	31.0994

6. Conclusion

A virtual instrument software for online measurement of fatigue crack growth has been presented. The developed software is user friendly and flexible enough to be developed for most type of test specimens. The developed software also has options of operator pre-set data logging intervals and operator pre-set failure criteria that can be based on elapsed cycles, crack length achievements or other machine control parameters. The developed software was implemented on Instron make 100 kN servo-hydraulic test machine by using National Instruments' PCI-6014 A/D-D/A card. The comparison of compliance based measurement results with visual measurement has been presented, which are in excellent agreement, that validates the use of developed software.

References

1. ASTM- E 647, Standard Test Method for Measurement of Fatigue Crack Growth Rates, Annual Book of ASTM Standards, Vol. 03. 01.
2. Ewalds H.L., Wanhill R. J. H.: *Fracture Mechanics*. Edward Arnold (Publishers) Ltd., 1985.
3. FCS controls systems SmarTEST ONE software.
4. Lincoln A.P., May B.J.: *Use of computers to control and monitor fatigue tests*. Measurement & Fatigue-EIS'86, 19-31, March 1986.
5. Lincoln.A.P., Parramore D.: *Automated crack measurement using ACFM*. Measurement & Fatigue-EIS'86, 19-31, March 1986.
6. MTL-Windows 7.0 FCP software, BiSS Research, 2002.
7. Parramore D., Lincoln A.P.: *Automated crack measurement using ACFM*. Measurement & Fatigue-EIS'86, March 1986.
8. Paris P. C., Bucci V. J., Wessel E. T.: *Extensive Study of Low Cycle Fatigue Crack Growth Rates in A533 and Alloys*. ASTM STP 513, 141-176, 1972.

VELOCITY AND STATIC ANALYSIS OF PARALLEL MANIPULATORS

Bouanane K., Khaldi A.

Abstract: The kinematic problem of platform type parallel manipulators can be studied by modeling each leg as a serial type mechanism. By analyzing this mechanism, one can derive the complete kinematic model of these manipulators. Based on this model, this work presents a study of both the velocity and the static problems of this manipulator. Solutions to both of these problems have been developed by deriving the manipulator Jacobean matrix. Although this work is based on the study of the Stewart Platform, the solutions developed are applicable to all platform type parallel manipulators. To show this, an example has been solved by using both the equations developed in this work and an alternative method.

1. Introduction

The velocity analysis, often referred to as "instantaneous kinematic analysis", is the study of the relationship between actuator rates and end effector velocity components. This analysis was pioneered by Waldron in 1966. Several methods have been presented to solve this problem since then. The classical method of solution makes use of the manipulator Jacobean. One method of determining the Jacobean is by differentiating the kinematic equations obtained from the homogeneous transformations to define both the orientation and the position of the manipulator end effector. Using this method, Gardner, Kumar, and Ho studied the instantaneous kinematics of hybrid serial/parallel systems in 1989 [4]. For platform type parallel manipulators, a more convenient way to derive the Jacobean matrix is through the use of line coordinates, such as "Plucker coordinates". In this coordinate system each joint axis (line) is defined by six variables. The Jacobean is simply the matrix formed by the variables derived from the joint axes. Fichter in 1984 [3] determined the leg lengths, directions, and moments of the Stewart platform built at the Oregon State University using the normalized Plucker coordinates. Further, he studied the leg velocities using the instantaneous screws.

The screw theory is one of the best methods that can be used to study the instantaneous kinematic problem of parallel manipulators. This can be justified by a theorem mentioned by Mohammed and Duffy in 1985 [6]. A general discussion of in-parallel-actuated robot arms was presented by Hunt in 1983 [5]. The velocity analysis of a Stewart platform was also analyzed using Motor algebra (Sugimoto, 1987) [7]. The static analysis is the study of the relationship between the force and moment acting on the end effector and the forces and torques supplied by the actuators at a static equilibrium condition. Using the virtual work procedure to derive the static equation of a manipulator, one can easily see that the Jacobean matrix is used again in the solution.

Using the virtual work method mentioned above, Gardner, Kumar, and Ho in 1989 [4] derived the relationship between actuator torques and the end effector force and moment for a three-degree-of-freedom planar parallel manipulator. Cwikala and Langrana in 1987 [2] discussed the relationship between the static load and the working volume in which the forces transmitted to the actuators remain less than a maximum specified value. This analysis was conducted for a Stewart platform. The present work is an effort to generate general solutions to the velocity and the static problems of the platform type parallel manipulators using the classical Jacobean method. The solution developed is applicable to all platform type parallel manipulators.

2. Velocity analysis

The velocity analysis is concerned with the relationship between the manipulator actuator rates and manipulator end effector velocity components. This analysis, which is often referred to as the instantaneous kinematic analysis of the robot, can be classified as the forward velocity analysis, which provides the linear and angular velocities of the end effector as functions of the Joint rates, and the inverse velocity analysis which determines the joint rates required to generate specified end effector linear and angular velocities.

In this work both the forward and the inverse velocity equations are derived by obtaining the manipulator's Jacobean matrix, the matrix that relates the actuator differential kinematic changes to the differential kinematic changes of the end effector. Although this analysis is applicable to all platform type parallel manipulators it is based on the study of the Stewart Platform. For the Steward Platform (Fig.1), by modeling each leg as a serial type manipulator (Fig. 2), the position of the end effector with respect to the base is determined by the vector $[P^*_{x2}, P^*_{y2}, P^*_{z2}]$, determined in our previous work [1], and defined as:

$$\begin{cases} P^*_{x2} = \cos\alpha_b P^*_x + \sin\alpha_b P^*_y \\ P^*_{y2} = -\sin\alpha_b P^*_x + \cos\alpha_b P^*_y \\ P^*_{z2} = P^*_z \end{cases} \quad (1)$$

where:

$$P^*_x = P_x - pt_x \;;\; P^*_y = P_y - pt_y + b \;;\; P^*_z = P_z - pt_z \quad (2)$$

$$\begin{cases} P_x = c_1 s_2 d_3 \\ P_y = s_1 s_2 d_3 \\ P_z = c_2 d_3 \\ t_x = -(c_1 c_2 c_4 c_5 s_6) + (s_1 s_4 c_5 s_6) + (c_1 s_2 s_5 s_6) - \\ \quad (c_1 c_2 c_4 c_6) - (s_1 c_4 c_6) \\ t_y = -(s_1 c_2 c_4 c_5 s_6) - (c_1 s_4 c_5 s_6) + (s_1 s_2 s_5 s_6) - \\ \quad (s_1 c_2 s_4 c_6) + (c_1 c_4 c_6) \\ t_z = (s_2 c_4 c_5 c_6) + (c_2 s_5 s_6) + (s_2 s_4 c_6) \end{cases} \quad (3)$$

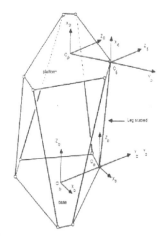
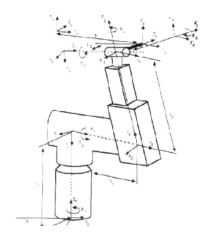

Figure 1. The Stewart Platform

Figure 2. The 5R1P model of the manipulator leg

Substituting equations (2) and (3) in equation (1) yields:

$$\begin{cases} P^*{}_x = \cos\alpha_b \{c_1 s_2 d_3 - p[-(c_1 c_2 c_4 c_5 s_6) + (s_1 s_4 c_5 s_6) + (c_1 s_2 s_5 s_6) - \\ (c_1 c_2 s_4 c_6) - (s_1 c_4 c_6)]\} + \sin\alpha_b \{s_1 s_2 d_3 - p[-(s_1 c_2 c_4 c_5 s_6) - \\ (c_1 s_4 c_5 s_6) + (s_1 s_2 s_5 s_6) - (s_1 c_2 s_4 c_6) + (c_1 c_4 c_6)] + b\} \\ P^*{}_y = -\sin\alpha_b \{c_1 s_2 d_3 - p[-(c_1 c_2 c_4 c_5 c_6) + (s_1 s_4 c_5 s_6) + (c_1 s_2 s_5 s_6) - \\ (c_1 c_2 s_4 c_6) - (s_1 c_4 c_6)]\} + \cos\alpha_b \{s_1 s_2 d_3 - p[-(s_1 c_2 c_4 c_5 s_6) - \\ (c_1 s_4 c_5 s_6) + (s_1 s_2 s_5 s_6) - (s_1 c_2 s_4 c_6) + (c_1 c_4 c_6)] + b\} \\ P^*{}_z = c_2 d_3 - p[(s_2 c_4 c_5 s_6) + (c_2 s_5 s_6) + (s_2 s_4 c_6)] \end{cases} \quad (4)$$

The general form of the Jacobean matrix of the leg model described in Fig. 2 is:

$$J = \begin{vmatrix} j_{11} & j_{12} & j_{13} & j_{14} & j_{15} & j_{16} \\ j_{21} & j_{22} & j_{23} & j_{24} & j_{25} & j_{26} \\ j_{31} & j_{32} & j_{33} & j_{34} & j_{35} & j_{36} \\ j_{41} & j_{42} & j_{43} & j_{44} & j_{45} & j_{46} \\ j_{51} & j_{52} & j_{53} & j_{54} & j_{55} & j_{56} \\ j_{61} & j_{62} & j_{63} & j_{64} & j_{65} & j_{66} \end{vmatrix} \quad (5)$$

Some of the "J_{ik}" terms of the above matrix can be determined by differentiating equations (5), yielding (6). Let $\vec{z}_i^b = [z_{ix} \; z_{iy} \; z_{iz}]^T$ be the unit vector parallel to the z-axis of frame "i" defined in the base co-ordinate frame, i.e. z_{ix}, z_{iy} and z_{iz} are elements of the third column of matrix A_i^b, where $A_i^b = A_0^b A_1^0 A_2^1 ... A_i^{i-1}$. Therefore, we have the following (7).

$$\begin{cases}
j_{11} = \dfrac{\partial P^*_{x2}}{\partial \theta_1} = c_\delta \{-s_1 s_2 d_3 - p[(s_1 c_2 c_4 c_5 s_6) + (c_1 s_4 c_5 s_6) - (s_1 s_2 s_5 s_6) + \\
\qquad (s_1 c_2 s_4 c_6) - (c_1 c_4 c_6)]\} + s_\delta \{c_1 s_2 d_3 - p[-(c_1 c_2 c_4 c_5 s_6) + \\
\qquad (s_1 s_4 c_5 s_6) + (c_1 s_2 s_5 s_6) - (c_1 c_2 s_4 c_6) - (s_1 c_4 c_6)]\} \\[4pt]
j_{12} = \dfrac{\partial P^*_{x2}}{\partial \theta_2} = \cos \alpha_\delta \{c_1 c_2 d_3 - p[(c_1 s_2 c_4 c_5 s_6) + (c_1 c_2 s_5 s_6) + (c_1 s_2 s_4 c_6)]\} + \\
\qquad \sin \alpha_\delta \{s_1 c_2 d_3 - p[(s_1 s_2 c_4 c_5 s_6) + (s_1 c_2 s_5 s_6) + (s_1 s_2 s_4 c_6)]\} \\[4pt]
j_{13} = \dfrac{\partial P^*_{x2}}{\partial d_3} = \cos \alpha_\delta \{c_1 s_2\} + \sin \alpha_\delta \{s_1 s_2\} \\[4pt]
j_{14} = \dfrac{\partial P^*_{x2}}{\partial \theta_4} = \cos \alpha_\delta \{-p[(c_1 c_2 s_4 c_5 s_6) + (s_1 c_4 c_5 s_6) - (c_1 c_2 c_4 c_6) + (s_1 s_4 c_6)]\} + \\
\qquad \sin \alpha_\delta \{-p[(s_1 c_2 s_4 c_5 s_6) - (c_1 c_4 c_5 s_6) - (s_1 c_2 c_4 c_6) - (c_1 s_4 c_6)]\} \\[4pt]
j_{15} = \dfrac{\partial P^*_{x2}}{\partial \theta_5} = \cos \alpha_\delta \{-p[(c_1 c_2 c_4 s_5 s_6) - (s_1 s_4 s_5 s_6) + (c_1 s_2 c_5 s_6)]\} + \\
\qquad \sin \alpha_\delta \{-p[(s_1 c_2 c_4 s_5 s_6) + (c_1 s_4 s_5 s_6) + (s_1 s_2 c_5 s_6)]\} \\[4pt]
j_{16} = \dfrac{\partial P^*_{x2}}{\partial \theta_6} = \cos \alpha_\delta \{-p[-(c_1 c_2 c_4 c_5 c_6) + (s_1 s_4 c_5 c_6) + (c_1 s_2 s_5 c_6) + (c_1 c_2 s_4 s_6) \\
\qquad (s_1 c_4 s_6)]\} + \sin \alpha_\delta \{-p[-(s_1 c_2 c_4 c_5 c_6) - (c_1 s_4 c_5 c_6) + \\
\qquad (s_1 s_2 s_5 c_6) + (s_1 c_2 s_4 s_6) - (c_1 c_4 s_6)]\} \\[4pt]
j_{21} = \dfrac{\partial P^*_{y2}}{\partial \theta_1} = -s_\delta \{-s_1 s_2 d_3 - p[(s_1 c_2 c_4 c_5 s_6) + (c_1 s_4 c_5 s_6) - (s_1 s_2 s_5 s_6) + \\
\qquad (s_1 c_2 s_4 c_6) - (c_1 c_4 c_6)]\} + c_\delta \{c_1 s_2 d_3 - p[-(c_1 c_2 c_4 c_5 s_6) + \\
\qquad (s_1 s_4 c_5 s_6) + (c_1 s_2 s_5 s_6) - (c_1 c_2 s_4 c_6) - (s_1 c_4 c_6)]\}
\end{cases} \qquad (6)$$

$$\begin{cases}
j_{22} = \dfrac{\partial P^*_{y2}}{\partial \theta_2} = -\sin \alpha_\delta \{c_1 c_2 d_3 - p[(c_1 s_2 c_4 c_5 s_6) + (c_1 c_2 s_5 s_6) + (c_1 s_2 s_4 c_6)]\} + \\
\qquad \cos \alpha_\delta \{s_1 c_2 d_3 - p[(s_1 s_2 c_4 c_5 s_6) + (s_1 c_2 s_5 s_6) + (s_1 s_2 s_4 c_6)]\} \\[4pt]
j_{23} = \dfrac{\partial P^*_{y2}}{\partial d_3} = -\sin \alpha_\delta \{c_1 s_2\} + \cos \alpha_\delta \{s_1 s_2\} \\[4pt]
j_{24} = \dfrac{\partial P^*_{y2}}{\partial \theta_4} = -\sin \alpha_\delta \{-p[(c_1 c_2 s_4 c_5 s_6) + (s_1 c_4 c_5 s_6) - (c_1 c_2 c_4 c_6) + (s_1 s_4 c_6)]\} + \\
\qquad \cos \alpha_\delta \{-p[(s_1 c_2 s_4 c_5 s_6) - (c_1 c_4 c_5 s_6) - (s_1 c_2 c_4 c_6) - (c_1 s_4 c_6)]\} \\[4pt]
j_{25} = \dfrac{\partial P^*_{y2}}{\partial \theta_5} = -\sin \alpha_\delta \{-p[(c_1 c_2 c_4 s_5 s_6) - (s_1 s_4 s_5 s_6) + (c_1 s_2 c_5 s_6)]\} + \\
\qquad \cos \alpha_\delta \{-p[(s_1 c_2 c_4 s_5 s_6) + (c_1 s_4 s_5 s_6) + (s_1 s_2 c_5 s_6)]\} \\[4pt]
j_{26} = \dfrac{\partial P^*_{y2}}{\partial \theta_6} = -\sin \alpha_\delta \{-p[-(c_1 c_2 c_4 c_5 c_6) + (s_1 s_4 c_5 c_6) + (c_1 s_2 s_5 c_6) + (c_1 c_2 s_4 s_6) \\
\qquad (s_1 c_4 s_6)]\} + \cos \alpha_\delta \{-p[-(s_1 c_2 c_4 c_5 c_6) - (c_1 s_4 c_5 c_6) + \\
\qquad (s_1 s_2 s_5 c_6) + (s_1 c_2 s_4 s_6) - (c_1 c_4 s_6)]\} \\[4pt]
j_{31} = \dfrac{\partial P^*_{z2}}{\partial \theta_1} = 0 \\[4pt]
j_{32} = \dfrac{\partial P^*_{z2}}{\partial \theta_2} = -s_2 d_3 - p[(c_2 c_4 c_5 s_6) - (s_2 s_5 s_6) + (c_2 s_4 c_6)] \\[4pt]
j_{33} = \dfrac{\partial P^*_{z2}}{\partial d_3} = c_2 \\[4pt]
j_{34} = \dfrac{\partial P^*_{z2}}{\partial \theta_4} = -p[-(s_2 s_4 c_5 s_6) + (s_2 c_4 c_6)] \\[4pt]
j_{35} = \dfrac{\partial P^*_{z2}}{\partial \theta_5} = -p[-(s_2 c_4 s_5 s_6) + (c_2 c_5 s_6)] \\[4pt]
j_{36} = \dfrac{\partial P^*_{z2}}{\partial \theta_6} = -p[(s_2 c_4 c_5 c_6) + (c_2 s_5 c_6) - (s_2 s_4 s_6)]
\end{cases} \qquad (7)$$

$$\begin{cases}
j_{41} = z_{0x}^{\delta} = & 0 \\
j_{42} = z_{1x}^{\delta} = & -s_1 \cos\alpha_b + c_1 \sin\alpha_b \\
j_{43} = & 0 \\
j_{44} = z_{2x}^{\delta} = & c_1 s_2 \cos\alpha_b + s_1 s_2 \sin\alpha_b \\
j_{45} = z_{4x}^{\delta} = & -c_1 c_2 s_4 \cos\alpha_b - s_1 c_4 \cos\alpha_b - s_1 c_2 s_4 \sin\alpha_b + c_1 c_4 \sin\alpha_b \\
j_{46} = z_{5x}^{\delta} = & c_1 c_2 c_4 s_5 \cos\alpha_b - s_1 s_4 s_5 \cos\alpha_b + c_1 s_2 c_5 \cos\alpha_b + s_1 c_2 c_4 s_5 \sin\alpha_b \cdot \\
& c_1 s_4 s_5 \sin\alpha_b + s_1 s_2 c_5 \sin\alpha_b \\
j_{51} = z_{0y}^{\delta} = & 0 \\
j_{52} = z_{1y}^{\delta} = & s_1 \sin\alpha_b + c_1 \cos\alpha_b \\
j_{53} = & 0 \\
j_{54} = z_{2y}^{\delta} = & -c_1 s_2 \sin\alpha_b + s_1 s_2 \cos\alpha_b \\
j_{55} = z_{4y}^{\delta} = & c_1 c_2 s_4 \sin\alpha_b + s_1 c_4 \sin\alpha_b - s_1 c_2 s_4 \cos\alpha_b + c_1 c_4 \cos\alpha_b \\
j_{56} = z_{5y}^{\delta} = & -c_1 c_2 c_4 s_5 \sin\alpha_b + s_1 s_4 s_5 \sin\alpha_b - c_1 s_2 c_5 \sin\alpha_b + s_1 c_2 c_4 s_5 \cos\alpha_b \\
& c_1 s_4 s_5 \cos\alpha_b + s_1 s_2 c_5 \cos\alpha_b \\
j_{61} = z_{0z}^{\delta} = & 1 \\
j_{62} = z_{1z}^{\delta} = & 0 \\
j_{63} = & 0 \\
j_{64} = z_{2z}^{\delta} = & c_2 \\
j_{65} = z_{4z}^{\delta} = & s_2 s_4 \\
j_{66} = z_{5z}^{\delta} = & -s_2 c_4 s_5 + c_2 c_5 \\
& \frac{d\theta_4}{} \\
j_{35} = & \frac{\partial P^*_{z2}}{\partial \theta_5} = -p[-(s_2 c_4 s_5 s_6) + (c_2 c_5 s_6)] \\
j_{36} = & \frac{\partial P^*_{z2}}{\partial \theta_6} = -p[(s_2 c_4 c_5 c_6) + (c_2 s_5 c_6) - (s_2 s_4 s_6)]
\end{cases} \quad (8)$$

From equations (7) and (8), one can obtain six matrices "J_i" (with i = 1,2,3,4,5,6) for the six legs of the manipulator. Let J_i be the Jacobean matrix of leg "i", then we have the following:

$$\begin{vmatrix} V_e^b \\ \omega_e^b \end{vmatrix} = J_i \begin{vmatrix} \dot{\theta}_{1i} \\ \dot{\theta}_{2i} \\ \dot{d}_{3i} \\ \dot{\theta}_{4i} \\ \dot{\theta}_{5i} \\ \dot{\theta}_{6i} \end{vmatrix} \qquad \begin{vmatrix} \dot{\theta}_{1i} \\ \dot{\theta}_{2i} \\ \dot{d}_{3i} \\ \dot{\theta}_{4i} \\ \dot{\theta}_{5i} \\ \dot{\theta}_{6i} \end{vmatrix} = J_i^{-1} \begin{vmatrix} V_e^b \\ \omega_e^b \end{vmatrix} \qquad J_i^{-1} = \frac{1}{\det[J_i]} \begin{vmatrix} C_{11}^i & C_{21}^i & C_{31}^i & C_{41}^i & C_{51}^i & C_{61}^i \\ C_{12}^i & C_{22}^i & C_{32}^i & C_{42}^i & C_{52}^i & C_{62}^i \\ C_{13}^i & C_{23}^i & C_{33}^i & C_{43}^i & C_{53}^i & C_{63}^i \\ C_{14}^i & C_{24}^i & C_{34}^i & C_{44}^i & C_{54}^i & C_{64}^i \\ C_{15}^i & C_{25}^i & C_{35}^i & C_{45}^i & C_{55}^i & C_{65}^i \\ C_{16}^i & C_{26}^i & C_{36}^i & C_{46}^i & C_{56}^i & C_{66}^i \end{vmatrix}$$

and C_{jk}^i is the cofactor of element j_{jk} of matrix J_i. The third row of above equation is:

$$\dot{d}_{3i} = \frac{1}{\det[J_i]} \left[C_{13}^i V_{ex}^b + C_{23}^i V_{ey}^b + C_{33}^i V_{ez}^b + C_{43}^i \omega_{ex}^b + C_{53}^i \omega_{ey}^b + C_{e3}^i \omega_{ez}^b \right]$$

Combining the above equation with respect to all six legs of the manipulator yields:

$$\begin{vmatrix} \dot{d}_{31} \\ \dot{d}_{32} \\ \dot{d}_{33} \\ \dot{d}_{34} \\ \dot{d}_{35} \\ \dot{d}_{36} \end{vmatrix} = [J^*]^{-1} \begin{vmatrix} V_{ex}^b \\ V_{ey}^b \\ V_{ez}^b \\ \omega_{ex}^b \\ \omega_{ey}^b \\ \omega_{ez}^b \end{vmatrix} \qquad [J^*]^{-1} = \begin{vmatrix} \frac{C_{13}^1}{\det[J_1]} & \frac{C_{23}^1}{\det[J_1]} & \frac{C_{33}^1}{\det[J_1]} & \frac{C_{43}^1}{\det[J_1]} & \frac{C_{53}^1}{\det[J_1]} & \frac{C_{63}^1}{\det[J_1]} \\ \frac{C_{13}^2}{\det[J_2]} & \frac{C_{23}^2}{\det[J_2]} & \frac{C_{33}^2}{\det[J_2]} & \frac{C_{43}^2}{\det[J_2]} & \frac{C_{53}^2}{\det[J_2]} & \frac{C_{63}^2}{\det[J_2]} \\ \frac{C_{13}^3}{\det[J_3]} & \frac{C_{23}^3}{\det[J_3]} & \frac{C_{33}^3}{\det[J_3]} & \frac{C_{43}^3}{\det[J_3]} & \frac{C_{53}^3}{\det[J_3]} & \frac{C_{63}^3}{\det[J_3]} \\ \frac{C_{13}^4}{\det[J_4]} & \frac{C_{23}^4}{\det[J_4]} & \frac{C_{33}^4}{\det[J_4]} & \frac{C_{43}^4}{\det[J_4]} & \frac{C_{53}^4}{\det[J_4]} & \frac{C_{63}^4}{\det[J_4]} \\ \frac{C_{13}^5}{\det[J_5]} & \frac{C_{23}^5}{\det[J_5]} & \frac{C_{33}^5}{\det[J_5]} & \frac{C_{43}^5}{\det[J_5]} & \frac{C_{53}^5}{\det[J_5]} & \frac{C_{63}^5}{\det[J_5]} \\ \frac{C_{13}^6}{\det[J_6]} & \frac{C_{23}^6}{\det[J_6]} & \frac{C_{33}^6}{\det[J_6]} & \frac{C_{43}^6}{\det[J_6]} & \frac{C_{53}^6}{\det[J_6]} & \frac{C_{63}^6}{\det[J_6]} \end{vmatrix} \begin{vmatrix} V_{ex}^b \\ V_{ey}^b \\ V_{ez}^b \\ \omega_{ex}^b \\ \omega_{ey}^b \\ \omega_{ez}^b \end{vmatrix} = J^* \begin{vmatrix} \dot{d}_{31} \\ \dot{d}_{32} \\ \dot{d}_{33} \\ \dot{d}_{34} \\ \dot{d}_{35} \\ \dot{d}_{36} \end{vmatrix}$$

(9)

where: C_{jk}^i is the co-factor of j_{jk}^i of matrix $[J^*]$,

J^* is the six by six matrix relating the linear and angular velocity components of the manipulator end effector to the actuator rates. Therefore, J^* is the Jacobean of the parallel manipulator, and above equations are the inverse and forward velocity equations.

3. Static analysis

The static analysis determines the relationship between the force and moment applied by the manipulator's end effector to the environment and the forces and torques supplied by the manipulator's actuators in a static equilibrium state. By using the virtual work procedure to derive the manipulator's static equations from its instantaneous kinematic ones, one can see that the transpose of the manipulator Jacobean is the matrix that relates the end effector forces and moments to the actuator torques and/or forces.

The virtual work exerted by the manipulator's end effector is equal to the virtual work done by its actuator motors. Therefore, one has:

$$\delta W_e = \delta W_m \qquad (10)$$

where: $\qquad \delta W_e = [\vec{f}_e]^T \delta \vec{P} + [\vec{g}_e]^T \delta \vec{\phi} \qquad \delta W_m = [\vec{\tau}]^T \delta \vec{q} \qquad (11)$

$[\vec{f}_e]$ is the force exerted by the end effector defined with respect to the base co-ordinate frame,

$\vec{P}$ is the linear displacement of the end effector,

$[\vec{g}_e]$ is the torque exerted by the end effector defined with respect to the base co-ordinate frame,

$\vec{\phi}$ is the angular displacement of the end effector,

$[\vec{\tau}]$ is the torque (or force) vector exerted by the actuators,

$\vec{q}$ is the actuator angular (or linear) displacement vector.

Substituting equations (11) in equations (10) yields:

$$[\vec{\tau}]^T \delta\vec{q} = \begin{pmatrix} \vec{f}_e \\ \vec{g}_e \end{pmatrix}^T \begin{pmatrix} \delta\vec{P} \\ \delta\vec{\phi} \end{pmatrix} \text{, since } [\delta\vec{P}\delta\vec{\phi}]^T = J^*\delta\vec{q} \text{, one gets: } [\vec{\tau}]^T \delta\vec{q} = \begin{pmatrix} \vec{f}_e \\ \vec{g}_e \end{pmatrix}^T J^*\delta\vec{q}$$

Simplifying and taking the transpose of both sides yields:

$$(\vec{\tau}) = [J^*]^T \begin{pmatrix} \vec{f}_e \\ \vec{g}_e \end{pmatrix}$$

Writing the above equation explicitly becomes:

$$\begin{vmatrix} \tau_1 \\ \tau_2 \\ \tau_3 \\ \tau_4 \\ \tau_5 \\ \tau_6 \end{vmatrix} = [J^*]^T \begin{vmatrix} f_{ex} \\ f_{ey} \\ f_{ez} \\ g_{ex} \\ g_{ey} \\ g_{ez} \end{vmatrix} \quad \begin{vmatrix} f_{ex} \\ f_{ey} \\ f_{ez} \\ g_{ex} \\ g_{ey} \\ g_{ez} \end{vmatrix} = \left[(J^*)^{-1}\right]^T \begin{vmatrix} \tau_1 \\ \tau_2 \\ \tau_3 \\ \tau_4 \\ \tau_5 \\ \tau_6 \end{vmatrix} \tag{12}$$

where J" is the manipulator Jacobean matrix derived in section 2.

Equations (12) are the inverse and the forward static equations of the manipulator. Due to the fact that the method developed in section 2 to derive the Jacobean matrix is applicable to all platform type parallel manipulators, the procedure followed in this section is also valid for solving the static problem of all platform type parallel manipulators. The computation is illustrated in the following example.

4. Example: a three degree-of-freedom, planar, platform type, parallel manipulator

4.1. Derivation of the manipulator static equation based on the method developed in sections 2 and 3.

For the leg model shown in Fig. 2, let $\theta_2 = \theta_4 = \theta_6 = 90°$ and θ_1, d_3, θ_5 be variables. By using a mechanism with three equispaced legs having the above specifications, the parallel manipulator reduces to a planar, three degree-of-freedom, platform type, parallel manipulator shown in Fig.3. For leg "i" of this manipulator, the Jacobean matrix reduces to:

$$J_i = \begin{vmatrix} j_{11} & j_{13} & j_{15} \\ j_{21} & j_{23} & j_{25} \\ j_{61} & j_{63} & j_{65} \end{vmatrix} \tag{13}$$

For leg one, $\alpha_b = 0$. Substituting this value into equations (7) and (8) yields:

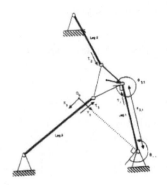

Figure 3. The three degree-of-freedom, planar, platform type manipulator

$$j_{11} = -s_1 d_3 - p[c_1 c_5 - s_1 s_5] = -s_1 d_3 - p\cos(\theta_1 + \theta_5)$$
$$j_{13} = c_1$$
$$j_{15} = p[s_1 s_5 - c_1 c_5] = -p\cos(\theta_1 + \theta_5)$$
$$j_{21} = c_1 d_3 - p[s_1 c_5 + c_1 s_5] = c_1 d_3 - p\sin(\theta_1 + \theta_5)$$
$$j_{23} = s_1$$
$$j_{25} = -p[c_1 s_5 + s_1 c_5] = -p\sin(\theta_1 + \theta_5)$$
$$j_{61} = 1$$
$$j_{63} = 0$$
$$j_{65} = 1 \qquad (14)$$

Substituting the above equations in equation (13) yields:

$$J_1 = \begin{vmatrix} -s_1 d_3 - p\cos(\theta_1 + \theta_5) & c_1 & -p\cos(\theta_1 + \theta_5) \\ c_1 d_3 - p\sin(\theta_1 + \theta_5) & s_1 & -p\sin(\theta_1 + \theta_5) \\ 1 & 0 & 1 \end{vmatrix} \qquad \begin{cases} \det(J_1) = -d_{3,1} \\ C_{12}^1 = -c_1 d_{3,1} \\ C_{22}^1 = -s_1 d_{3,1} \\ C_{32}^1 = -d_{3,1} p c_{5,1} \end{cases} \qquad (15)$$

Similarly, by setting the value of α_b to = 120° and = 240° for legs two and three simultaneously, we get:

$$\begin{cases} \det(J_2) = -d_{3,2} \\ C_{12}^2 = -\left(\dfrac{\sqrt{3}}{2}\right) s_{1,2} d_{3,2} + \left(\dfrac{1}{2}\right) c_{1,2} d_{3,2} \\ C_{22}^2 = \left(\dfrac{1}{2}\right) s_{1,2} d_{3,2} + \left(\dfrac{\sqrt{3}}{2}\right) c_{1,2} d_{3,2} \\ C_{32}^2 = -d_{3,2} p \cos\theta_{5,2} \end{cases} \qquad \begin{cases} \det(J_3) = -d_{3,3} \\ C_{12}^3 = \left(\dfrac{\sqrt{3}}{2}\right) s_{1,3} d_{3,3} + \left(\dfrac{1}{2}\right) c_{1,3} d_{3,3} \\ C_{22}^3 = \left(\dfrac{1}{2}\right) s_{1,3} d_{3,3} - \left(\dfrac{\sqrt{3}}{2}\right) c_{1,3} d_{3,3} \\ C_{32}^3 = -d_{3,3} p \cos\theta_{5,3} \end{cases} \qquad (16)$$

Velocity and static analysis of parallel manipulators 355

For this manipulator, the forward static equation reduces to:

$$\begin{vmatrix} f_x \\ f_y \\ g_z \end{vmatrix} = \left[(J^*)^{-1} \right]' \begin{vmatrix} \tau_1 \\ \tau_2 \\ \tau_3 \end{vmatrix} \qquad [J^*]^{-1} = \begin{vmatrix} \frac{C_{12}^1}{\det[J_1]} & \frac{C_{22}^1}{\det[J_1]} & \frac{C_{32}^1}{\det[J_1]} \\ \frac{C_{12}^2}{\det[J_2]} & \frac{C_{22}^2}{\det[J_2]} & \frac{C_{32}^2}{\det[J_2]} \\ \frac{C_{12}^3}{\det[J_3]} & \frac{C_{22}^3}{\det[J_3]} & \frac{C_{32}^3}{\det[J_3]} \end{vmatrix} \qquad (17)$$

Substituting equations (15), (16) and (17) into the above equation yields:

$$[J^*]^{-1} = \begin{vmatrix} \cos(\theta_{1,1}) & & \sin(\theta_{1,1}) \\ \left(\frac{\sqrt{3}}{2}\right)\sin(\theta_{1,2}) - (1/2)\cos(\theta_{1,2}) & -(1/2)\sin(\theta_{1,2}) - \left(\frac{\sqrt{3}}{2}\right)\cos(\theta_{1,2}) \\ -\left(\frac{\sqrt{3}}{2}\right)\sin(\theta_{1,3}) - (1/2)\cos(\theta_{1,3}) & -(1/2)\sin(\theta_{1,3}) + \left(\frac{\sqrt{3}}{2}\right)\cos(\theta_{1,3}) \\ & p \cos(\theta_{s,1}) \\ & p \cos(\theta_{s,2}) \\ & p \cos(\theta_{s,3}) \end{vmatrix} \qquad (18)$$

4.2. Alternative method to derive the manipulator static equation

For the manipulator shown in Fig.4, consider only the force of the actuator of leg one. This force results in the following end effector forces and torque:

$$\begin{aligned} f_{x,1} &= -\tau_1 \sin\left[(3/2)\pi - \theta_{1,1}\right] = \tau_1 \cos\theta_{1,1} \\ f_{y,1} &= -\tau_1 \cos\left[(3/2)\pi - \theta_{1,1}\right] = \tau_1 \sin\theta_{1,1} \\ g_{z,1} &= -p\tau_1 \sin\left[(3/2)\pi - \theta_{s,1}\right] = p\tau_1 \cos\theta_{s,1} \end{aligned} \qquad (19)$$

Likewise, the force and torque components of the manipulator end effector resulting from the actuator forces of legs two and three are:

$$\begin{vmatrix} f_{x,3} \\ f_{y,3} \\ g_{z,3} \end{vmatrix} = \begin{vmatrix} -(1/2)\tau_3 \cos(\theta_{1,3}) + \left(\frac{\sqrt{3}}{2}\right)\tau_3 \sin(\theta_{1,3}) \\ \left(\frac{\sqrt{3}}{2}\right)\tau_3 \cos(\theta_{1,2}) - (1/2)\tau_3 \sin(\theta_{1,3}) \\ p\tau_3 \cos\theta_{s,3} \end{vmatrix} \qquad \begin{vmatrix} f_{x,3} \\ f_{y,3} \\ g_{z,3} \end{vmatrix} = \begin{vmatrix} -(1/2)\tau_3 \cos(\theta_{1,3}) + \left(\frac{\sqrt{3}}{2}\right)\tau_3 \sin(\theta_{1,3}) \\ \left(\frac{\sqrt{3}}{2}\right)\tau_3 \cos(\theta_{1,2}) - (1/2)\tau_3 \sin(\theta_{1,3}) \\ p\tau_3 \cos\theta_{s,3} \end{vmatrix} \qquad (20)$$

By adding the above equations, the total end effector force components and torque are:

$$\begin{vmatrix} f_x \\ f_y \\ g_z \end{vmatrix} = \begin{vmatrix} \cos(\theta_{1,1}) & \left(\frac{\sqrt{3}}{2}\right)\sin(\theta_{1,2}) - \left(\frac{1}{2}\right)\cos(\theta_{1,2}) & -\left(\frac{\sqrt{3}}{2}\right)\sin(\theta_{1,3}) - \left(\frac{1}{2}\right)\cos(\theta_{1,3}) \\ \sin(\theta_{1,1}) & -\left(\frac{1}{2}\right)\sin(\theta_{1,2}) - \left(\frac{\sqrt{3}}{2}\right)\cos(\theta_{1,2}) & -\left(\frac{1}{2}\right)\sin(\theta_{1,3}) + \left(\frac{\sqrt{3}}{2}\right)\cos(\theta_{1,3}) \\ p\cos(\theta_{5,2}) & p\cos(\theta_{5,2}) & p\cos(\theta_{5,3}) \end{vmatrix} \begin{vmatrix} \tau_1 \\ \tau_2 \\ \tau_3 \end{vmatrix} \qquad (21)$$

Recalling that the forward static equation of the manipulator is:

$$\begin{vmatrix} f_x \\ f_y \\ g_z \end{vmatrix} = \left[(J^*)^{-1} \right]^T \begin{vmatrix} \tau_1 \\ \tau_2 \\ \tau_3 \end{vmatrix} \qquad (22)$$

Comparing the above two equations yields

$$\left[(J^*)^{-1} \right]^T = \begin{vmatrix} \cos(\theta_{1,1}) & \left(\frac{\sqrt{3}}{2}\right)\sin(\theta_{1,2}) - \left(\frac{1}{2}\right)\cos(\theta_{1,2}) & -\left(\frac{\sqrt{3}}{2}\right)\sin(\theta_{1,3}) - \left(\frac{1}{2}\right)\cos(\theta_{1,3}) \\ \sin(\theta_{1,1}) & -\left(\frac{1}{2}\right)\sin(\theta_{1,2}) - \left(\frac{\sqrt{3}}{2}\right)\cos(\theta_{1,2}) & -\left(\frac{1}{2}\right)\sin(\theta_{1,3}) + \left(\frac{\sqrt{3}}{2}\right)\cos(\theta_{1,3}) \\ p\cos(\theta_{5,1}) & p\cos(\theta_{5,2}) & p\cos(\theta_{5,3}) \end{vmatrix} \qquad (23)$$

and the transpose of the above equation yields equation (18). This indicates that the velocity and the static equilibrium equations developed in sections 2 and 3 are correct.

5. Conclusion

In previous literature the static equilibrium equations of platform type parallel manipulators have been derived based on direct geometrical methods. Although these methods are simple and useful for simple structure manipulators, they are complicated and non applicable for manipulators with complicated geometries such as the Stewart Platform.

In this work, based on the kinematic model described in Fig. 2 and using the kinematic equations developed in our previous work [I], the Stewart Platform end effector position vector was derived. By differentiating the components of this position vector with respect to time, both the velocity and the static equilibrium problems of the Stewart platform were solved. The result is a set of velocity and static equilibrium equations applicable to any platform type parallel manipulator. An example (a three degree-of-freedom planar, platform type manipulator) has been solved using these equations and the results were then checked using a geometrical method.

References

1. Bouanane K., Fenton R.G.: *Kinematic Analysis of Parallel Manipulators*. 3rd Int. Workshop on Advances in Robot Kinematics, September 1992.
2. Cwiakala M., Langrana N.A.: *Static Load Optimization in a Platform Manipulator Design*. 10th Applied Mechanisms Conference, pp. 1-7, 1987.
3. Fichter E.F.: *A Stewart Platform-Based Manipulator: General Theory and Practical Construction*. Int. 3 of Robotics Research, Vol. 5, No 2. pp. 165-190, 1986.
4. Gardner J.F., Kumar, Ho J.H.: *Kinernatics and Control of Redundantly Actuated Closed Chains*. IEEE Conf. on Robotics and Automation, Vol. 1, pp. 418-423, 1989.
5. Hunt K.H.: *Structural Kinematics of In-Parallel-Actuated Robot Arms*. Trans. of ASME, J. of Mechanisms, Transmissions and Automation in Design, Vol. 105, pp. 705-712, 1983.
6. Mohammed M.G., Duffy J.: *A direct Determination of the Instantaneous Kinematics of Fully Parallel Robot Manipulators*. J. of Mechanisms, Transmissions and Automation in Design, Vol. 107, pp. 225-229, 1985.
7. Sugimoto K.: *Kinematic and Dynamic Analysis of Parallel Manipulator by Means of Motor Algebra*. Journal of Mechanisms, Transmissions and Automation in Design, Vol. 109, pp. 3-7, 1987.

DEVELOPMENT OF SOFTNESS EQUIPMENT AND SOFTNESS MEASURING METHOD

Nagao M.M., Sakai Y., Yokota O.

Abstract: In the conventional testing equipments, the indenters or the needles are made in diamonds, steel balls, etc. With the deformation given to the soft material using the hard material, the hardness can be measured from the deformation volume. However, it has never done that the hard measured object using the soft indenter material is measured. Then, the softness of the measured object is examined using indenter material which is softer than the hardness of the measured object. Here, the measurement principle and measuring method are proposed. The testing equipment based on this is manufactured, and the research is carried out. It is possible to measure the material which is softer than the indenter. And, it is possible to confirm the performance of the softness equipment.

1. Introduction

The hardness of an object defines the resistance force of the object, when a indenter pressed the measured objects [1]. In the hardness test method in metal and ceramics, there are the static indentation method like Brinell hardness, Vickers hardness and Rockwell hardness, the shocking indentation method like Shore hardness and the scratch test method [2] and the hardness tester of high polymer material such as rubber and synthetic resin pushes the indenter (needle) into the measured object, and it is transformed, and it measures the penetration depth of the needle, and the hardness is obtained by the method for digitizing it [3]. There are electrical methods, magnetic methods, ultrasonic wave methods, etc. for indirectly measuring the hardness. The hardness shows the value which differs by property, shapes and dimensions, degrees of causing deformation, measuring method of resistance force and presentation method etc.. Therefore, the type is very abounding for the hardness tester, and the relations between the results are complicated.

In conventional hardness tester equipment, the indenter or the needle which made diamonds, steel ball, etc. is harder than the measured object. However, the test method which measures hard object using the soft indenter material has not been done until now.

Then, the following are proposed using indenter material which is softer than the hardness of the measured object: Measurement principle which examines the softness of object and the measuring method. The testing equipment based on these was manufactured, and the research was carried out. It is possible to measure the material which is harder than the indenter, and it is possible that the performance and advantage of the tester are confirmed.

2. Measurement principle

The measurement principle of this study [4] is shown in Figure 1. The balloon is used as a soft indenter, and the measured object which is harder than the indenter is measured. The measured object is made to contact the balloon. The contact plane becomes concave surface of balloon, when the pressure of the object is maintained lower than the pressure of measured object (Figure 1(a)). The object is statically pressurized, while it contacted in respect of two materials each other. The contact plane is almost flattened (Figure 1(b)). In addition, the contacting plane changes to convex surface of the object from the flat surface, when the balloon pressure rises further than the pressure of the measured object (Figure 1(c)). The pressure of the balloon in flattening the contact plane is defined as softness (kPa) of the measured object.

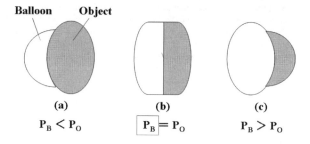

Figure 1. Principle of measurement

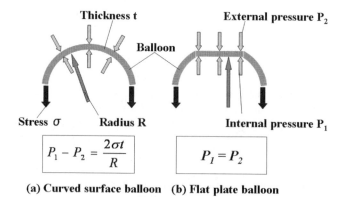

(a) Curved surface balloon (b) Flat plate balloon

Figure 2. Membrane theory of the balloon

In the meantime, the contact surface is put on in respect of the balloon pressure which is higher than the resistance force of measured object. The softness can be measured, when the internal pressure of the balloon is decompressed in order to flatten the contact plane from convex in the balloon surface (The measuring sequence is Figure 1(c)→(b)→(a)).

The membrane theory of the balloon is shown in Figure 2. The P_1 and P_2 are respectively applied in internal pressure and external pressure of the balloon, and the condition for $P_1 > P_2$ is given. The stress which works in the balloon shows σ, the radius of the balloon is R, thickness of the balloon is t:

$$\pi R^2 (P_1 - P_2) = 2\pi R t \sigma \tag{1}$$

$$P_1 - P_2 = 2\sigma t / R \tag{2}$$

The contact plane of the balloon and measured object is flatten as shown in Fig. 2(b), when the working fluid was injected in the balloon:

$$R = \infty, \quad 2Tt/R \: P \: 0 \tag{3}$$

$$P_1 = P_2 \tag{4}$$

When contact plane of the balloon is almost flattened, the pressure in the balloon becomes equal the pressure of the outside. That is to say, the pressure in the balloon will measure the softness of the measured object.

3. Measuring equipment

The schematic drawing of the equipment which measures the softness of the measured object is shown in Figure 3. The equipment has been composed of the balloon which contacts measured object, compressor which pressurizes the balloon, regulator which decompresses the balloon, the pressure sensor which measures the pressure in the balloon, control section and display which measure the softness of measured object. The balloon can be gradually pressed to the measured object. Under low pressure and minute pressure changes, the possibility in which the measured object is damaged can be reduced. The observation of the contact condition is possible from the balloon inside using the fiberscope. Water and salt solution and so on can be utilized for the working fluid except for the air.

4. Experimental results

4.1. Shape of the balloon

It is measurable, when the shape of the balloon is flat and the shape of the measured object is convex in Figure 4(a). The measurement becomes impossible the case in which the measured object is even and concave. It is possible to measure the softness that the balloon shape of the hemisphere does not influence the surface shape of measured object. The example is shown in Figure 4. Therefore, balloon shape used for this experiment was made to be the hemisphere. The dimension of the rubber was made in the silicone rubber of 7.15 mm inner diameter and 0.4 mm film thickness, and the hardness of the rubber adopted the materials of 20°, 30° and 50°. The balloon can be easily exchanged.

4.2. Color of the balloon

The effect of the light income was examined by the color of the measured object. Here, the balloon of which the color differed was produced. The balloon colors used were the 3 types (white, red, and black), and the light reception waveform by the difference between those colors was examined. The result obtained is shown in Figure 5.

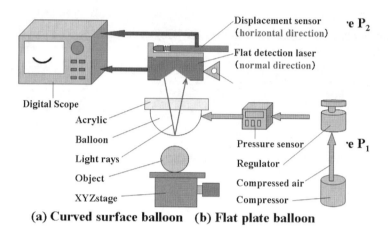

Figure 3. Schematic drawing of equipment

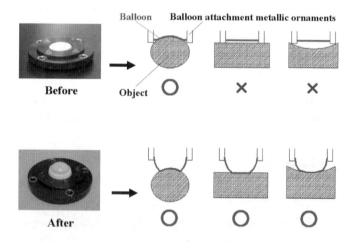

Figure 4. Balloon shapes and measured object shape for the measurement

From Figure 5, the change of the light income by the white balloon shows clearly semicircle shape. However, the light waveform can not be clearly shown in black and red of the balloon and the resolution is also inferior. Still, light reception waveform from reflection plane shows clarified planar shape in red, yellow and white. However, the surface shape with the ruggedness was shown in black and blue.

4.3. Surface quality of the balloon

In balloons with the luster and with the ruggedness surface which produced by the metal mold, the balloon shape between both surfaces by the laser beam was examined. However, there was not large difference in hemispherical state by the laser beam. In the meantime, the balloon which conducted the blasting exploded early than the balloon without the blasting, when the internal pressure in the balloons was raised. The burst pressure of blasting have became 7.0kPa. And the burst pressure without the blast became 11.5 kPa. Therefore, the adoptive balloon was white, and was made to be the surface with the luster.

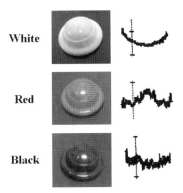

Figure 5. Balloon colors used and these light reception waveform

4.4. Measurement

The pencil-form rubber with hollow used as a measured object is shown in Figure 6. The dimensions of the pencil-form rubber are the diameter of 6 mm, the wall thicknesses of 0.3 mm, and the length of 30 mm. This measuring example is shown in Figure 7. In the measurement of the light reception waveforms, the balloon plane contacted with the measured object is vertically convex. The pressure of the balloon was taken each 0.5 kPa and the maximum pressure was taken 7.0 kPa. Afterward, internal pressures in the balloon were decompressed at 0 kPa from 7.0 kPa. The result is shown in Figure 7. By pressurizing in the balloon , the light reception waveform from the laser beam on contact plane hanges to the mountain which becomes flat from the hemisphere. It is almost flattened at 4.5 kPa. In addition, it rises of the pressure of balloon higher than the hardness of the measured object, when the balloon pressure is taken at 7.0 kPa. The height of the concave surface decreases, when the balloon pressure is decompressed from 7.0 kPa. The shape of contact plane becomes a concave surface of measured object. At 3.5 kPa, it is almost flattened or flattened. The shape of contact plane becomes a concave surface of measured object. The measurement is comparatively easy, because it can be pressurized until the contact plane is flattened.

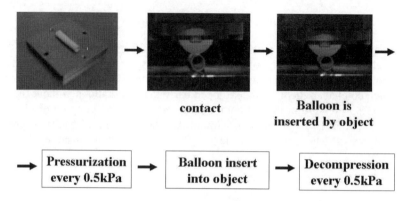

Figure 6. Pencil-form rubber with hollow used as measured object

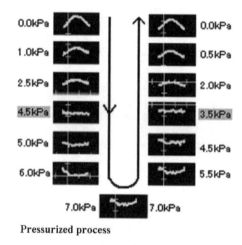

Figure 7. Contact surface of the balloon rubber measured by pressurization process and decompression process

As a method for measuring the softness of the measured object, in contacting the balloon of the pressure which is lower than measured object to the measured object each other, it is pressurized (pressurization stroke). And, there is a method for measuring the softness of the measured object with the decompression of the balloon pressure after the pressure which is higher than the measured object contacted the balloon (decompression stroke).

5. Conclusions

In order to describe the above, softness testing equipment using the indenter material which was softer than hardness of the measured object was proposed. That effectiveness has been confirming this equipment. This measuring device has the advantage of the following:

1. The softness of the measured object can be measured, after the balloon of the pressure which is lower than measured object is made to contact the measured object.
2. The balloon can be gradually pressed to the measured object.
3. Under low pressure and minute pressure changes, the possibility in which the measured object is damaged can be reduced.
4. The measurement is comparatively easy, because it can be pressurized until the contact plane is flattened.
5. The observation of the contact condition is possible from the balloon inside using the fiberscope.
6. The balloon can be easily exchanged.
7. Water and salt solution can be utilized for the working fluid except for the air.

References

1. JSME: *Handbook of mechanical engineering, a fundamentals*. A4::strength of materials (1990) p.148.
2. Komatsu K.: *Talk of the rubber*. JMA, 2002, p.115, ISBN 4-542-90141-6.
3. Hardness measurement method and measuring device of the rubber, JIS K 6253, 1997.
4. NIHON Univ.: *Softness measuring method and measuring device*. JP 2004-77182, 2004.

DESIGN, CONSTRUCTION, CONTROL AND TESTS OF AN EXPERIMENTAL DIESEL FUEL HIGH PRESSURE BURNER CANNON TYPE

Tibaquirá G.J.E., Burbano J.J.C., Holguín L.G.A.

Abstract: This paper shows the obtained results in the project: "Design, construction, control and tests of an experimental diesel fuel high pressure burner cannon type" . One by one are explained the development steps for the design, components selection and the burner construction, tests and control in order to install this device in the "Universidad Tecnológica de Pereira" in Colombia. This type of devices is important in countries like Colombia, coffee producers and where the drying process plays an important role. This study proposes a design and construction methodology economically viable of burners for small producers.

1. Introduction

The research project: "Design, construction, control and tests of an experimental diesel fuel high pressure burner cannon type, for the Thermal Lab", is a project developed by Thermal Systems and Mechanical Power Research Group from the Faculty of Mechanical Engineer at the "Universidad Tecnológica de Pereira" (Colombia). This article will show the following stages of the project:

- design and selection of components of the fuel injection system,
- design and selection of components of the air feed system,
- design and selection of components of the burner ignition system,
- design and construction of the combustion chamber,
- design and implementation of the burner control system.

A diesel fuel high pressure burner cannon type is a device used to burn liquid fuels and it is composed by: Fuel system, Air feed system, Ignition system and Control system [1].

2. Design and selection of components of the fuel injection system

2.1. Pump

It requires a pressure between 100 and 150 psig and a flow of 3 GPH. In this way, a positive displacement internal gear pump is selected. See figure 1. The characteristics of this pump are listed below:

- Manufacturer: SUNTEC
- Model: A1VA-7112
- Speed: 1725 RPM
- Pressure: 100- 150 psig

Figure 1. Selected Pump

With the previous conditions, was selected the electrical motor for the pump, figure 2, with the following characteristics:

- Manufacturer: Siemens
- Model: 1LA7 070-4Y
- Frecuency: 60 Hz
- Power: 0.4 HP (300 W)
- Speed: 1640 RPM
- Power factor: 0.77
- Voltage: 220 V, trifase.

Figure 2. Mounted motor and pump

2.2. Type of nozzle

The different types of nozzle used in the atomization of the fuel presents a large variety in flow ranks and functioning principles. The common nozzles that are used on the burners of the high pressure cannon are whirl type because of its cheap price and mainly because it doesn't requiere other mediums to produce the atomization.

2.3. Fuel injection circuit design

The fuel injection circuit will be composed of the following: A fuel deposit (Figure 3), a filter (Figure 4), a fuel pump, a valve that interupts the flow, Nozzle and pipes for atomization. Pressure and flow gauges will also be installed. A deposit tank will also be used as a source of fuel.

Figure 3. Fuel deposit

Figure 4. Filter

3. Design and selection of components of the air feed system

3.1. Ventilator

The maximum quantity of burnt fuel is 3 GPH, and according to the obtained results it is necessary around 85 CFM (cubic feet per minute) of air to burn the fuel completely ($C_{12}H_{26)}$, with the 100 % theoretical air. A ventilator that gives up to 200 CFM to try it out with 200% of the theoretical air will be designed. Analysing the different regulations that the burner has, and to bear in mind that for burners that have under 100 kg/h of fuel flow, the losts in pressure of these devices doesn't outdo the 5 mm.c.a.[1] , taking into account that the pipe that presents the biggest losts is the round one, that has an internal diameter of 10 cm, and a SP of 0.114 in c.a. The SP for the ventilator will be of 0.114 in.c.a. plus 0.2 in.c.a, so in total it will be 0.32 in.c.a that correspond to the addition of the wastage of the pipes with bigger wastages. For this reason and taking into account the over size, for future innovations in the design and experimentation, a 0.5 in c.a of SP fan was selected. The acquired ventilator (Figure 5) has the following characteristics:

- Manufacturer: Niche Fans
- Nominal power: 135 W, supplied by an incorporated electrical motor.
- Frecuecy: 60 Hz.
- Rotation Speed: 1350 RPM
- Flow: 0 - 266 CFM
- Static Head (SP): 0 – 0.8 pul c.a

Figure 5. Ventilator

3.2. Air Feed Pipe

The chosen air feed pipe is a 5'' HR steel tube. The unload path of the ventilator has a rectangular section, that is why the join will be designed and constructed, so the two parts can fit together (Figure 6).

4. Design and selection of components of the burner ignition system

The generation of an electric arc (About 10000V), in the air and fuel mixture, for an adequate amount of time, so that the combustion process will uphold itself is necessary for the ignition system. This arc is generated by using two 9/16"X6" electrodes, see figure 7, that are feed by a transformer that raises from 120 V to 10000V.

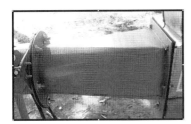
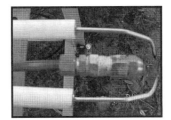

Figure 6. Transition   Figure 7. Electrodes and Fuel nozzle

5. Combustion chamber

The dimensions recommended for the combustion chamber for burners (fuel volume as much as 3 GPH), where determinated keeping in mind the dimensions specified for combustion chambers for diesel fuel, by the rules of reference [2], and according to the references in the Beckett catalogue (reference 3). The main dimensions for this aplication are: cannon diameter Dc= 12 cm, total angle of the diffuser: 70°, internal chamber diameter: Di= 40 cm, chamber length: L=50cm, product vent diameter: Dp=20 cm, chamber nozzle total angle: 70°. The selection of the adequate materials is the second part of the design. Several factors such as, maximum work temperature, work conditions, lessen the heat watsed through the walls of the chamber, and the economic posibilities where taken into account for the selection of the right materials. The maximum working temperature is related inherently with the adiabatic flame temperature; due to it is the highest temperature that can be reached, that is why this temperature will be used for the design. The ratio between the temperature at the surface of heat transfer and the adiabatic flame temperature will be defined with the τ symbol.

For industrial burners this ratio does not exceed a value of 0.6, that is to say that the temperature at the surface of heat transfer, does not exceed the 60% of the adiabatic flame temperature, that is, if it was possible that this idealized case occured. Knowing the adiabatic flame temperature for diesel fuel, references [4] and [5], and assuming the maximum working temperature of the 60% of the adiabatic makes it possible to choose a refractory brick or the apropiate cement for the case. Using catalogues of refractory materials, CONCRAX 1500 cement is selected to be used, with a recomended maximum working temperature of 1540°C and a rupture module of 15 MPa.

In figure 8 the designed and constructed chamber is shown, two holes with an internal diameter of 1-3/8" can be observed. One of them is used to install a visor that will allow the observer to see the flame inside the chamber. The other one may be used to introduce elements such as instruments to measure temperature or sounding to monitor the gases inside the chamber.

Figure 8. Combustion chamber and Structure

The visor, which allows the vizualisation of the flame inside the chamber through the 1-3/8 holes, consist of a 32 mm diameter glass lens made of boron silicate Pyrex®. Pyrex may be used up to 232°C with out a problem, it also resist thermal impacts even up to 120°C. The threaded end of a 2" galvanized iron pipe with its plug was used as the chassis to install the lens.

6. Control system

The implemented control system has two options for its operation: manual operation from a control board and automatic operation from a computer. The events controlled by the manual operation are: Turning the burner on and off, Preliminar gas sweep at the moment of ignition of the burner, Time of ignition of the electrodes, All of them are controled using a simple contact logic.

The operation from the computer is possible thanks to the National Instruments acquisition data card, wich captures the conditioned signs of the following elements:

- fuel pressure sensor: the selected pressure sensor (see figure 9 and 10) is the KOBOLD SEN 3296, it works with an inside diaphragm technology, which is ideal for the highest industrial demands. Its compact design decreases the cost of installation and makes easier the assembly. Its stainless stell is perfect for measurements of corrosive fluids and supports high mechanical charges,

Figure 9. Selected Presssure gauge   Figure 10. Installed Presssure gauge

- flow fuel sensor: The selected sensor is a Kobold DPM, which uses the principle of Pelton wheel to meassure the water flow and other fluids with low viscosity. The fluid goes through a nozzle at the body entrance of the sensor and it is exactly guided to the blades of the Pelton turbine. This will turn the turbine at an angular velocity proportional to the flow. The turbine momevent is detected through an optic sensor. The sign that comes out of the optical sensor is procesed as an amplified pulse, or as a 4-20 mA sign, both proportional to the flow. See figures 11 and 12,

Figure 11. Selected flow sensor   Figure 12. Installed flow sensor

Figure 13. Not installed flam sensor   Figure 14. Flame sensor installed in the burner.

- flame Sensor: For the burner the selected flame detector has been a C 554A Honeywell (see figures 5 and 6). It consists on a photocell of cadmium sulphide and electrodes that are attached to it, to transmit an electrical signal to the primary control that operates the burner. In darkness the cadmium sulphide has a high electric resistance which blocks the way of the current. In lightness the cell has a low current resistance allowing its way for the current, creating an electrical signal which is in charge of the detention of the burner. See figures 13 and 14,

- temperature Sensor: The temperature of the combustion products is measured with a type K thermocouple,
- potentiometer of the butterfly valve: this has a linear resitance that may vary in one revolution between 0 y 100 kΩ, a 13mm long and 1/8" diameter axis.

A program created in Lab View is used to control the following actions from the computer: Turning the burner on and off. Gas sweep before turning on the burner: To know at any moment where the butterfly valve is, a signal is given by the potentiometer. This information is processed in the computer program. To obtain an effective gas sweep when the burner is turned on, a step by step engine joined to the axis of the valve, that opens the valve completly, is also turned on; the computer program is in charge of all this process. Time for the Ignition Electrodes to turn on: The electrode circuit is opened or closed by using the switch. Butterfly valve regulator of air Opening: This valve is opened by the stepping motor.

Figure 15. Not installed potentiometer

Figure 16. Installed potentiometer

Figure 17. Stepping motor

7. Stability and elasticity tests

The tests that are made to the burner are: elasticity and stability. The elasticity is the capacity of the burner to allow excess and deficiency of theoretical air, with the source of ignition ignited permanently of such form that does not appear extinction of the flame. The stability is the capacity of the burner to allow excess and deficiency of theoretical air, of such form that appears support of the flame after extinguished the ignition source.

It is possible to write down that the previous tests are affected by external environment because are made without a combustion chamber, since the air that enters combustion is not only provided by the ventilator but also the ambient air. The temperature, the ambient humidity as well as the wind speed also affect the tests. The elasticity of the experimental burner is of the total range of air flow that can be handled.

In this way the obtained burner elasticity goes from 15 CFM to 266 CFM. This is equivalent to different percentages of air excess using several nozzles. In order to give a range of elasticity, the nozzle with greater volume flow rate is taken (3 GPH) and is obtained and air excess equivalent. This represents burner elasticity from 15.7% to 260% of theoretical air. In the case the nozzles with smaller volume flow rate (1 GPH) is obtained burner elasticity from 46.97% to 780% of theoretical air. According with described above, it can be concluded that the elasticity increases by the superior limit.

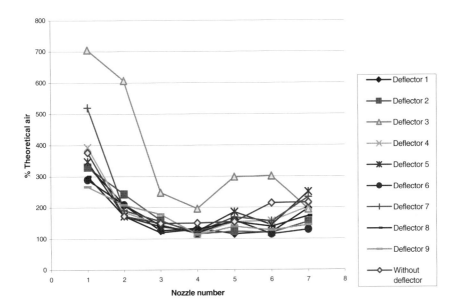

Figure 18. Percentage of theoretical air versus nozzle number for different flame deflectors

For the stability case, is not posible to give a fix range that characterized it. However, is possible to determine the stability for burner operation point, like in industrial installations. In the figure 18 can be seen that the best performance flame deflector is number 3 that has 12 blades and 30 degrees of inclination.

8. Conclusions

An experimental diesel cannon burner of high pressure was designed and built. It has the following descriptions:
- flujo de combustible: 0 – 3 GPH,
- fuel flow: 0-3GPH,
- air flow: 0-250 CFM,
- fuel pressure: 100-150 psig.

The door is opened to continue developing proyects that involve combustion and air quality, in the Universidad Tecnologica de Pereira (Colombia) and specifically in the Mechanical Power and Termichal Systems by this project.

Acknowledgements

The authors want to thank to next students who have participated in this project: Mauricio Carmona, Andres Acosta, Jorge Alzate, Hector Calvo, Julian Maya and Andres Muñoz.

References

1. Salvi G.: *La combustión, teoría y aplicaciones*. Madrid: Editorial Dossat., 1975.
2. Grupo combustión industrial del Perú. Available in Internet:
3. www.combustionindustrial .com/combustibles.html
4. Instruction manual model: Oil Burner. R.W. BECKETT Corporation
5. Kuo K. K.: *Principles of combustion*. John Wiley & Sons Inc., 1986.
6. Turns S.R.: *An introduction to combustión*. Singapore, McGraw Hill, 1996.

REDUCING THE SAMPLING-RATE OF DIGITAL DRIVERS FOR A STEPPING-MOTOR BY FREQUENCY-RESPONSE RESHAPING

Fukuzawa M., Hori N., Sejimo I.

Abstract: In this paper, how the frequency response properties of the digital driver for stepping-motors change as the sampling interval increases is observed and this knowledge is used to modify the control parameters. In particular, the sampling interval of the so-called Plant-Input-Mapping method, which guarantees the stability for any non-pathological sampling interval, is increased from 250μ seconds, which is used previously, to 400 μ seconds, without sacrificing the performance.

1. Introduction

Analog controllers, which have long been used in a wide range of industrial control applications, are being replaced with digital devices thanks to advances in digital technologies. Whereas physical components need to be changed to vary the parameters of an analog control system, software changes usually suffice to alter those of a digital control system. In spite of the cost reduction of microprocessors per computation speed, the need for control algorithms that can perform good control actions using less-expensive processors remains in industrial applications, especially those dealing with fast dynamics. This holds true, for example, in electronic driver circuits for stepping-motors. Presently, the cost of the processors required to implement the digital control algorithm designed using the Tustin method (Franklin, Powell, and Workman, 1998), which is one of the most popular discretization methods, would be several times higher than the entire analog driver device. This method does not give a performance close to the analog original even with the sampling interval of 10 microseconds, which is the minimum possible using one of the most powerful DSP available commercially at present.

The Plant-Input-Mapping (PIM) method proposed in (Markazi and Hori, 1992) is a digital redesign method for converting an analog control system into a digital one with guaranteed stability for any non-pathological sampling interval and good performances even for large sampling intervals. This method is successfully applied to the stepping-motor driver using the sampling interval of 250 micro-seconds in (Shimamura and Hori, 2002). Although the cost of processors is lowered using this method, it is still high for commercial production and the demand for a technique that allows a larger sampling interval still remains. Unfortunately, at such a large interval, the digital control system even designed with the standard PIM method does not give a good transient response, although the stability is achieved.

In the present paper, a method is proposed for the PIM digital redesign method that can increase the sampling interval to 400 microseconds and give a performance close to the original analog driver even in the transient stage. This is achieved by observing how the frequency response characteristics of the transfer function from the reference input to the plant input, called the plant-input transfer-function (PITF), changes as the sampling interval increases. The use of a PITF is very usefull not only for digital controllers but for analog ones as well (Kawasumi, Hori, and Fukuzawa, 2003). Knowing how discretization changes the frequency response characteristics, the target analog frequency response characteristics can be modified by taking into account such changes.

2. The plant-input-mapping (PIM) method

In this section, the so-called the Plant-Input-Mapping (PIM) method, which is used for designing a digital control system, is briefly reviewed. Consider the analog control system shown in Figure 1, where the plant and analog controller blocks are given by

$$\overline{G}(s) = \frac{\overline{n}_G(s)}{\overline{d}_G(s)}, \overline{A}(s) = \frac{\overline{n}_A(s)}{\overline{d}_A(s)}, \overline{B}(s) = \frac{\overline{n}_B(s)}{\overline{d}_B(s)}, \overline{C}(s) = \frac{\overline{n}_C(s)}{\overline{d}_C(s)}, \quad (1)$$

which are assumed to satisy all the design specifications, such as good transient response, steady-state performance, disturbance rejection property and robustness.

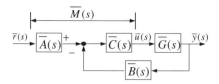

Figure 1. The continuous-time control system

The continuous-time (CT), Plant-Input Transfer-Function (PITF) is defined as the transfer-function from the reference input $\overline{r}(s)$ to the plant input $\overline{u}(s)$, and is given by:

$$\overline{M}(s) \triangleq \frac{\overline{u}(s)}{\overline{r}(s)} = \frac{\overline{n}_A(s)}{\overline{d}_A(s)} \frac{\overline{d}_B(s)\overline{n}_C(s)\overline{d}_G(s)}{\overline{n}_B(s)\overline{n}_C(s)\overline{n}_G(s) + \overline{d}_B(s)\overline{d}_C(s)\overline{d}_G(s)}, \quad (2)$$

which can also be written, according to the plant input pole and zero principles (Sain and Schrader, 1990), as:

$$\overline{M}(s) = \frac{\overline{n}_M(s)\overline{d}_G(s)}{\overline{d}_M(s)} \quad (3)$$

where $\overline{n}_M(s)$ represents the portion that does not correspond to the plant and $\overline{d}_M(s)$ is the characteristic polynomial of the closed-loop transfer function.

It should be noted that all that is required for the digital controller design is the CT PITF and the analog control system does not have to be designed using the PIM method. For instance, when the desired closed-loop transfer function can be determined directly from the design specifications, the CT PITF can be calculated using equations (1) and (3). The method of designing a PIM digital control system assuming that an analog control system is available is called the PIM digital redesign. The PIM digital design, on the other hand, refers to the case where the CT PITF is obtained directly from the design specification without designing an analog control system.

Let the Step-Invariant Model (SIM) of the plant (Franklin, Powell, and Workman, 1998) and the Matched-Pole-Zero (MPZ) model (Markazi and Hori, 1992) of the CT-PITF be given by

$$G(\varepsilon) = \frac{n_G(\varepsilon)}{d_G(\varepsilon)}, M(\varepsilon) = \frac{n_M(\varepsilon)d_G(\varepsilon)}{d_M(\varepsilon)}, \qquad (4)$$

where: ε is the discrete-time operator defined as $\varepsilon = (z-1)/T$, z is the usual z operator, and T the sampling interval.

The SIM is used to represent, as closely as possible, the A/D-plant-D/A combination while the MPZ model is used to retain the plant input zero principle. This discrete-time (DT) PITF is the target PITF to be realized by the DT system shown in Figure 2, where the controller blocks are chosen as:

$$A(\varepsilon) = \frac{n_A(\varepsilon)}{\lambda(\varepsilon)}, B(\varepsilon) = \frac{n_B(\varepsilon)}{\lambda(\varepsilon)}, C(\varepsilon) = \frac{\lambda(\varepsilon)}{d_C(\varepsilon)} \qquad (5)$$

with $\lambda(\varepsilon)$ being an arbitrary stable polynomial of appropriate degree.

The resulting DT-PITF is given by:

$$M(\varepsilon) = \frac{n_A(\varepsilon)d_G(\varepsilon)}{n_B(\varepsilon)n_G(\varepsilon) + d_C(\varepsilon)d_G(\varepsilon)} \qquad (6)$$

where: $n_G(\varepsilon)$ and $d_G(\varepsilon)$ are known from equation (4).

By comparing the denominator and numerators of the DT PITFs in equations (4) and (6), $n_A(\varepsilon), n_B(\varepsilon), d_C(\varepsilon)$ can be determined (Markazi and Hori, 1992).

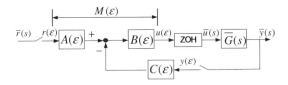

Figure 2. A discrete-time control system

3. PIM digital redesign of the driver

In the present study, the analog control system is the current regulator portion of the driver (Shimamura and Hori, 2002) and the plant for the PIM design purpose is that from the actuation voltage signal to the detected current (from the PWM circuit to the Stepping Motor). In this section, PIM digital redesign of the analog driver at the sampling interval of 400 microseconds is attempted. The analog control system under consideration is given (Sejimo and Hori, 2004) as

$$\bar{G}(s) = \frac{50900(s+280)}{(s+3300)(s+3600)}, \bar{A}(s) = 0.502, \bar{B}(s) = 1.83, \bar{C}(s) = \frac{10^4}{s}, \quad (7)$$

where: the plant transfer function $\bar{G}(s)$ is that of the equilibrium point at the hold mode.

The CT-PITF of the analog control system is given by

$$\bar{M}(s) = \frac{5020(s+3600)(s+3300)}{(s+277)(s^2+6623s+9.415\times10^8)}. \quad (8)$$

The MPZ model of this PITF at the sampling interval of 400 microseconds is denoted as $M_{400}(\varepsilon)$ and is calculated to be

$$M_{400}(\varepsilon) = \frac{61.5223(\varepsilon+1908)(\varepsilon+1832)}{(\varepsilon+262.2)(\varepsilon^2+3758\varepsilon+3.586\times10^6)}. \quad (9)$$

The digital control system that achieves this DT PITF $M_{400}(\varepsilon)$ is denoted as PIM_{400}. The design polynomial $\lambda(\varepsilon)$ is chosen in the same manner as in (Shimamura and Hori, 2003), whose pole λ_p at the sampling interval T of 250 microseconds is selected to be $\lambda_p = -1.2/T$.

Experiments have been carried out using the PIM_{400} controller. Shown in Figure 3 is the Drive Voltage (DV), which is the voltage that actually drives the motor and is closely related to the actuation signal, which is the output of the PITF, and is a more convenient signal to assess the controller performance. The motor is at the hold-mode initially, is commanded to rotate at 1000 pulse-per-seconds for 0.2 seconds, and goes back to the hold-mode. For comparison, the results are also shown for the PIM_{250} controller, which is the PIM digital control system running at the sampling interval of 250 microseconds. It can be seen from the figure that the performance of PIM_{250} is very close to that of the analog controller, while the performance of PIM_{400} is different, with slower response, a phase delay, and no overshoot. The change in the time response of PIM_{400} from the analog controller as evident in Figure 3 is looked at from the viewpoint of the frequency response characteristics of the PITF. Shown in Figure 4 are the frequency responses of the PITFs, $\bar{M}(s)$ and $M_{400}(\varepsilon)$. Shown also is the PITF $M_{250}(\varepsilon)$ at the sampling interval of 250 microseconds. It can be seen that the frequency response of $M_{250}(\varepsilon)$ is closer to $\bar{M}(s)$ than $M_{400}(\varepsilon)$; the gain of $M_{400}(\varepsilon)$ monotonically decreases with no peak and is noticeably lower above 1000 rad/sec.

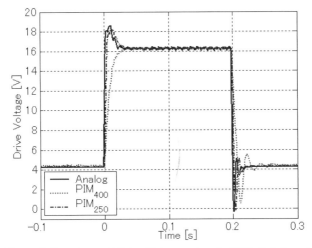

Figure 3. Experimental Drive Voltage using the PIM digital controller

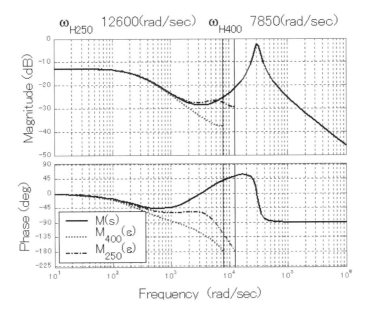

Figure 4. Bode plots of the PITFs

4. Reshaping frequency response of PITF

It was judged from Figure 3 that the performance of PIM_{250} controller is acceptable, whereas that of PIM_{400} is not. It was felt that, as long as the digital redesign approach is used, there is not much room left for improvement using the sampling interval of 400 microseconds. Therefore, $M_{400}(\varepsilon)$ will be modified in this section in an attempt to obtain a satisfactory digital control system. It should be noted that there is a change in the design approach here, from digital redesign to digital design. Controllers PIM_{250} and PIM_{400} designed in Section 4 are obtained from the CT-PITF $\overline{M}(s)$. In this section, this CT-PITF is modified to $\overline{M}_{\mathrm{mod}400}(s)$ such that the DT-PITF at 400 microseconds, denoted as $M_{\mathrm{mod}400}(\varepsilon)$, leads to a PIM controller $PIM_{\mathrm{mod}400}$.

Since the real pole $s=-277$ of the CT-PITF $\overline{M}(s)$ given by (8) is slow and the sampling interval of 400 microseconds is already small enough, only a modification to the complex conjugate poles is considered. Denoting the un-damped natural frequency and the damping ratio of the complex conjugate part, respectively, as ω_n [rad/sec] and ξ, define the modified CT-PITF at the sampling interval of 400μ sec. as

$$\overline{M}_{\mathrm{mod}400}(s) = \frac{328.8792(s+3600)(s+3300)}{(s+277)(s^2+2\xi\omega_n s+\omega_n^2)}. \tag{10}$$

The MPZ model sampled with $T=400[\mu\text{ sec}]$ is denoted as $M_{\mathrm{mod}400}(\varepsilon)$.

It was found from extensive simulations and experiments that the DV signal is closely related to the frequency response of the PITF such that the speed of DV response is fastest by choosing $\omega_n = \omega_{H400} = \pi/T$ and that the amount of overshoot in DV can be made adjusted by suitably selecting the damping ratio ξ. To see the gain change as the damping ratio changes, let the gain of $\overline{M}_{\mathrm{mod}400}(s)$ at ω_{H400} be $g_{CT}[dB]$ and that of $M_{\mathrm{mod}400}(\varepsilon)$ be $g_{DT}[dB]$. They are given by

$$g_{CT} = 20\log_{10}\left|\overline{M}_{\mathrm{mod}400}(j\omega_{H400})\right|, g_{DT} = 20\log_{10}\left|M_{\mathrm{mod}400}\left(\frac{e^{j\omega T_S}-1}{T_S}\right)\right|. \tag{11}$$

Shown in Figure 5 are the peak gains $g_{CT}[dB]$ and $g_{DT}[dB]$ for the range of damping ratio ξ from 0.05 to 0.5. It can be seen from the figure that for a given peak gain value, the DT damping ratio is much larger than the CT damping ratio, meaning that when the peak gains are the same, the DT response will have smaller overshoot or no overshoot and slower response, than the CT response. Thus, in general, the process of PIM discretization implies a reduced overshoot and this becomes more prominent as the sampling interval increases. Looking that this discretization effect from a different angle, it canbe said that having the peak gain value of DT-PITF $M_{\mathrm{mod}400}(\varepsilon)$ with $\xi=0.33$ is equivalent to modifying the CT-PITF $\overline{M}_{\mathrm{mod}400}(s)$ to have the damping ratio of $\xi=0.17$.

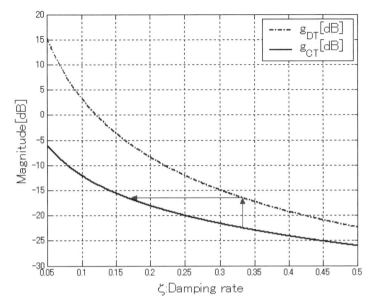

Figure 5. Peak gain values versus damping ratio

Simulations using a second-order plant model with known physical properties, such as a dead-zone in the PWM circuit and anti-aliasing filters, have shown that decreasing the damping ratio further causes excessive overshoots. However, the overshoot did not become too excessive in the experiments, due to known causes. Therefore, the value of the damping ratio was determined by trials-and-errors in experiments and was chosen to be 0.33. Increasing the damping ratio did not change the transient DV response much, but the hold-mode DV signal became oscillatory. The MPZ model sampled at 400 microseconds, $M_{\mathrm{mod}400}(\varepsilon)$, is given by

$$M_{\mathrm{mod}400}(\varepsilon) = \frac{195.5583(\varepsilon+1908)(\varepsilon+1832)}{(\varepsilon+262.2)(\varepsilon^2+6746\varepsilon+1.14\times10^7)}. \tag{12}$$

Figure 6 shows the frequency response of $\overline{M}_{\mathrm{mod}400}(s)$, $M_{\mathrm{mod}400}(\varepsilon)$, and $M_{400}(\varepsilon)$. It can be seen that the gain of $M_{\mathrm{mod}400}(\varepsilon)$ is increased around $\omega_{H\,400}$. Let the PIM controller that achieves the desired DT-PITF $M_{\mathrm{mod}400}(\varepsilon)$ be denoted by $PIM_{\mathrm{mod}400}$. Figure 7 shows the experimental results of the DV signal for analog and $PIM_{\mathrm{mod}400}$ controllers under the same conditions and design parameter λ_p, as in Figure 3. It can be observed from Figure 7 that, except for a slightly larger overshoot, the performance of the $PIM_{\mathrm{mod}400}$ controller is almost the same as that of the original analog controller, and perhaps even better than the PIM controller at 250 microsecond sampling interval PIM_{250} shown in Figure 3.

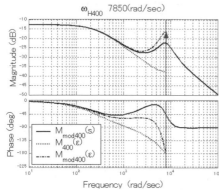
Figure 6. Frequency response of PITFs

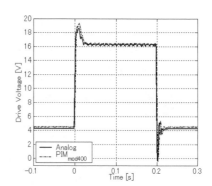
Figure 7. DV signal under analog and digital controls

5. Conclusions

The issue of the performance degradation of control systems due to discretization was looked at from the viewpoint of the frequency response of the transfer-function from the reference input to the plant input. Rather than reducing the sampling interval, the frequency response characteristics of the CT-PITF can be modified to achieve the desired performances at a given sampling interval. Based on this approach, a digital driver that performs well with the sampling interval of 400 microseconds was successfully designed for a stepping motor. Considering the fact that the popular Tustin method cannot give a satisfactory performance even with the 10 microsecond interval, the use of such a large sampling interval is beneficial since it enables the commercial production using the presently available inexpensive microprocessors, rather than wait for future, significantly faster, and yet less costly, microprocessor. However, it is desired to establish general guidelines for modifying the CT-PITF.

References

1. Franklin G. F., Powell J. D., Workman M.: *Digital Control of Dynamic Systems*. Addison Wesley, 1998.
2. Hori N., Markazi A. H. D.: *A new approach to the design of digital model reference control systems*. Trans. JSME Series-C 1995; 38: 712-718.
3. Kawasumi T., Hori N., Fukuzawa M.: *Selecting pole-locations under constraints on plant-inputs*. IEEE Int. Conf. on Methods and Models in Automation and Robotics, 2003; 609-614.
4. Markazi A. H. D., Hori, N.: *A new method with guaranteed stability for discretization of continuous time control systems*. Proc. American Control Conference, 1992; 2: 1397-1402.
5. Shimamura, Hori N.: *Digital Redesign of a Stepping Motor Driver*. Proc. SICE Annual Conf., 2002; 1333-1338.
6. Sain M. K., Schrader C. B.: *The role of zeros in the performance of multiinput, multioutput feedback systems*. IEEE Trans. Education 1990; 33: 244-257.
7. Sejimo I., Hori N.: *Models for PIM digital redesign of a stepping motor current regulator*. Proc. SICE Annual Conf., 2004; 355-360.

MONITORING OF LASER ALIGNMENT BASED ON INDUCED ULTRASONIC WAVES IN LASER-ASSISTED JET ELECTROCHEMICAL MACHINING

Dąbrowski J., Pająk T.P., Harrison D.K., Szpytko J.

Abstract: A relatively small power (375mW) Nd:YAG laser has been used to enhance electrochemical dissolution in Laser Assisted Jet Electrochemical Machining. This paper investigates into a possible monitoring system of a LAJECM process which emanates from the laser's ability to deposit large quantities of energy in the thin surface layer of the workpiece, resulting in the acoustic emission (AE).

1. Introduction

A number of different physical processes may take place when a solid surface is illuminated by a laser. The absorption of the incident laser beam over a short time duration (typically of the order of 10ns) is followed by a local thermal expansion as well as by melting and vaporization of the material.

The generation of sound with a laser is based upon heating of the ground (Scruby and Drain, 1990). However, it must be noted that pulsed lasers, with satisfactorily short pulses can locally but extensively raise the temperature. Otherwise, the phenomenon of thermal diffusion takes over and it limits the maximum temperature which is essential to cause expansion of the ground resulting in an acoustic pulse. Low power laser waves arise due to thermoelastic stresses, while for high power lasers acoustic emission is caused by the material ablation. The amplitude of an ultrasonic wave depends on the laser power.

These ultrasounds can be measured either with piezoelectric sensors or with laser interferometers. The latter method is much more accurate and has a higher spatial resolution as well as a wider bandwidth. However, although the cost of piezoelectric transducers makes them more popular for research purposes they are very brittle and can be easily damaged, especially in an industrial environment.

Measurements of laser induced waves proved to be an efficient method of monitoring various processes under the non-destructive regime. Much research has been undertaken in this field, although an attempt to monitor acoustic waves in the laser assisted jet electrochemical machining has not been reported so far.

The aim of this project was to measure acoustic waves induced by a laser in a process of laser assisted jet electrochemical machining in order to verify proper laser alignment in the jet cell. The experiments described below were carried out where the electrolyte was not present, i.e. there was only the laser's impact on the specimen.

2. LAJECM

Laser assisted electrochemical jet machining has been developed in order to improve both the productivity and accuracy of the process of electrochemical or jet electrochemical machining.

McGeough (1974) gives a detailed study of ECM, which is regarded as the major alternative to conventional methods for machining difficult-to-cut materials of and/or generating complex contours, without inducing residual stress and tool wear. It is applied in a variety of fields, for example: aviation (cooling holes in jet turbine blades), space, automobile, electronics and computers (printed circuit boards), medical (surgical implants) and many others (Sen&Shan, 2005, De Silva et al, 2004). Kozak et al (2004) emphasize the importance of miniaturization and the challenges it brings. In order to meet the requirements of miniaturisation some improvements must be done.

As ECM is a process of dissolving an anode when an electrolyte flows between the anode and a cathode tool, it strongly depends on the distribution of electric field. In order to ensure a precise dissolving of the material, the electric field must be concentrated exactly in the desired machining area. De Silva et al (2004) mention tested methods aimed at improving ECM accuracy by usage of: smaller inter-electrode gaps, pulsed power, low-concentration electrolytes etc.

In LAJECM (Figure 1) besides the electrolyte jet, emerging from the nozzle, there is a laser beam, focused by a lens before entering the jet cell. In experiments carried out the nozzle and laser beam diameters were 0.3mm and 0.11mm respectively.

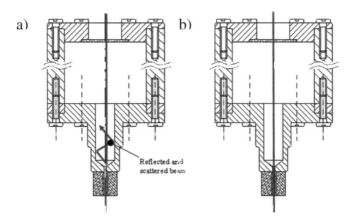

Figure 1. A diagram of LAJECM (jet cell): a) bad alignment i.e. the laser beam is reflected and scattered, b) desired alignment – the laser beam passes freely

It is essential to properly align the laser beam with a nozzle in order to assure that the process can use the laser's entire available power i.e. 375mW. The laser, because of its relatively low power, does not remove the material from the workpiece but it thermally activates the surface layer, which accelerates the electrochemical dissolution. De Silva et al (2004) thoroughly investigate the laser impact as well as giving a theoretical explanation of the whole process of LAJECM. They point out that it is relatively difficult to obtain a proper alignment of the laser beam with the electrolyte jet and that during processing this alignment can be violated by bubbles, electrolyte disturbances or just defocusing of the beam on the workpiece surface as the material is being removed and a hole is drilled.

Therefore a system to monitor (and ideally correct) the alignment could guarantee the fulfilment of the abovementioned needs. It has been explained that a laser can induce ultrasonic waves in a specimen. The amplitude of ultrasonic wave depends on the laser power. For that reason measurements of these waves as a possible solution will be investigated.

3. Methodology

The measurements carried out are based on acoustic emission, which takes place when a laser beam is incident on a material surface. In order to measure the induced ultrasonic waves in materials, a piezoelectric sensor was used. Figure 2 shows a schematic drawing of the setup during measurements of laser induced ultrasonic waves (the setup with a nozzle was already presented in Figure 1). The synchronous output of the laser controller was used to trigger the signal acquisition in the oscilloscope - LeCroy LT342. The signal from a piezoelectric transducer was amplified 10 times via high impedance probes connected to the scope. The ultrasonic waves were displayed, digitized and stored in a binary format supported by the scope. Then files were transferred to a PC and converted to ASCII files for further processing. The typical sampling rate was 500MS/s (mega samples per second).

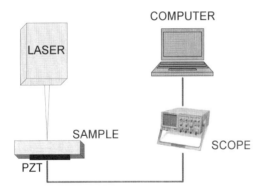

Figure 2. A schematic diagram of the experimental setup

The transducers used were disk-shaped, with the diameter of 7mm and 0.2mm thick. First, wires were attached to the transducer electrodes, and then the PZTs have been stuck, with epoxy adhesive[1], to different types of metals (Aluminium, Hastelloy and Titanium).

Before the laser beam was incident on the surface it was focused with a lens, thus the laser spot was approximately 0.11mm in diameter. Experiments were carried out first without

---

[1] Bison Combi-Super

and then with the nozzle, in order to verify effects of the latter. Different sensor locations relative to the laser beam were investigated as well. Matlab Digital Signal Processing Toolbox was used for analyzing and filtering signals.

4. Results and discussion

For the purposes of this project only signals corresponding to a single laser shot (pulse) are presented and compared. Figure 3 shows a typical long timescale signal obtained from a PZT. Its duration is approximately 200 ms, however, such a signal does not allow a detailed analysis because of the signal-to-noise ratio. Although it was possible to filter and analyse signals in the whole time scale, the initial signals proved to be more indicative of processes described above.

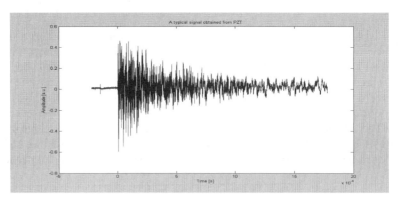

Figure 3. A typical long timescale signal obtained from a PZT, blue line corresponds to an original signal, the red one to a filtered one (averaged)

In the first part of measurements, the laser was operated in a low energy mode i.e. 5mJ and signals were recorded. It has been confirmed that the sensor response was about 60ns after the data acquisition was triggered. It is in a very good agreement with the laser specification, which informs that the actual laser firing takes place 60ns after the trigger was released. The waveforms presented in Figure 4 are not the ultrasonic waves and no ultrasonic wave is induced in such a low energy mode. This leads to a conclusion that even when the laser is operated in full power, but it is very badly aligned ultrasounds are not induced. These waves correspond to the laser electromagnetic radiation. This is due to the change in momentum of the photons as they are reflected from or absorbed by the surface. This radiation pressure is much smaller than this from the thermoelastic effects and actually does not depend on laser energy. Small differences in amplitudes are based on it's intrinsicmaterial properties. It has also been checked that these waves are not the subject of the laser beam-sensor relative position. Even when the beam hits the material in an away (few cm) location from the sensor, the signal is not changed. It must be indicated that in a narrow timescale signals were not filtered, graphs are based on 'raw' data obtained from sensors.

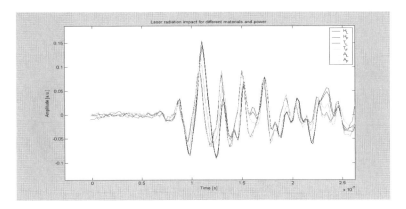

Figure 4. Electromagnetic radiation waves due to change in momentum of photons as they are reflected or absorbed by material. Letters H, T, A correspond to Hastelloy, Titanium and Aluminium respectively, superscripts L and F, to low and full laser power

Having obtained signals in a low energy mode, the full pulse laser energy was used – 25mJ. Not only characteristic signals were recorded for different materials: Aluminium, Hastelloy and Titanium, but also the effect of sensor positioning was examined. Figure 5 shows different waves induced in Hastelloy for different sensor positions: epicentre, off-epicentre and when the laser beam was not focused above the sensor (off-sensor position). The positioning of the sensor against the laser beam is shown in Figure 6.

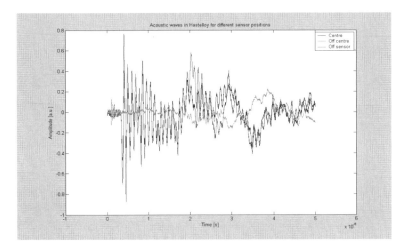

Figure 5. Laser induced ultrasonic waves in Hastelloy for different sensor positions

The first part of the signal is due to the abovementioned radiation pressure. Frequencies of signals in off-epicentric (red signal) and epicentric (blue signal) positions are equal to the sensor's resonating frequency i.e. 10.7MHz which explains why the amplitudes are much bigger than the amplitude for off-sensor position (green signal). Similar results were found for Titanium and Aluminium.

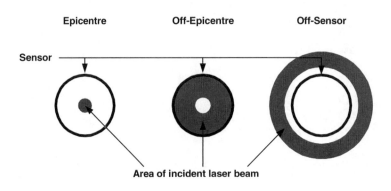

Figure 6. Piezoelectric sensor positioning against the striking laser beam

The next graph (Figure 7) demonstrates differences in the amplitudes for analysed materials. Amplitudes in Aluminium are bigger than in Hastelloy and Titanium and they arrive faster because the speed sound is higher.

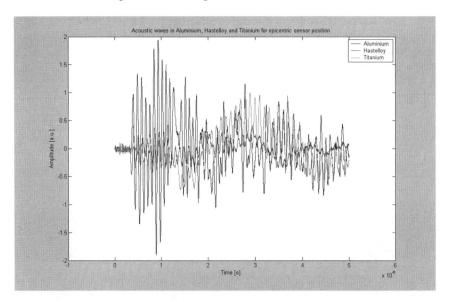

Figure 7. A comparison of waves' amplitudes in Aluminium, Hastelloy and Titanium for epicentre sensor position

The most important and complex part of the experiments is now presented. That time the nozzle was used. The laser beam passed through a hole of 0.3mm in diameter. As explained earlier it is essential to ascertain that the entire laser power is available in LAJECM i.e. laser beam (0.11mm in diameter) can freely pass through the nozzle. Therefore, the relative laser beam position to the nozzle and for different setups will be investigated and the amplitudes of ultrasonic waves will be measured. When the laser beam alignment was correct and the beam was passing the nozzle unobstructed it resulted in high power density on the surface, which in turn causes bigger amplitudes of these waves. In this type of setup the piezoelectric sensor was placed exactly (within experimental error) below the incident beam. The time axis was narrowed to display only the first wave fully; subsequent waves are the result of signal reverberation through the sample and are not taken into account. The thermal diffusion seems to be a major cause of the initial positive spike of the epicentral waveform, which is visible just before the first big negative peak. Figures 8 and 9 explicitly illustrate that the amplitude depends on the laser alignment - the better the alignment the bigger the amplitude. Amplitudes differ for various materials due to their physical properties, however, the main factor that drives the amplitude for a given material is the laser power, which strongly depends on the alignment factor.

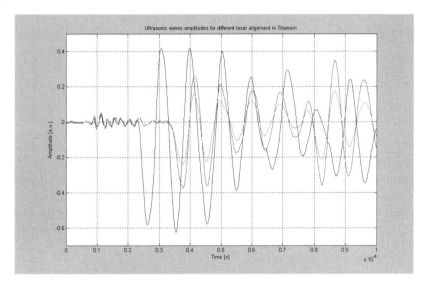

Figure 8. Changes in amplitude for different laser alignment in Titanium (the better the alignment the bigger the amplitude)

The maximum voltage level obtained from the PZT for these measurements was about 40mV, while the duration of the first wave was approximately 0.6µs. The values gained for other materials oscillated around these numbers.
It must be taken into account that optoacoustic signals are proportional to the surface reflectivity/absorbtivity.

5. Conclusion and future works

The experiments carried out proved that a monitoring of laser alignment based on ultrasound waves in a process of LAJECM is feasible and that it is a relatively cheap and easy method. The results obtained are highly consistent and clearly differentiate good and bad alignment. As a part of non-destructive testing the material's quality measurements of sound speed were carried out too.

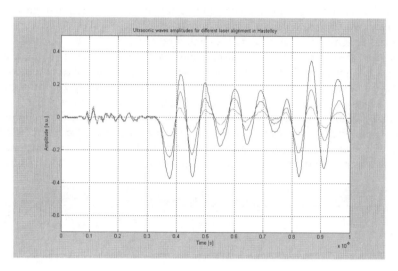

Figure 9. Changes in amplitude for different laser alignment in Hastelloy (the better the alignment the bigger the amplitude)

Future works could include a replacement of the oscilloscope by a personal computer equipped with an analogue data acquisition card. Ideally this card could also drive motors used to position the nozzle relatively to the laser beam which is stationary, using the feedback signal from the piezoelectric sensor. Other possible works could consist of creating a finite elements model, which would regard the three-dimensional nature of the waves' propagation. Moreover using a known force or a laser interferometer the actual surface displacements could be evaluated.

References

1. De Silva A.K.M, Pajak P.T., Harrison D.K., McGeough J.A. (2004): *Modelling and Experimental Investigation of Laser Assisted Jet Electrochemical Machining.* Annals of the CIRP, 53, pp. 179-182.
2. Kozak J., Kamlakar P., Rajurkar, Makkarb Y. (2004): *Selected problems of micro-electrochemical machining.* Journal of Materials Processing Technology 149, pp. 426–431.
3. McGeough J.A. (1974): *Principles of Electrochemical Machining.* Chapman and Hall Ltd., London.
4. Scruby C.B., Drain L.E. (1990): *Laser Ultrasonics: Techniques and Application.* Adam Hilder, New York.
5. Sen M., Shan H.S. (2005): *A review of electrochemical macro- to micro-hole drilling processes.* International Journal of Machine Tools&Manufacture, 45, pp. 137–152.

THE INTEGRATION OF VISION BASED MEASUREMENT SYSTEM AND MODAL ANALYSIS FOR DETECTION AND LOCALIZATION OF DAMAGE

Kohut P., Kurowski P.

Abstract: The document covers the application of vision techniques applying modal analysis of damage detection. The main concept of the solution suggested involves vibration registering by means of visual system; executing modal experiment and applying the energy method. The key feature of this new solution is that global mode shapes are obtained by partial mode shapes gathered together. Developed algorithms (for image processing and modal experiment executing) were implemented and tested in MATLAB programming environment, in which the modal experiment was carried out by means of – VIOMA –software developed in KRiDM AGH.

1. Introduction

The contemporary diagnostics lays great emphasis on detection and localization of damage at the earliest possible stage. During such processes modal model parameters change very often. Besides the stage of modal model creation is very expensive and time-consuming. Alternative techniques are investigated for non-contact vibration measurements. There are two categories of systems for vision techniques used for 3-D object geometry measurements: active and passive. In depth measurements active ones make use of additional devices (e.g. lasers, LCD projectors), which generate structured light [1, 12]. The latter systems make a depth measurements based on provided image sequences from one or more cameras. Non-contact signal-registration techniques have complied with a new tendency of construction designing and give the possibility to fulfill the contemporary modal analysis requirements. The basics of modal analysis requirements can be defined as: test accuracy increase, broadening of frequency bandwidth, measuring points number increase, reducing testing preparation-time and results analysis time, facilitating analyses and tests. Another very important aspect is a damage detection process. It is generally known from experience that simple methods based on frequency or modal-damping-factor changes in an analyzed construction prove to be ineffective. This document presents the method involving the analysis of deflection-energy changes in mode-shapes in relation with propagated damage. The crucial part in this method plays the knowledge about mode shapes of the research object. Earlier researches by the authors indicate that the determination of smooth and well mapped global mode shapes by means of vision techniques is not a trivial task.

2. Methodolody

The connection of visual technology, used as a measuring device, with a modal tool of modal analysis gives the possibility to estimate the modal parameters (modal shapes, damping, frequencies).

The great challenge for vision-system designers is to create systems, processing three-dimensional scene, based solely on obtained images and external illumination. This category of systems, which does not require a structural lighting, is known as passive technique. The following issues were developed as parts of the technique: methodology, algorithms and procedures for automatic geometry mapping and measurement-points location (key of classical image-processing methods); and vibrations measurements of specified construction points. Thus the coordinates of marker centroids at the selected construction points could be computed. By the application of developed calibration module relationship was settled between the calculated centroids coordinates of the analyzed objects, shown in the form of pixels (on the image plane), and real coordinates defined in millimeters. For this purpose a calibration pattern, in the form of circle with known diameter, was attached to the analyzed construction. Image Processing Toolbox [3] embedded in MATLAB programming environment was used to implement image pre-processing and analysis procedures. The developed image analysis algorithm was based on the region-oriented segmentation technique [4, 5, 6]. The scheme of procedures carried out is shown in figure 1. The developed methodology and algorithms, in the form of software cooperating with computer-aided modal experimental tool – VIOMA – [8] embedded in Matlab environment, allowed to obtain the desired modal parameters (shape modes, damping, frequency) and reduced the process time.

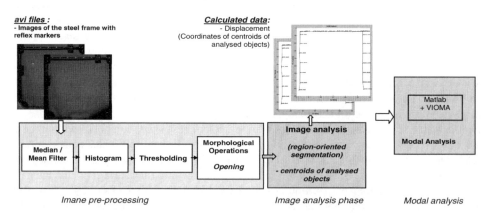

Figure 1. Methodology of the automatic geometry mapping and modal test realization

The method of damage detection and localization by means of the energetic coefficient was applied on the basis of the obtained results (mode shapes). Proposed method involves comparing modal shapes of a model - damage free - object and modal shapes of the construction which sustained damage. This method is based on the deflection energy changes and the modal shapes knowledge as a result of damage. The crucial issue in this method is modal-shapes examination - particularly IInd derivative of the modal shapes with respect to coordinates. It means that the deformation energy is changing depending on the specific modal shape and the specific construction damage.

The damage coefficient applied in an experimental method can be defined as:

$$\beta_{ik} = \frac{(\overline{U}_{ik}+\overline{U}_k)U_k}{(U_{ik}+U_k)\overline{U}_k} \qquad (1)$$

where:

$\overline{U}_{ik}$ the deformation energy at ith point for the kth mode shapes of damaged construction,
$\overline{U}_k$ the total deformation energy for kth mode shapes of damaged construction,
U_{ik} the deformation energy at ith point for the kth mode shapes of damage-free construction,
U_k the total deformation energy for kth mode shapes of the damage-free construction.

Deformation energies of the given modal shape can be obtained as follows:

$$\overline{U}_{ik} = \int_a^b [\overline{\psi}_i(x)]^2 dx \;;\; \overline{U}_k = \int_0^L [\overline{\psi}(x)]^2 dx \;;\; U_{ik} = \int_a^b [\psi_i(x)]^2 dx \;;\; U_k = \int_0^L [\psi(x)]^2 dx \qquad (2)$$

where:

ψ modal vector for the damage-free construction,
$\overline{\psi}$ modal vector for the damaged construction,
a, b   area borders in which damage-development possibilities are examined,
L   beam length.

Implementation of the formulas above was performed in Matlab environment as a part of VIOMA software.

3. Experiments

The experimental setup for the modal test by the use of vision system was composed of:
1. The vision system:
- illumination (halogen -500 W),
- the X-Stream digital camera. Image sequences acquisition were performed by means of the X-Stream digital camera and saved in the form of 'avi' files. The camera enables image acquisition with the rate of above 30.000 frames per sec. [www.idtpiv.com],
- software tools developed in Matlab environment.
2. The laboratory object:
- a steel frame with attached reflex markers serving as objects for the image analysis (figures. 2 and 3). The centroids of the reflex markers were calculated and expressed in millimeters after the calibration process,
3. The shaker - random excitation.

The analyzed objects were represented by the images of reflex markers obtained from the digital camera. The images acquisition was executed with 400 rates by 1260x116 image resolution. The 'avi' files generated by the digital camera were automatically saved in a computer's memory. These files were input-data of developed and implemented algorithms and procedures for image pre-processing and analysis in Matlab software (figure 2). The centroids' coordinates of analyzed objects were calculated for each image frame on image plane. Transformation of obtained data onto the image plane (from pixels into millimeters) was carried out by means of developed calibration module.

By the use of the developed software analyzed construction geometry was obtained and modal analyses were carried out. Results achieved by the vision system were processed by the software tool VIOMA [8]. Subsequently the input data were subject to the following processes:
- measurements of absolute displacements normalization,
- computation of correlation functions with respect to selected reference points,
- correlation functions averaging for all measurement sessions (there were 48 measurement sessions).

The key feature of the described vision-based measurement is multiple partial-experiments data availability and lack of common global references points. Such a situation makes it impossible to obtain global results of estimation. The objective of this particular experiment was to receive smooth and well-mapped mode shapes. They formed the bases for the damage-coefficient calculation related to fault. Prior to the assessments the following assumptions were made:
1. Test sessions associated with individual damage causes were divided into groups including at least one common point.
2. The common points were adapted as reference points in specified analysis.
3. Modal parameters were estimated for each group.
4. Global mode shapes were glued together by means of partial shapes. The bases of the gluing process were common points situated on boundaries of partial shapes.
5. Resulting global mode shapes were used to detect and localize the fault.

To fulfill the above postulations series of tests were performed in order to detect and localize damage appeared in the construction. The first measurement was carried out on the damage-free construction. The three consecutive series of measurements were executed on damaged construction. The damage was represented by three notches on the horizontal beam of steel frame (see Figure 3). Depth of the notches was respectively: 5mm (12,5%), 13mm (32,5%), 20mm (50%), the thickness was 1mm. The construction was excited by random signal. Then the vision system was applied for measurements. The main concept of the test involved carrying out a few measurements for the selected construction areas. Next selected areas (fields of view) were merged during the analysis. The analyzed construction was divided into four areas. Each measurement had at least one common point with the next one. Registered images were processed according to the procedure presented in figure 1. Displacement amplitudes of the investigated construction points as a function of time were calculated based on the image analysis process. The obtained results were used to compute correlation functions. Then the correlation functions were utilized to modal analysis process. Since an input force was not measured, the operation modal analysis was executed. The BR (Balance Realization) algorithm was used. The test was carried out in full measurement bandwidth (0-200Hz). Modal models for stabilization-diagram requirements were estimated from 2 to 50.

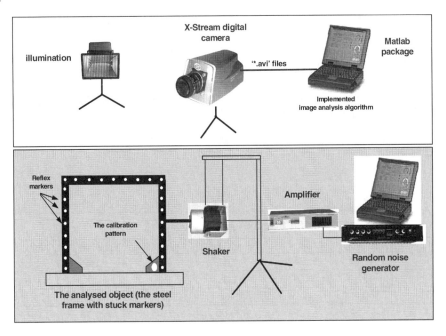

Figure 2. The vision based measurement experimental setup

Measurement 1	Measurement 2	Measurement 3	Measurement 4
○ ○ ○ ○ ○ ○ ○ ○	○ ○ ○ ○ ○ ○ ○ ○	○ ○ ○ ○ ○ ○ ○ ○	○ ○ ○ ○ ○ ○ ○ ○
1 2 3 4 5 6 7 8	9 10 11 12 13 14 15 16 17 18	19 20 21 22 23 24 25 26 27	28 29 30 31 32 33

Figure 3. The scheme of the construction division corresponding to the estimation process

In standard measurement methods applying classical vibration sensors one of the sensors has been assigned the role of the reference. Its position would not change in time of the test session. In cases when vision measurement test was carried out the adaptation of one measurement point as a reference point hasn't been satisfying. This situation results in that mode shapes are obtained by a few partial analyses. It is necessary for modal shapes to "glue together" within the whole analyzed object on the basis of the several partial modal shapes. The crucial fact is that the modal shapes can differ from one another in respect of both amplitude and phase.

The proposed solution was based on calculated correlation functions (obtained through amplitude of displacements). There were four selected areas (figure 3) subject to analysis and measurement processes: area 1): points 1-10; area 2): points 9-18; area 3): points 17-26; area 4): points 25-33 .Two independent sets of correlation functions were evaluated. One was drawn from measurements: No 1 and 2. The point No 9 was adapted as the reference point. The other set was calculated from measurements: No 3 and 4 with the reference point No. 25. Modal analyses were applied to both sets of the correlation functions. Modal shapes obtained from two modal analyses contained two common points (No. 18 and 17). That point was used to merge modal shapes with use of partial modal shapes

Collected data was converted into format compatible to VIOMA software by means of then vision system. The Balanced Realization algorithm [7] was applied to estimation process. There were two analyses performed for the damage-free case and the damaged construction case. The partial modes were stuck together in order to obtain the global mode shapes. The natural frequencies and damping factors related to the first part of the mode shape were taken as the pole parameters. The example mode shapes and selected gluing results are presented in table1 and table 2.

Table 1. The mode shapes

No.	Damage-free construction		Damaged construction 12.5%		Damaged construction 50%	
	Freq [Hz]	Damp [%]	Freq [Hz]	Damp [%]	Freq [Hz]	Damp [%]
1	11,01	2,42	11,04	2,23	10,99	1,86
2	43,96	0,67	44,02	0,70	44,06	0,69
3	60,09	1,94	59,86	2,18	59,76	2,24
4	122,48	0,54	123,00	0,45	122,88	0,43
5	161,50	0,51	162,02	0,47	160,90	0,35

Table 2. The selected gluing results

ModeNo	Damage-free constr.	Damaged constr. 12.5%	Damaged constr. 50%
2			
3			
4			

The obtained global mode shapes were used to evaluate the damage coefficient (1) and thus it made way for fault detection. Resulting data are presented on figure 4.

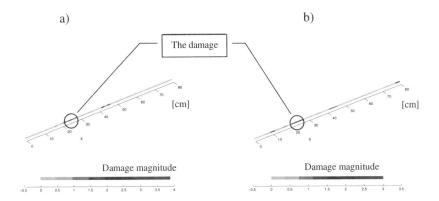

Figure 4. The results of the damage coefficient evaluation for: a) construction damaged in 12.5%; b) construction damaged in 50%.

The place where the construction was cut features the highest value of the magnitude of damage coefficient. As it can be noticed (figure4) the highest magnitude of the damage coefficient corresponds to the place 25cm away from the origin of the x-axis. It indicates the location of the notch.

4. Summary

Detection and localization of the expanded fault in constructions can be done through the application of the vision system to vibration measurements. However, in the experiments occurred a problem of low image space resolution. During preliminary measurements, images received from the vision system failed to yield satisfactory results after the analyzing process, since the image space resolution was too low. As a result the mapping of vibration amplitude was not sufficiently dense. The drawback was removed by the increase of image space resolution, at the cost of narrowing the field of view. As a result the area of the analyzed construction had to be divided into four parts (the number of experiments increased), as well as it was necessary to glue together mode shapes. On the other hand the increase of image space resolution proved to be favorable since it allowed the mode shapes smoothing.

The fact that this method proves to be useful with large-dimensions objects is the additional great advantage. The new algorithm for the mode shapes sticking was developed. The application of the energetic method based on the global mode shape knowledge enabled to detect and localize damages. Research described in this document proves that it is possible to detect and localize faults that feature more than 12% of the whole damage extent.

Acknowledgements

This work was supported by a grant from KBN State Committee for Scientific Research (grant No: 4 T07B 057 26).

References

1. Freymann R., Honsberg W., Winter F., Steinbichler H.: *Holographic modal analysis*. Laser in Research and Engineering, Springer Verlag Berlin, , 1996, 530-542.
2. Heylen W., Lammens S., Sas P.: *Modal Analysis Theory and Testing*. Department of Mechanical Engineering, Katholieke Universiteit Leuven, Leuven, Belgium, 1995.
3. Image Processing Toolbox for use with MATLAB, The MathWorks Inc., 2002.
4. Jahne B.: *Digital image processing: concepts, algorithms, and scientific application*. Springer-Verlag, Berlin, 1995.
5. Klette R., Zamperoni P.: *Handbook of image processing operators*. JohnWiley&Sons Ltd., New York, 1996.
6. Kohut P., Uhl T.: *The rapid prototyping of the visual servoing on Matlab/Simulink/dSPACE environment*. Proc. of the 7th IEEE International Conference on Methods and Models in Automation and Robotics, Międzyzdroje, 28-31 August 2001, 672-677.
7. Kurowski P.: *Identification of mechanical constructions modal models on the basis of exploitational measurements*. (in Polish), PhD Thesis, AGH, Krakow 2001.
8. Kurowski P., Uhl T.: *VIOMA 2.0 – users guide*. Dept of Robotics and Machine Dynamics AGH, Krakow, 2003.
9. Shapiro L., Stockman G.: Textbook: *Computer vision*. Prentice-Hall, 2001.
10. Tadeusiewicz R.: *Vision systems of industrial robots* (In Polish). WNT, Warszwa, 1992
11. Uhl T.: *Computer aided identification of mechanical constructions models* (in Polish). WNT, Warszawa, 1997.
12. Van Der Auweraer H., Steinbichler H., Vanlanduit S., Haberstok C., Freymann R., Storer D., Linet V.: *Application of Stroboscopic and Pulsed-Laser ESPI to Modal Analysis Problems*. Measurement Science and Technology, Vol. 13 (4), Apr. 2002, 451-463.

TESTING MOTIONAL ACCURACY OF A MANUFACTURING MACHINE - A TASK IMPOSED ON MODERN MAINTENANCE

Blacharski W.

Abstract: This paper concerns maintenance of manufacturing machines and its tasks connected with drives. Some possible software solution dedicated for maintenance and their location within subsystems of CIM was briefly overviewed and discussed. Importance of maintenance of the drives was underlined. Different ways of data collecting in the researches of motional accuracy of a manufacturing machine were described and some examples connected with testing motional accuracy of a servo-drive were included.

1. Introduction

Keeping manufacturing machines in their good repair is one of the preconditions to satisfy if a computer aided manufacturing system has to operate efficiently. This is an obvious fact, however hardly ever maintenance or wider management of the stock of machines tends to be presented as a separate area in graphical illustrations of the CIM structures. Nevertheless there are some exceptions, for instant in the thematic publications by Siemens, where maintenance is presented as a separate subsystem within the framework of CAM [5]. On the other hand great variety of the graphical models of CIM, that are so often published in many works, give only a simplified picture. More genuine picture can be obtained by analysis capabilities of the software tools that are offered on the market and adopted in manufacturing plants. This leads to a conclusion that there is a variety of ways to apply computer aiding for maintenance.

The related to maintenance elements can be implemented within subsystems that serve other tasks, such as processes and machines control, visualization, measurements, data acquisition, data base, etc. In this case HMI/SCADA and other cooperating software tools for industrial automation can be adopted. However, it can involve a necessity to implement a new application for solving a maintenance problem in near every case. Another way is applying a great variety of software tools that are dedicated especially for solving particular maintenance tasks and are offered by manufacturers of different machines or subassemblies. Leading manufacturers tend to introduce tools for solving very detailed maintenance tasks remotely by a network. An example can be SIMATIC PDM [4] – a program dedicated to remote maintenance of a very wide group of devices manufactured not only by Siemens. The next way can be using universal software dedicated especially for the maintenance. This kind of software usually operates in a network and can aid management of the stock of machines on the level of the whole enterprise. The typical tasks in this case can be connected with recording and planning details of maintenance processes, workflow, inventory, downtime, costs etc. Some programs of the group, for example popular MP2 [3], are dedicated for aiding so named assets lifecycle management. In this case computer assisting concerns whole history of the assets and helps to rationalize different technical, logistic and economical decisions between others connected with maintenance and repairs.

2. Maintenance within a framework of CIM

Computer aided maintenance in every case can be composed of different tasks. Possible location for some of the tasks within a framework of CIM has been illustrated in the Figure 1, which was prepared accordingly to a very popularly presented graphical model. It can be concluded that in such presentation every particular task of maintenance can be allocated for some different subsystems. The Figure 2 illustrates scope of the related to maintenance groups of programs within a framework of CAM. Also in this case the maintenance-oriented elements can be seen as allocated for some subsystems. On the other hand it should be noticed, that in fact the allocation depends on the software tools that an enterprise has adopted and can be not comparable with any simplified graphical model. In many cases, especially when well developed programs dedicated for maintenance are in use, the management of the stock of machines tend to be seen as a separate subsystem of computer aided manufacturing.

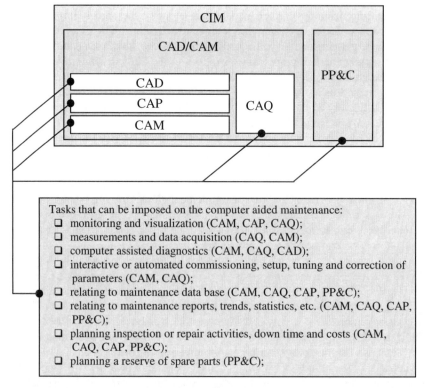

Figure 1. Location for the computer aided maintenance within a framework of CIM (according to a popularly presented model of CIM)

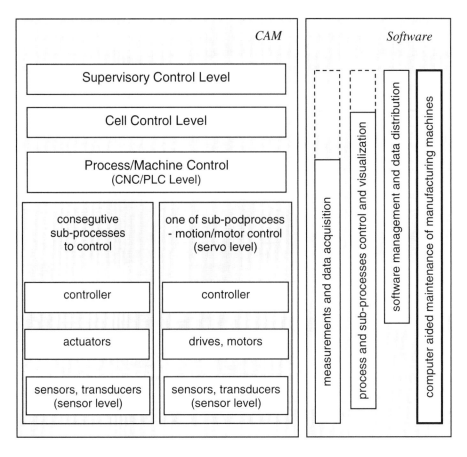

Figure 2. Scope of the related to maintenance software components within a framework of CAM

3. Testing motional accuracy of a machine

Tasks connected with maintenance of controls and drives built in machines play particular role, because of their importance for efficiency and quality of manufacturing processes. Accuracy of a machine depends on its drives capabilities what in turn impose on maintenance some additional tasks connected with monitoring, troubleshooting and diagnostics. Setting up and tuning adjustable parameters during commissioning and periodically repeated acceptance testing play a special role. Users of machines and drives can have at disposal a variety of well developed and very specialized software tools and measuring equipment for these purposes, however, there is a lack of fully universal tools. This results in needs to develop methods and tools that allow effective maintenance also in not typical cases.

Figure 3 illustrates some groups of methods that can be used for data collecting during testing of a servo-drive. Groups "C" and "D" are commonly used for acceptance testing purposes. The tests "A" are very simple to realise when appropriate tools from manufacturer of the drive are at disposal and is commonly applied during commissioning and retuning parameters. Methods "B" and "E" demand some measuring equipment and a DAQ application [1,2]. They can be effectively used also in not typical tasks. Idea of such measurements has been presented in the Figure 4. In practice parameters of a drive can be corrected in some ways what depends on design of the drive and on its parameters system.

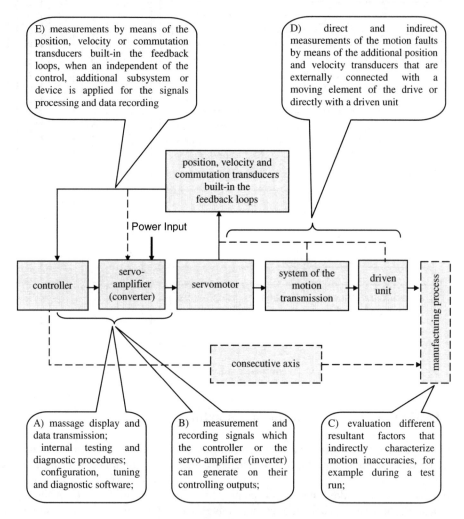

Figure 3. Illustration of the methods used for data collecting in the diagnostic researches of a servo-drive

Figure 5 illustrates principles of the ways. An inaccuracies correction can be done by means of a controller program or by means of a drive parameters system.

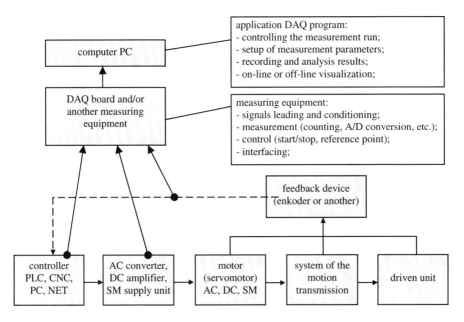

Figure 4. Illustration of the computer-assisted measurements used for data collecting in the diagnostic researches of automated drives

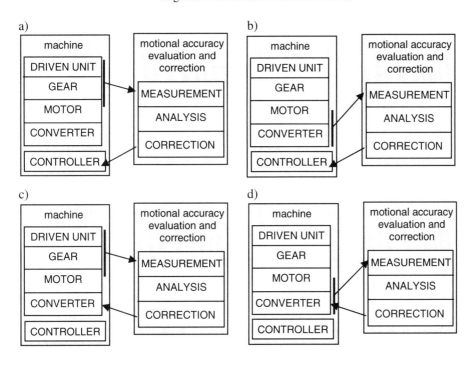

Figure 5. Ways of computer assisted testing of the motional inaccuracy and correction an automated drive: a,b – retuning the parameters inside of the controller; c,d – retuning the drive adjusting parameters

4. Testing a servo-drive by independent measurements

Independent measurements of the controlling signals, both from feedback loops and from outputs of a converter or a controller (see Fig.3 and Fig.4) have some essential advantages. A standard DAQ board and software can be used what makes the methods simple to prepare, flexible and costs effective. It is possible to research drives operating in one axis or in several axes at a time, during an idle test run and under load as well. However, the most important is possibility to apply them in the not typical and complex cases, when other solution are not obtainable, not effective or too costly. Finally the methods can be used as a complement to the others, especially when influence of the different factors on the observed faults has to be separated.

Examples of the tests have been presented in the Figures 6 and 7. Diagrams in the Figure 6 show some results of a test when a reason why an AC servo-drive operates unstably had to be found. The test was carried out by measurement of an analogue voltage signal on a programmable output of the AC converter. As it can be seen this

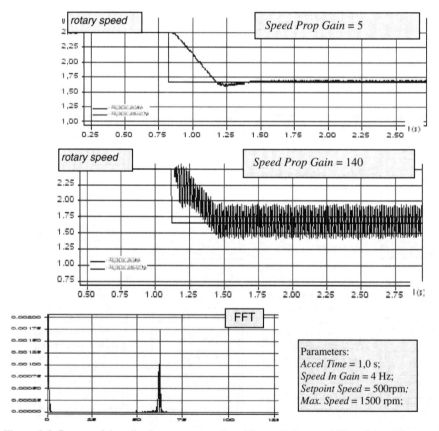

Figure 6. Influence of the adjusting parameter *Speed Prop. Gain* on stability of the AC servo-drive; example of testing a drive by measurement of a controlling signal on an output of the AC converter proportional to speed signal turned out to be quite useful also for dynamic tests

Some results of another test carried out by a similar way, were presented in the Figure 7. A main purpose was to find reasons of inaccuracies observed during reversal motion. The diagrams present experimentally obtained acceleration. Some inaccuracies are visible, for instant a difference between acceleration and deceleration or a perturbation when the rotary speed is very close to zero during the reversal.

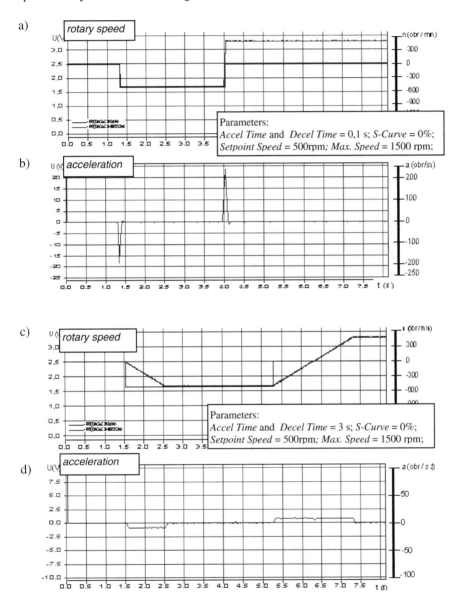

Figure 7. Inaccuracies of the acceleration realized by the AC servo-drive during reversal motion; example of testing a drive by measurement of a controlling signal on an output of the AC converter

5. Final remarks

Simplified graphical models of CIM that are published in many modifications usually do not give enough consideration to maintenance-oriented subsystems. In the author's opinion their existence should be clearly pointed out as a part of CAM taking into account their importance and great variety of maintenance oriented software tools that are in use in practice.

Testing of automated drives to find out whether the motion faults exist and what their possible reasons are tends to be one of the most essential parts of the modern maintenance. However, the situation sometimes can be met where in a plant a computer aided preventive maintenance system has been adopted, but drives as subassemblies of a machine stay in long use with only poor or without any inspection till failure and replacement. On the other hand adopting the general rules of preventive maintenance also for the drives can help keeping their capabilities and also performances of a whole machine. This is often applied for CNC machine tools in connection with requirements of quality certification. Nevertheless, there is no obstacle to adopt the solution wider.

The motion faults can be caused by performances of a drive, by properties of the being worn mechanical elements, by incorrect pre-setting of the adjustable parameters of a control and by different factors connected with a load. This determines a related to the drives set of tasks for maintenance that can include:
- evaluation of the control quality when the motion in one axis is taken into account and analysed separately from the other axes,
- evaluation of the control quality when the motion is realised in more then one axis at the same time and when the analysed criterions are connected with accuracy of a resultant trajectory or with a relative velocity,
- a search for reasons for the motion inaccuracies,
- correcting and retuning faulty pre-settings of adjustable parameters,
- eliminating other reasons of the observed faults.

References

1. Blacharski W.: *Digital measurements of the motion faults in the diagnostics of servo-drives*. International Conference MECHATRONICS 2000, Warsaw University of Technology, Warsaw, 2000, p.169-172.
2. Blacharski W.: *Computer assisted measurements in the servo-drives diagnostics based on the feedback signals evaluation*. Proc. ICCC'2002 – 3 rd International Carpatian Control Conference, Ostrava -Malenovice, May 27-30, 2002, Vysoka Skola Banska – Technicka Univ., CSVTS KAKI, 2002.
3. Datastream Co.: *MP2 Brochure*.
4. Leidel S.: *Totally Integrated Automation SIMATIC PDM*. Siemens. Automation & Drives.
5. Rembold U., Njani B., Storr A.: *Computer Integrated Manufacturing and Engineering*. Addison -Wesley Publishing Company, 1993.

The International Journal of **INGENIUM** 2005 (4)

ENGINEERING ACHIEVEMENTS ACROSS THE GLOBAL VILLAGE

edited by

Janusz SZPYTKO

Transport Systems and Devices

Cracow - Glasgow - Radom, 2005

TABLE OF CONTENTS   page

6. Transport Systems and Devices **407**

6.1. Modeling of bridge cranes for dimensioning needs of their load-carrying structures,
Chmurawa M.P., Gąska D. 409

6.2. The rapid prototyping of a crane intelligent control system,
Smoczek J., Szpytko J. 415

6.3. Customized mass-manufacturing: low cost, radio-frequency, node based indoor tracking networks,
Diegel O., Bright G., Potgieter J. 423

6.4. Decision support systems in electric power supply in liberalized electricity market,
Kunicina N., Levchenkov A., Greidane S. 431

6.5. Low cost laser guidance system for de-mining robots,
Potgieter J., Bright G., Diegel O., Tinnelly M.J. 439

All rights reserved. No part of this book may be reproduced, stored in a retrieval system, or transmitted, in any form or by any means, without prior written permission from the Publisher.

The International Journal of INGENIUM
Chief Editor: Professor David K. Harrison, Glasgow Caledonian University, UK

© GCU Glasgow

ISSN 1363-514x

A CIP catalogue record for this publication is available from British Library

Publishing cooperation: Instytut Technologii Eksploatacji – PIB w Radomiu

MODELING OF BRIDGE CRANES FOR DIMENSIONING NEEDS OF THEIR LOAD-CARRYING STRUCTURES

Chmurawa M.P., Gąska D.

Abstract: *Dimensioning of load-carrying structures is very important at the stage of cranes design. This dimensioning relates to defining the last form and constructional features, the most often steel load-carrying structure, able for safety taking over the suitably combined service loads (Markusik, 2001; prEN 13001-2), connected with hoisting capacity, the working environment and conditions of use. This paper presents current strength, durability and stability conditions of load-carrying structures and problematic of modeling the real structure with bridge crane as a example.*

1. Introduction

The dimensioning of crane load-carrying structure, which own mass comes to 500 tones – about 70% of whole cranes mass – despite lot of development works (Grabowski, 2004; Chmurawa, 2004; Smolnicki, 2004) and access to FEM (finite element method) software (Rusiński, 2000) stays still an open problem. The design engineer, who designs the structure on the basis of computer aided design, analytical calculations and his own experience, decides about construction and figure of crane load-carrying structure, therefore about its mass, costs and labor consumption. Simultaneously the new European Normative Acts will be in force for the second part of 2005 year, which for the sake of safety considerations should be applied in cranes dimensioning and design.

The presented problem of crane's load-carrying structures dimensioning, especially in the light of European Norms, requires a work out of numerical or numerically – experimental method for relatively quick prototyping (rapid prototyping) of steel structures which will fulfill current strength, durability and stability conditions that are required by European Normative Acts.

2. New principles of cranes load-carrying systems dimensioning in aspects of European norms

In calculation of load-carrying cranes structures (fig. 1) according to European Norms limit design stresses method should be applied, however also allowable stress method is possible (Markusik, 2001; prEN 13001-3.1; prEN 13001-1).

The individual structure members (like: girder, bearings, buffer beams, fig. 2-6) are subjects to the proof of: static strength, fatigue strength, strength of connections (pin, welded, bolt) and stability of components and special elements.

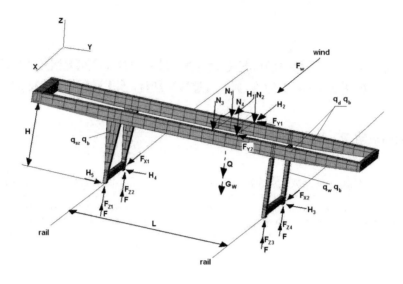

where:
- F_{zi}, F, N_i — wheel forces and forces caused by traveling over a step and over a gap,
- H_i — forces caused by skewing and wind,
- q_d, q_{sz}, q_w — gravitational forces of load-carrying structure,
- q_b — inertial forces,
- F_{xi}, F_{yi} — forces caused by acceleration of traveling drives,
- Q, G_w — gravitational forces of load and hoisting winch,
- F_w — wind force,
- L — bearings span.

Figure 1. Geometrical model of bridge crane with potential state of loads

2.1. Proof of Static Strength

For the structural member to be designed it shall be proven that:

$$\sigma_{Sd,x} \leq f_{Rd,x}; \quad \sigma_{Sd,y} \leq f_{Rd,y}; \quad \tau_{Sd} \leq f_{Rd} \tag{1}$$

The load-carrying crane structure is a thin-welled shell construction (fig. 2-5), therefore a plane state of stresses will apply and it shall additionally be proven that:

$$\left(\frac{\sigma_{Sd,x}}{f_{Rd,x}}\right)^2 + \left(\frac{\sigma_{Sd,y}}{f_{Rd,y}}\right)^2 - \frac{\sigma_{Sd,x} \cdot \sigma_{Sd,y}}{f_{Rd,x} \cdot f_{Rd,y}} + \left(\frac{\tau_{Sd}}{f_{Rd}}\right)^2 \leq 1,0 \tag{2}$$

where:
- $\sigma_{Sd,x}$, $\sigma_{Sd,y}$, τ_{Sd} — normal design stresses in "x" and "y" directions and shear stress determined by applying the loads, load combinations with taking into

x, y	consideration the dynamic factors ϕ_i and partial safety factors γ_p according to Table 10 of EN 13001-2, the orthogonal directions of stresses,
$f_{Rd,y}$, $f_{Rd,x}$, f_{Rd}	the corresponding limit design stresses in "x" and "y" directions and shear stress, adequate for the type of material.

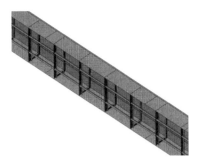

Figure 2. Shell construction of crane main girder with stiffenings (unequal angles)

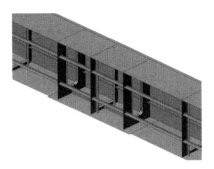

Figure 3. Node where main girder with stiffenings joints the fixed bearing

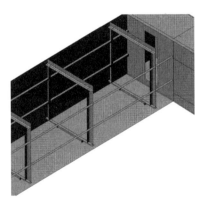

Figure 4. Node where main girder joints the buffer beam

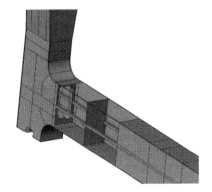

Figure 5. Bottom part of fixed bearing

The conditions (1) and (2) have the same figure both for material of structure, and for the welded joints. Difference concerns the meanings and value of suitable design stresses and limit design stresses, different for material and different for joints. Condition (2) for welded joints is in form of:

$$\left(\frac{\sigma_{Sd,x}}{f_{Rd,x}}\right)^2 + \left(\frac{\sigma_{Sd,y}}{f_{Rd,y}}\right)^2 - \frac{\sigma_{Sd,x} \cdot \sigma_{Sd,y}}{f_{Rd,x} \cdot f_{Rd,y}} + \left(\frac{\tau_{Sd}}{f_{Rd}}\right)^2 \leq 1,1 \qquad (2a)$$

2.2. Proof of fatigue strength

For the detail under consideration it shall be proven that:

$$\Delta\sigma_{Sd,x} \leq \Delta\sigma_{Rd,x}; \quad \Delta\sigma_{Sd,y} \leq \Delta\sigma_{Rd,y}; \quad \Delta\tau_{Sd} \leq \Delta\tau_{Rd}; \tag{3}$$

$$\Delta\sigma_{Sd,x} = \max\sigma_x - \min\sigma_x, \tag{4}$$

$$\Delta\sigma_{Sd,y} = \max\sigma_y - \min\sigma_y, \tag{5}$$

$$\Delta\tau_{Sd} = \max\tau - \min\tau, \tag{6}$$

where:

$\Delta\sigma_{Sd,x}, \Delta\sigma_{Sd,y}, \Delta\tau_{Sd}$ — calculated maximum range of design stresses in "x" and "y" directions and shear stress,

$\max\sigma_x, \max\sigma_y, \max\tau, \min\sigma_x, \min\sigma_y, \min\tau$ — extreme values of design stresses resulting from load combination A according to Table 10 of EN 13001-2, by applying $\gamma_p = 1$ (compression stresses with negative sign),

$\Delta\sigma_{Rd,x}, \Delta\sigma_{Rd,y}, \Delta\tau_{Rd}$ — permissible stress range for each material (or welded joint) by taking the stress history parameter $s_{(m)}$ and inverse slope of Wöhler-curve (σ/N-curve) m into consideration, adequate for the "x" or "y" directions and type of stress (normal or shear).

Additionally the action of independently varying ranges of normal and shear stresses shall be considered by:

$$\sqrt[m]{\left(\frac{\Delta\sigma_{Sd,x}}{\Delta\sigma_{c,x}}\right)^{m_x} \cdot S_{x(m_x)} + \left(\frac{\Delta\sigma_{Sd,y}}{\Delta\sigma_{c,y}}\right)^{m_y} \cdot S_{y(m_y)} + \left(\frac{\Delta\tau_{Sd}}{\Delta\tau_c}\right)^{m_\tau} \cdot S_{\tau(m_\tau)}} \leq \frac{1,0}{\gamma_{Mf}} \tag{7}$$

where:

$\Delta\sigma_{c,x}, \Delta\sigma_{c,y}, \Delta\tau_c$ — are the characteristic values of stress range in "x" and "y" directions and shear stress for constructional detail. It represents the fatigue strength under $2 \cdot 10^6$ constant stress range cycles and a probability of survival of P = 97,7% (mean value minus double standard deviation). Values of $\Delta\sigma_{c,x}, \Delta\sigma_{c,y}, \Delta\tau_c$ for the characteristic constructional details are announced in Annex A and Annex E of prEN 13001-3.1,

m — inverse slope of Wöhler-curve (σ/N-curve) announced for the characteristic constructional details in Annex A and Annex E of prEN 13001-3.1,

m_x, m_y, m_τ — inverse slope of Wöhler-curve (σ/N-curve) adequate for the "x" or "y" directions and type of stress (normal or shear),

$S_{x(m)}, S_{y(m)}, S_{\tau(m)}$ — stress history parameter calculated according to the formula in point 4.4.4 of prEN 13001-1 adequate for the "x" or "y" directions and type of stress (normal or shear),

γ_{Mf} — fatigue strength specific resistance factor according to Table 10 of prEN 13001-3.1.

The permissible stress ranges $\Delta\sigma_{Rd}$ for the detail under consideration shall be determined either by direct use of stress history parameter $s_{(m)}$ or simplified by use of class S (Markusik, 2001; prEN 13001-3.1).

2.3. Proof of static strength of hollow section girder joints

Considering the similarity of cranes main girders construction (fig. 2-6) to conventional steel girders used in building engineering, the proof shall be executed according to the rules of Eurocode 3 (prEN 1993-1-1).

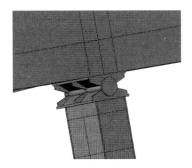

Figure 6. Articulated joint of pivot bearing

2.4. Proof of elastic stability of structural members and special elements

The proof of static stability prevents structural members from the loss of stability by lateral deformation (e.g. buckling). Considering the similarity of this members and elements to typical steel construction, the proof shall be executed according to the rules of above-mentioned Eurocode 3 (prEN 1993-1-1).

3. Modeling problems of the real load-carrying structure with bridge crane as an example

Creation of geometrical model is the first step of modeling a load-carrying structure. Figure 1 shows such a structures geometrical model of large bridge crane with hoisting capacity Q = 45 tones, made with CAD software. The most suitable model is a shell model, which faithfully imitates de facto the shell construction of structure. It also allows to identify the state of effort in constructional kinematical pairs with the help of FEM (finite element method) software. Such image is much more complicated to be made and in principle is impossible in case of beam or solid model.

In next stage on the basis of this geometrical model the FEM one should be worked out with reference to analysis using finite element method. The digitization of geometrical model should be made with use of shell finite elements e.g. "thin shell linear triangles" or "thin shell linear quadrilateral". The finite elements, mentioned above, have 6 degrees of freedom in knots, which assure good agreement of boundary conditions.

The next stage of numerical dimensioning is the definition and combination of loads according to prEN 13001-2 as well as loading it as external loads to FEM load-carrying structure model. In this way with the help of FEM software it is possible to determine suitable components of the design stresses e.g. $\sigma_{Sd,x}$, $\sigma_{Sd,y}$, τ_{Sd} in individual knots of structure, such as the ones shown on fig. 2-4 and then to check the strength conditions (1) and (2) that ensure correct dimensioning of cranes load-carrying structure being under consideration.

4. Conclusions

The load-carrying structures of cranes are complex thin-walled structures which in stage of design should be checked according to European Norms or other recommended ones, for strength, durability and stability conditions.

FEM is possible and should be applied in dimensioning that structures. However the criterion of such approach is preparing the adequate numerical model of real structure which in FEM preprocessors software is unusually labor-consuming. According to this, there is a need to work out a relatively fast and numerically coherent transformation of geometrical model of structure to numerical model, which would make the digitization fast and take less time for prototyping of new load-carrying crane structure. As a result of working out of this method, at the stage of design, the almost exact determination of labor-, material-consuming and costs of crane manufacture will be possible.

References

1. Chmurawa M, Gąska D: *Dimensioning problems of overhead traveling cranes load carrying structures* (in polish). Transport Przemysłowy nr 2/2004, pp. 6-9.
2. Grabowski E, Kulig J: *Durability in dimensioning of cranes load carrying structures* (in polish). Transport Przemysłowy nr 1/2004.
3. Markusik S: *New cranes classification according to European Norms* (in polish). Transport Przemysłowy nr 1(3)/2001, pp. 32-36.
4. Markusik S: *Determination of loads in cranes load carrying structures according to European Norms* (in polish). Transport Przemysłowy nr 2(4)/2001, pp. 22-29.
5. Markusik S: *Loads due to wind in cranes load carrying structures according to European Norms* (in polish). Transport Przemysłowy nr 3(5)/2001, pp. 32-35.
6. Rusiński E., Czmochowski J., Smolnicki T.: *Zaawansowana metoda elementów skończonych w ustrojach nośnych maszyn*. Oficyna Wyd. Pol. Wroc., Wrocław 2000.
7. Smolnicki T., Rusiński E., Czmochowski J.: *Dimensioning problems of surface mining machines load carrying structures* (in polish). Przegląd Mechaniczny nr 1/2004.
8. prEN 13001-1:2003. Cranes – General design – Part 1: General principles and requirements.
9. prEN 13001-2:2003. Cranes – General design – Part 2: Load effects.
10. CEN/TS 13001-3.1:2004. Cranes – General design – Part 3.1: Limit states and proof of competence of steel structures.
11. prEN 1993-1-1:2003. Eurocode 3: Design of steel structures – Part 1-1: General rules and rules for buildings.

THE RAPID PROTOTYPING OF A CRANE INTELLIGENT CONTROL SYSTEM

Smoczek J., Szpytko J.

Abstract: The paper presents methodology and results of researches conducted on the real device, the two-spare overhead crane to elaborate intelligent control system based on fuzzy controller with Takagi-Sugeno-Kang inference system. Applied programming environment based on Matlab program and hardware's solutions gave possibility to built control system using rapid prototyping method. The proposed control system was elaborated, built and optimized during simulations conducted on mathematical models of the device and experiments conducted on the real device and next the control algorithm was implemented on the final control device, programmable logic controller PLC.

1. Introduction

In manufacturing industrial systems the greater and greater requirements are put before operation quality and precision of automated transportation systems. One of the important elements of internal transportation systems are the overhead cranes. The issues of minimizing transport time, precision of positioning with swing angle of the load minimization are the more and more significant in practice as well as exploitation quality of a device (Szpytko, 2004; Szpytko et al., 2003). Complexity of phenomenon occurred during loads transportation process using cranes as a result of wide change exploitation parameters causes that in crane's control systems are required tools taking into consideration complexity of such systems characterized by uncertainty, imprecision and subjectivity of parameters. One of that mathematical tool is fuzzy logic which is a nonlinear system that convert a crisp input vector into a crisp output vector. Fuzzy models are experts systems in which control strategy is expressed in form of IF-THEN rules built using linguistic terms that gives simple and legible interface in process of controllers' realization. In the researches works the most solutions of those problems are based on fuzzy approaches with Mamdani models. Compound mathematical mechanism Mamdani's inference system cause that it is difficult to put into practice control algorithm in industrial application (e.g. on PLC controllers). Also, the most solutions of intelligent control systems based on fuzzy logic or artificial neural network are realized only during simulations conducted on mathematical models of a device and seldom tests on laboratory models (Mahfouf et al., 2000; Mendez et al., 1999). Implementation of the control system on the real device is time-consuming process that can be shorten by using prototyping methods and integrated software and hardware computers' tools enabling to elaborate control systems during simulations on the mathematical models and experiments conducted on the real devices.

The methodology and results of researches conducted on the real device for elaborating control system of crane's movement are shown in this paper. Intelligent control system was based on fuzzy controller with Takagi-Sugeno-Kang (TSK) inference system and elaborated using rapid prototyping method (Smoczek, 2003; Szpytko and Smoczek, 2004a;

Szpytko and Smoczek, 2004b). Using computer software's tools and equipment, PC computer with Matlab *MathWorks* program and multifunction data acquisition cards it was able to build compound control system, elaborating and optimizing control algorithm and testing property of working control and measurement circuit. The proposed research methodology considerably shortened time of building control system and its further implementation on the target control device, PLC controller.

2. Rapid prototyping of the crane's control system

Using rapid prototyping process shorten the time of control system designing: control algorithm optimization, sampling time selection, testing proper working measurement circuit. The prototyping process is performed in the result of testing virtual controller during computer simulations conducted on the mathematical models of the control object and than control system is verified and validated during experiments on the real device. In the result of experiments conducted on the device and its models controller's parameters can be adjusted (sampling time, measurement and control signals quantization) and mathematical models can be validated. As a result of built control system the ready control algorithm can be implemented on the target controller, e.g. PC computer, programmable logic controller PLC or other control device.

During researches conducted towards elaborating control system of the crane's movement, prototyping process was based on Matlab *MathWorks* program package equipped in Simulink and Fuzzy Logic Toolbox (FLT), tools enable building mathematical models of control object and fuzzy controllers as well as in Real Time Workshop (RTW), tool enables conducting real time experiments on the control object. Equipment's architecture of the measurement-control circuit was based on PC computer with two multifunction data acquisition cards produced by *Advantech* with symbol PCL 818HG and PCI 1720.

The aim of the researches conducted on the real object, the two-spars overhead crane with 12,5 tons hoisting capacity, 16 meters bridge length and 10 meters of hoisting high was building intelligent control system with fuzzy logic controller and finally implementation proposed control algorithm on programmable logic controller PLC. For this reason the methodology of prototyping control system was elaborated and realized in following steps (Figure 1):
1. Elaborating mathematical models of the control object in Matlab/Simulink program.
2. Elaborating control algorithm and its testing using simulation process realized in Mtlab/Simulink program.
3. Elaborating control system with virtual controller and developing it during experiments on the real object.
4. Control algorithm implementation on the PLC controller.

In the result of simulations conducted on mathematical models of the control object the control systems with conventional controller PI (Proportional-Integrative) and TSK fuzzy controller were built. Basis on proposed virtual controllers the measurement-control circuit was created in Simulink program using graphical blocks represented physical interface's cards inputs and outputs and S-functions, written in C program language, realized measurement signal conversion (Figure 2).

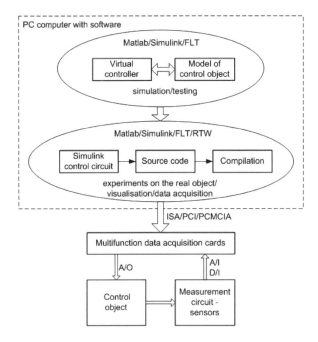

Figure 1. Rapid prototyping process of control system

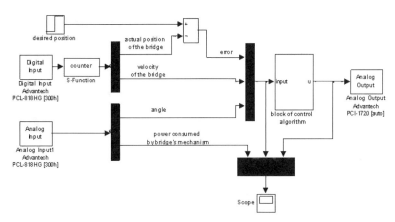

Figure 2. Simulink measurement-control circuit

Basis on control system created in Simulink program, presented at figure 2, the source code was generated and further compiled by using Real Time Workshop program. This automatically realized process enables conducting real time experiments on the control object. During tests conducted on the control object parameters of the PI and TSK controllers were adjusted, control algorithms were developed and sampling time and measurement signals quantization were selected. Employing software and hardware based on Matlab/Simulink/RTW programs and interface's cards enabled online visualization and data acquisition of measurement signals and control variables during conducting experiments.

Process of automatically generating of source code and its compilation considerably shortens the time of building control system, giving possibility of concentrating only on control system designing. Designer's work can fluently gone from the stage of computer simulation conducted on the models, preparing virtual controller to the stage of experiments on the control object, if need giving possibility of quickly returning to the phase of simulation (Figure 3).

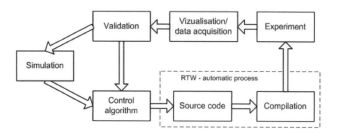

Figure 3. Rapid prototyping of control algorithm

3. Fuzzy control system

For research requirements three-mass model (mass: trolley m_1, bridge m_2) of the overhead crane was built as an electro-mechanical model of the device transporting the load m_3 suspended on the rope, Figure 4.

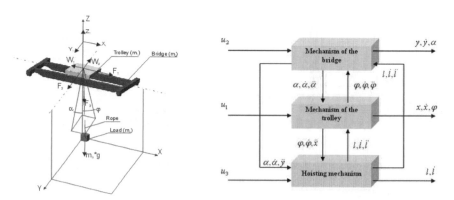

Figure 4. The overhead crane's model

The aim of the proposed control system was precision positioning of the transported load to the desired position in three-dimensional working space $OXYZ$ with minimization swing of the load after starting and during braking movements' mechanisms. The control algorithm was based on three TSK controllers, that control movements' mechanisms of the trolley, bridge and hoisting. The controllers' input signals were errors of mechanisms' positions, actual velocities of the mechanisms and swing angles of the load measured in direction of bridge and trolley movements. The TSK controllers output signals were desired velocity of individual crane's mechanism on which bases the final control signals were calculated as the increases of velocity error.

The researches conducted on the real object were oriented on controlling of the bridge movement. In the control algorithm of the bridge movement the TSK controller was used with inputs: error of bridge position $e_y = y_d - y$ (where: y_d - desired bridge's position, y - actual bridge's position), velocity of the bridge $\dot{y}$ and swing angle of the load measured in bridge movement direction α (Figure 5).

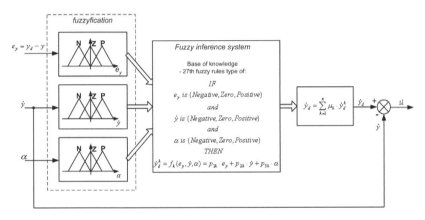

Figure 5. Fuzzy control of the bridge velocity

For the each input signal of TSK model the three fuzzy sets were assigned. Fuzzy sets were entitled using linguistic terms: *Negative*, *Zero* and *Positive*. As the membership function were used function type of triangular. Inference process was realized basis on base of knowledge expressed by 27 fuzzy rules type of IF-THEN. In the result of using single k-rule desired velocity of the bridge is counted $\dot{y}_d^k$.

$$\dot{y}_d = \sum_{k=1}^{n} \mu_k \cdot \dot{y}_d^k \qquad (1)$$

where:
μ_k – coefficient of weight rule [0, 1],
k – rule's number,
$n = 1, 2..., 27$

The output signal of the TSK model y_d is calculated as a sum of desired velocities calculated from each of the rule multiplied by rules' weight coefficients (1). The increase of signal control $\dot{u}$ is calculated as a difference of output signal from TSK model y_d (desired velocity of the bridge) and actual velocity of the bridge $\dot{y}$.

4. Results of experiments on the real object

Researches' works which results are presented in this paper were conducted on the real object - the two-spars overhead crane working in the workshop, with $Q = 12{,}5$ tons hoisting capacity and bridge width $L = 16$ meters. The object of the researches was modernized and frequency inverters were installed in driving subsystems: bridge, crane and hoisting motion mechanisms (Szpytko and Smoczek, 2004a; Szpytko and Smoczek 2004b). Installing frequency inverters and PLC controller type of *Melseca* FX2N 48MR, that assure control on motion mechanisms of the overhead crane, the smooth changing of motors' speeds and torques were obtained (Smoczek, 2003).

During experiments conducted on the real object the PI and TSK controllers were tested and verified on the bridge motion mechanism for assumptions: 12,3 [m] desired position of the bridge, with load about 10 tons suspended about 4 meters over the base. Elaborated control algorithms were verified and developed during computer simulations and tests conducted on the real object using Matlab/Simulink/FLT/RTW programs and interface's cards (Figure 6):
- PCL 818HG – multifunction data acquisition card with 16 analog inputs SE, 16 digital inputs, 16 digital outputs, 1 analog outputs, 12-bit resolution of analog-to-digital converters,
- PCI 1720 – 4 analog outputs.

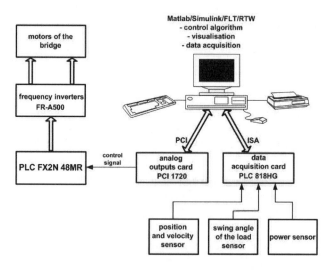

Figure 6. Control system

Proposed and tested during experiments control algorithm based on TSK controller was in the next step implemented on programmable logic controller PLC type of FX2N 48MR.

The results of experiments conducted using PI and TSK controllers were presented at the figures from 7 to 10.

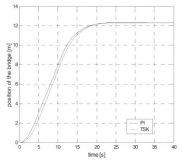

Figure 7. Position of the bridge

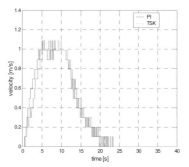

Figure 8. Velocity of the bridge

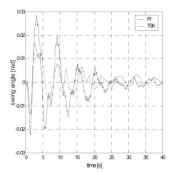

Figure 9. Swing angle of the load

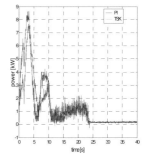

Figure 10. Power consumption of the bridge's motors

Using TSK controller the swing angle of the load was minimized better especially during starting (above 40%) and faster during braking the bridge mechanism (Figure 9). By using TSK controller overloads in the driving mechanism were decreased during starting above 12% in comparison with results obtained using PI controller (Figure 10). Basis of results obtained during experiments conducted on the real object it was stated using control system based on fuzzy logic techniques in movement mechanisms of the overhead crane the improvement of device's controlling quality is possible.

5. Conclusions

Realized researches works based on rapid prototyping process enabled to elaborate compound intelligent control algorithm during computer simulations and experiments on the real object. Applied software (Matlab/Simulink/RTW programs) and hardware (interface's cards) integrated tools gave possibility to shorten the time of control system designing and further implementation on target control device (PLC cotroller). Process of source code generation and compilation realized automatically shortens the time of gone from the simulations conducted on the mathematical models to the experiments on the real object. Prototyping process enabled to concentrate only on developing control system elements: control algorithm optimization, adjusting the control system parameters and testing used measurements circuits.

Presented results of experiments conducted on the overhead crane proved that employing fuzzy controller can be very useful in automated control systems of the cranes movement mechanisms. Using frequency inverters in crane's driving mechanisms and employing intelligent control system the accuracy of device's transportation tasks can be improved as well as exploitation of the device.

This paper has been financially supported by central budget on science for the years 2005-07 as the research project.

References

1. Mahfouf M., Kee C.H., Abbod M.F., Linkens D.A. (2000): *Fuzzy logic-based antisway control design for overhead cranes*. Neural Computating and Applications, Vol. 9, 2000, pp. 38-43.
2. Mendez J.A., Acosta L., Moreno L., Torres S., Marichal G.N. (1999): *An application of a neural self controller to an overhead crane*. Neural Computing and Applications, 8, 1999, pp. 143-150.
3. Smoczek J. (2003): *Fuzzy logic approach to the overhead crane movements mechanisms*. PhD Thesis, AGH Krakow.
4. Szpytko J. (2004): *Integrated decision making supporting the exploitation and control of transport devices*. Monographs, UWND AGH, Krakow.
5. Szpytko J., Schab J., Smoczek J. (2003): *Exploitation indicato of traveling cranes modeling through simulation*. 9th IEEE International Conference on Methods and Models in Automation and Robotics. 2003, Miedzyzdroje, Poland, pp. 1409-1414.
6. Szpytko J., Smoczek J. (2004a): *Fuzzy logic approach implementation in transport devices controlling*. Intrnational Carpathian Control Conference ICCC'2004, Zakopane, Poland, May 25-28, 2004, pp. 271-276.
7. Szpytko J., Smoczek J. (2004b): *Fuzzy logic control improve transport devices quality*. 10th IEEE International Conference on Methods and Models in Automation and Robotics. 2004, Miedzyzdroje, Poland, pp. 1391-1396.

CUSTOMIZED MASS-MANUFACTURING: LOW COST, RADIO-FREQUENCY, NODE BASED INDOOR TRACKING NETWORKS

Diegel O., Bright G., Potgieter J.

Abstract: The ability to track the position of goods throughout a production environment can be of great importance to agile mass-manufacturing systems. Once a controlling computer system knows where a particular product is within its manufacturing cycle, and what further operations are required, it can make smart decisions as to how this product can be customized, without affecting the production rate. This paper outlines the development of a Radio Frequency node based tracking system in which a RF tag, loaded with product specific information, is attached to a product. This allows the product to be tracked through the manufacturing cycle so customized operations can be carried out on it as part of the mass manufacturing process.

1. Introduction

Customized mass-manufacturing refers to the ability to mass-manufacture products in such a way that each particular product can be customized by the user to best meet their needs without affecting the flow of production and, more importantly, the rate of production (Goebel, 2004). If a product is manufactured in one continuous process, (such as in cell manufacturing), then it is relatively easy to keep track of simply by counting it. Many manufacturing environments, however, may use batch manufacturing systems, or even hybrid systems such as a combination of batch manufacturing and cell manufacturing, making it much more difficult to keep track of any particular product that may need special operations performed on it.

A particular manufacturing operation, such as chrome plating for example, may be done as a batch process after which all products in the batch are placed in storage until they are ready for the next process. It may not be possible to keep track of a particular product within that batch unless it is specifically tagged in some manner. To be of use in a customized mass-manufacturing system, each product must therefore be uniquely identifiable in some manner (Potgieter, Bright, 2002). A logical way of achieving this is through the use of an Indoor Position System, (IPS), to track and position each product in the indoor environment. There are many technologies used for IPS systems, and the most common are usually based on the technologies such as Infrared, Radio-frequency, DC Electromagnetic, Ultrasound, etc. (Harle et al, 2003).

2. Current tracking technologies

Tracking technologies are those that allow the control computer to know where any particular product is within a production environment and what processes it needs to have performed on it. Some of the commonly available technologies that are suitable for tracking networks include:

Infrared: Due to their ubiquitous deployment infrared (IR) transceivers are inexpensive, compact, and low power. IR propagation is fast but effective bandwidth is limited by interference from ambient light and from other IR devices in the environment. IR signals reflect off most interior surfaces but diffracts around few. Typical range is up to 5 meters. They are also restricted by line-of-sight limitations.

Radio frequency: Radio frequency (RF) signals offer several benefits over IR. RF signals diffract around and pass through common building materials. RF signals compare favorably to IR in propagation speed, bandwidth, and cost. Since the RF spectrum is heavily regulated, typical systems operate at 900MHz or 2.45GHz and comply with Part 15 FCC regulations so as not to require licensing. Transmission range of 10m-30m indoors is common.

RFID (Radio Frequency Identification) is a commonly available system which uses either low-cost passive Radio tags, or higher cost active tags. An RFID system comprises a reader, its associated antenna and the transponders (Tags/ RFID Cards) that carry the data. The reader transmits a low-power radio signal, through its antenna, that the tag receives via its own antenna to power an integrated circuit. Using the energy it gets from the signal when it enters the radio field, the tag will briefly converse with the reader for verification and the exchange of data. Once that data is received by the reader it can be sent to a controlling computer for processing and management.

DC Electromagnetic: DC electromagnetic fields have been used in many high-precision positioning systems. While the signal propagation speed is high range is limited to 1m-3m. These signals are very sensitive to environmental interference from a variety of sources including the earth's magnetic field, CRTs, and even metal in the area. Thus, systems based on these signals need precise calibration in a controlled environment.

Ultrasound: Ultrasound signals are becoming more common in positioning systems the relatively show propagation speed of sound (343m/s) allows for precise measurement at low clock rates, making ultrasound based-systems relatively simple and inexpensive. The signal frequency is limited by human hearing on the low end and by short range on the high end. A keen human can hear 20KHz sounds. Typical systems use a 40KHz signal. Conveniently, standard sound cards have a 48KHz sampling rate, sufficient for ~ 1cm resolution distance measurements. Environmental factors have substantial but not prohibitive effects on ultrasound propagation, particularly speed. Humidity can slow ultrasound by up to 0.3%. More drastically, a temperature rise from 0 °C to 30 °C alters the speed of sound by 3%. Finally, ultrasound reflects off most indoor surfaces. Empirical studies show that 40KHz ultrasound signals reverberate at detectable levels for at most about 20ms.

AT&T has developed a system called the bat ultrasonic system. The bat system involves the use of an ultra-sonic tracking technology developed at AT&T Laboratories in Cambridge. A small device carried by the users emits ultrasonic beeps, thus allowing a network grid of ultrasonic receivers to track which node the user is nearest (Harter, Hopper, Steggles, Ward, Webster, 1999).
Global Positioning System (GPS): Developed by the US military, GPS has been in consumer use in the last five years with the availability of affordable navigation tools.

These devices usually include GPS receivers to locate the user and a map database to give context such as streets and surroundings. Sometimes the device can also compute the best route from a source to a source to a destination, or store these planned trips for later retrieval. GPS features positioning accuracy of roughly 10m. For it to function, the receiver must be in line of sight of four satellites above, or be able to receive a supplementary correction signals from a ground station. Due to these limitations, GPS is not a useful tool for indoor or underground navigation. GPS cannot distinguish adjacent levels or floors of buildings.

3. RF ubiquitous networks

A ubiquitous network is one that is present everywhere throughout an area and has the ability to detect where users or items are within the network and, based on that information, meet their needs. A simple example of the use of a ubiquitous network could be that of working on a computer and then moving to a different part of the building to another computer and having the document automatically follow you to that computer allowing you to seamlessly continue your work.

The overall concept for the RF ubiquitous network was as follows: Small RF transceiver modules were attached to the ceiling of the environment to form a grid of linked modules, as shown in figure 1. The modules were spaced 2m apart and each module was set to have a range of 2.4 metres. The grid was connected to a computer via a low cost RS485 network and ran software to deal with the information received via the network. This software included the ability to track the product over a map of the environment, and display the product-specific data contained in the tag.

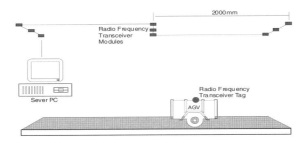

Figure 1. Radio Frequency Node Based Ubiquitous Network

A RF Enabled Tag was loaded with the product-specific data, and it had a range of 2.4 m. As the tag was within a distance of 2000 mm from the floor, it was always within range of at least one receiver module but never in range of more than 4 receiver modules. It then became simple geometry to determine the location of the user relative to either one, two, three or four modules.

As any RF grid module picked up a signal from a tag device, it sent back the information contained in the device, as well as it's own node number to the central computer which then knew which node the information was coming from (and therefore the products position) as well as the products pertinent information. To keep data transmission to a minimum, the complete product-specific data file was only sent on the first node contact, after which time it was stored in the central computer database. Thereafter, only a header identifying the product needed to be sent at a regular interval. From the location information and the product data, the central computer was able to reconfigure the various devices in the vicinity of the product to best suit the operations to be carried out. This burst transmission system also had the advantage of minimizing the risk of co-channel interference (Papadopoulos and Sundberg, 1998).

4. Tracking system construction

A network of several Radio Frequency receiver nodes were attached to the ceiling of the environment to form a grid of linked modules. These nodes were spaced 2 meters apart and each module was set to have a range of 2.4 meters.

The radio frequency receiver module (rfRXD0420) received the radio signals and passed the valid signals to the PIC18F242. The PIC18F242 sampled these signals and picked up the useful data from that. The receiver node, which received the information from the radio frequency transmitter, then transferred its address and received valid information to a central control PC via an RS485 network.

An RS485-to-RS232 converter was placed between PC and the RS485 network because computer's serial port was using the RS232 protocol. The PC was running software to deal with the information received from the network and to control the devices controlled by the system. In the case of the sample system described in this paper, after analyzing the information received, the PC then sent the command to the lighting control box to switch on the appropriate colored light to represent a particular manufacturing process.

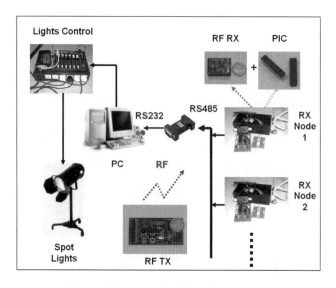

Figure 2. Diagram of test tracking network

4.1. Construction of the RF Tag

An rfPIC12F675 transmitter module was used to be the electronic transmission device. The rfPIC12F675 transmitter module contained:
- 2 push-button switches connected to GP3 and GP4,
- 2 potentiometers connected to GP0 and GP1,
- RF enable (RFenin) connected to GP5,
- Data ASK (DATAask) connected to GP2,
- optional 8-pin socked (U2) for In-Circuit Emulation (ICE) or inserting an 8-pin DIP package version of the PIC12F675.

The push-button switch GP3 was used as the main power switch. The push-button switch GP4 was removed and the Low Frequency Communication Circuit was linked to pin GP4. This Low Frequency Communication Circuit (LFCC) acted as an electronic switch with a very short range (typically, 20cm). If no low frequency signal was received by the rfPIC12F675 module, the module sent only the product ID every 1 second. If the rfPIC12F675 module received a low frequency command within the LFCC range, the pin GP4 was pulled-up and the module started to send all the information contained in the tag (Product ID, Operations to be performed, etc.).

Those potentiometers connected to GP0 and GP1 were not used. A power reduction resistor was added on, and the length of the antenna was shortened to decrease the transmission range to approximate 2.5 meters. This was in important step as, if the range of the transmitter were too large, it would communicate with too many receiver nodes, making it more difficult to pinpoint a precise location. The data transmitted from the rf12F675 module used its own code transmission format, in which there were four distinct parts to every code word transmission as follows: Preamble, Header, Data and Guard Time.

The preamble started the transmission and consisted of repeating low and high phases each of length Te representing the elemental time period. The header consisted of a low phase which had a length of 10*Te. Next came the data bits. The data bits were Pulse Width Modulated (PWM). A logic one was equivalent to a high of length Te, followed by a low of length 2*Te. A logic zero was equivalent to a high of length 2*Te, followed by a low of length Te. The final part of the code word transmission was the guard time which was the spacing before another code word was transmitted.

Figure 3. Transmitter Pulse Train

The encoding method used for the transmission was a $^1/_3$ $^2/_3$ PWM format with Te (basic pulse element).

The device used the following data format: the preamble was 10101010 (8-bits sequence), followed by a 0000 (4-bits) header. The data section contained the product ID, Description, and four different operations to be carried out. Each of these operations was represented by either a blue or red light depending on the configuration of each robot. The last section was the Guard Time which consisted of 8 bits 0.

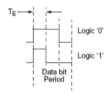

Figure 4. Encoding Method

4.2. Low frequency communication circuit

The power supply was a standard linear regulated main frequency unit with one difference: the mains earth was brought through to the circuit ground. This was necessary to provide the radio output with an earth reference, so it could function effectively as a transmitter (Agrawal, 2001).

The rest of the circuit consisted of two nearly identical halves, each producing one of the outputs. The first of the three op-amps was the oscillator, calibrated to the desired frequency. The output to the filter was taken not from the square wave output but from the timing capacitor, which had a continuous RC charge/discharge curve. This had fewer high frequency components than a square wave, which resulted in a cleaner output signal. As this output was easily loaded, a second op-amp was included as a 1:1 buffer. The signal was finally fed to a second order low-pass filter with a cut-off frequency identical to that of the oscillator. This removed the majority of the high-frequency components, resulting in an acceptable sine wave at the output.

4.3. Receiver node

The receiver Node consisted of two main elements: a radio frequency receiver module (rfRXD0420) and a PIC18F242. The receiver module received the radio frequency signals and passed them to the PIC18F242 for processing. The PIC18F242 picked up the useful information and sent them to the PC via the RS485 network.

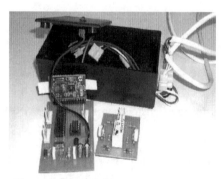

Figure 5. Tracking System Receiver Node

The assembly code for PIC18F242 performed the following tasks: Recognize incoming signal, pick up the data section of the incoming information package, process the data and send its receiver node address and useful receiving data to the PC via its USART port, Repeat.

A MAX485 Chip provided the half duplex RS-485 function with two data lines (A and B), a common ground line, and an extra control line. This control line was used to check whether the bus lines were free or not. The receiver node also had the in-circuit programming function for the PIC18F242. This function was used to easily re-program or upgrade the PIC18F242 code as the system was developed.

5. Customized mass-manufacturing setup

The system described in this paper used a ubiquitous network made up of RF nodes. The RF nodes placed throughout the environment formed a ubiquitous network which allowed for mass-customized production in which consumer products can be mass-manufactured while, at the same time, each product can be customized to a specific user. This allowed for direct integration between the end-consumer and the machines assembling products.

An example of such integration would be that of a fully automated computer assembly cell in which the customer could order his preferred computer configuration over the Internet. The process would begin by the customers logging in over the Internet and ordering a customized product. As the order leaves the order entry system, its order specific data is downloaded to the RF module on the AGV (Autonomous Guided Vehicle) which then begins its journey through the factory. At each automated workstation a robot sets itself up to, for example, insert 512Mb of Ram, until all the appropriate work is complete. At any stage the control computer system can tell the customer exactly where in the system his machine is, and can estimate when the customers' job will be finished. The product was accordingly assembled and dispatched to the customer.

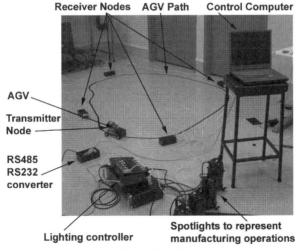

Figure 6. Tracking System Setup

The system setup, as shown in Figure 6, described in this paper was a simplified system which involved a vehicle with an RF tag moving through an environment along a predefined path. Around the path, pairs of spotlights were positioned to represent manufacturing processes. Each pair of spotlights included a blue and a red light to represent the reconfiguration of a particular assembly process. The RF tag on the vehicle was preprogrammed with a random combination of either red or blue. As it followed its path, the RF tag was detected by the RF receivers and the appropriate colored light was switched on.

6. Conclusion

Customized mass-manufacturing refers to the ability to mass-manufacture products in such a way that each particular product can be customized by the user to best meet their needs without affecting the flow of production or the rate of production. In order to achieve this, the control computer must have the ability to know the precise location of any particular product at any given time as well as what customized operations need to be performed on each product. This project successfully demonstrated the feasibility of a low-cost customized mass-manufacturing system using an RF node based tracking system.

The system consisted of a transmitter node, placed on an AGV, which transmitted product ID and required operations data at repeated intervals. Receiver nodes placed along the AGV path picked up this information as the AGV moved to within their receiving range, and transmitted this information to the control computer along an RS485 network. Based on this information, the computer was made aware of the precise location of the AGV and was then able to reconfigure the machines to perform the operation required by the product. In the case of the demonstration system, each machine was represented by two different colored lights to simulate different manufacturing operations.

References

1. Agrawal Jai P.: *Power electronic systems: theory and design*. Upper Saddle River, N.J., Prentice Hall, 2001.
2. Goebel P.: *Reconfigurable Manufacturing Systems*. Proceeds of the International Conference on Competitive Manufacturing, COMA'04, Stellenbosch, South Africa, 2004 pp. 69-79.
3. Harle R.K., Ward A., Hopper A.: Single *Reflection Spatial Voting: A Novel Method for Discovering Reflective Surfaces Using Indoor Positioning Systems*. MobiSys 2003, First International Conference on Mobile Systems, Applications and Services, San Francisco, 2003.
4. Harter A., Hopper A., Steggles P., Ward A., Webster P.: *The Anatomy of a Context-Aware Application*. Proceedings of the Fifth Annual ACM/IEEE International Conference on Mobile Computing and Networking, August, 1999.
5. Papadopoulos, H. C., Sundberg, C.-E. W., Reduction of mixed co-channel interference in microcellular shared time division duplexing (STDD) systems, IEEE Transactions on Vehicular. Technology, vol. 47, pp. 842-855, Aug. 1998
6. Potgieter J., Bright G.: *Modular Mechatronic Control System for Internet Manufacturing*. Proceeds of the 18 th International on CAD/CAM, Robotics and Factories of the future, Porto, 2002, pp. 529-536.
7. Kalpakjian S. Schmid S.R.: *Manufacturing Engineering and Technology*. Prentice-Hall, London, 2001.
8. Krar S., Arthur G.: *Exploring Advanced Manufacturing Technologies*. New York, Industrial Press, 2003.

DECISION SUPPORT SYSTEMS IN ELECTRIC POWER SUPPLY IN LIBERALISED ELECTRICITY MARKET

Kunicina N., Levchenkov A., Greidane S.

Abstract: Article presents the mathematical problem formulation of electric power supply in liberalised electricity market. The is methods of the problem solving, structure of problem solving algorithm is given in the article. The is experimental check of algorithm and main conclusions are given in the article. This research was supported by EU ESF postdoctoral grant (Nr 2004/0002/VPD1/ESF/PIAA/04/NP/3.2.3.1/0002/0007)

1. Introduction

By the present moment research of Latvian and foreign scientists were connected with the problems of decision making in the area of power engineering production, designing and maintenance. It includes also the investigations of those electric power supplies, for the solving of which the methods of operation investigation and the method of decision making in fuzzy conditions are applied, allowing obtaining the recommendations for an optimal (fairly) alternative choice. The methods of vector optimization allow analysing and making of an optimal alternative choice for solving of complicated problems, when an alternative together with contradictory features is considered.

Scientific novelty of the research covers the following aspects of the work: The problem of the development of the software agents modelling methods for the logistic systems of electric power supply in the situation of global networks has been formulated. This work considers the development of the algorithm of minimum expenses loop with losses for a server.

The purpose of the paper is to demonstrate how simple logistic problems Jenes and Bominik algorithms with the help of PHP and My SQL programming languages can realize. The application of effective commands is shown in the paper. The problem of transport intellectual systems is solved for the solving of logistic problems. The assessment and analysis of the algorithm are realized in the work applying a real server and solving the given informational logistic problems. A practical example of delivery for consumers with the lowest expenses is considered. The work is practically realized for the solutions in the global network. It is comfortably to use at any time as it is possible to input the necessary information on the road expenses. The development of the algorithm foresees searching for the shortest way with minimum payments for the delivery of raw materials to the electric energy producers, when a capacity of highway of a road and the distance between the raw materials seller and electric energy producer are given. For realization of this problem Jenesa and Bominik algorithm on the shortest way searching with losses is applied. The development of algorithm realization example with MySQL and PHP was completed to find the shortest way with the lowest expenses till the final point.

2. Mathematical problem setting

An oriented graph G=(V,A), with two separated peaks s and t, losing factor $h(v) \in Z^+$ for each peak $v \in V\setminus\{s,t\}$, capacity of road $c(a) \in Z^+$ for each loop $a \in A$ and requirements $R \in Z^+$ are given. In the article will consider algorithm for such function realisation $f: A \to Z_0^+$, with the following requirements:

$f(a) \leq c(a)$ for all $a \in A$;

for peaks $v \in V\setminus\{s,t\}$ the interrelation ia valid:

$$\sum_{(u,v) \in A} h(v) * f((u,v)) = \sum_{(v,u) \in A} f((v,u)) \qquad (1)$$

to peak f a flow comes not less than R.

If $f(v)=1$ for all $v \in V\setminus\{,t\}$, then the problem could be solved for the polinomial time with the standard theory of flows in the network.

Let us consider the following problem: searching for the shortest way of electric energy delivery to the consumers with minimum paste expenses. Road capacity and distance between post office and consumers are given. It is necessary to develop an algorithm for the shortest way searching till the final point. A practical example from the application of logistic algorithm for the post with the application of the Internet is suggested.

3. Method of the problem solving

For realisation of this problem, we will apply algorithm for the shortest way searching with losses. The nets with losses are the nets where the flow decreases outdoing from the loops. It means that input flow can differ from output. Let us consider the oriented graph, which contains the loops (i, j). from i to j flow f_{ij}, goes, value of expenses is c_{ij}. Permissible flow value, which can go around the loop (i, j) can be not less than $Emin_{ij}$ and not more than $Emax_{ij}$. Loop factor is marked with A_{ij} (Figure 1, Table 1).

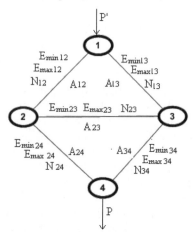

Figure1. Incoming data

Table 1. Sample table

Loup	Emin$_{ij}$	Emax$_{ij}$	N$_{ij}$	A$_{ij}$
(1,2)	0	12	2	1/3
(1,3)	0	16	20	½
(2,3)	0	16	20	½
(2,4)	0	4	12	¼
(3,4)	0	2	2	¼

At $A_{ij}>1$ the flow increases, at $A_{ij}<1$ the flow decreases.

The goal is achieved for the input flow F to be with minimum expenses and for the output flow F' to be with minimum ones. The problem with minimum expenses in the network with losses can be formulated as follows:

To minimise $\Sigma_i \Sigma_j c_{ij} f_{ij}$ under the condition that:

$\Sigma_j f_{ij} - \Sigma_j A_{ji} f_{ji} = 0$, $\quad\quad \Sigma_j f_{ij} - \Sigma_j A_{ji} f_{ji} = -F$, $\quad\quad 0 \leq f_{ij} \leq U_{ij}$ for all (i, j)

The algorithm of the problem solving:

step 1. Constructing of the flow:
- to find the minimum expenses loop which connects the source with the flow,
- to increase the minimum expenses loop, until one of the loops is saturated.

step 2. Marginal network construction: When the flow is given the direction along the scheme the output network is modified in order to reflect the current network condition and define the possible flow changes in the network. This network is called a marginal one. It contains all the saturated loops of the output network.

step 3. the process of the flow increasing: With the use of marginal network we will define a new flow way from the source to the flow and search for the maximum flow with the minimum expenses. At the end of the 3rd step the required flow is achieved, if not it is necessary to repeat steps 1 and 2.

3.1. Mathematical solution

Iteration 1

To find delivery expenses in node j we will apply the following formulas:

$V'_{ij} = (V_i + c_{ij})/A_{ij}$,

where: V_i- are the delivery expenses in node *i*, V'_{ij}- delivery expenses in node *j*

$V_j = \min[V'_{ij}] = \min[(V_i + c_{ij})/A_{ij}]$

The minimum expenses scheme contains loops (1,2), (2,3), (3,4). Maximum value of flow is:

$P_t = \min_{1 \le k < m}[U_k \Pi^m_{r=k} a_r]$,

where: m- is quantity of loops in the scheme

$P_t = \min[\frac{1}{2}, \frac{3}{4}, \frac{1}{2}] = 1/2$

$P_{ij}^{(I)}$- flow along the loop (i,j) in I.iteration

Going along the minimum expenses scheme from flows to the source we will get:

$I = 1: f_{34}^{(1)} = (1/2)/(1/4) = 2$, $f_{23}^{(1)} = 2/(1/2) = 4$, $f_{12}^{(1)} = 4/(1/3) = 12$

$f_{13}^{(1)} = f_{24}^{(1)} = 0 \Rightarrow F = 1/2, F' = 12$,

but the flow expenses, which is ½, is equal to ½*64=32.

Loops (1,2) and (3,4) - are saturated, thus we will switch them off from the output network. The reflected loops are (2,1), (3,2), (4,3) connected to the marginal scheme.

Iteration 2

$c^*_{ij} = c_{ij}$, $c^*_{ji} = -c_{ij}/A_{ij}$, $Emin^*_{ij} = Emin_{ij} - A_{ij}$, $A^*_{ij} = A_{ij}$, $A^*_{ji} = 1/A_{ij}$

The factor of acquisition g is:

$g = a_{34} * a_{32} * a_{24} = 2 * 1/4 * 4 = 2$; $P_h = 1/2 \min[1,1,1] = 1/2$

For the case of network generating, with the presence of cycle, the following formula should be applied:

$f'_k = P_h[g/(g-1)]/\Pi^m_{r=k} a_r$ $P_h = (g-1)/g\{\min_{1 \le k < m}[Emin_k \Pi^m_{r=k} a_r]\}$,

Increasing of the flow = ½ and after the second iteration P=1 P'=12. The expenses are 32 (from the 1 iteration)+80*½=72. The flow at the output is 1, flow expenses = 72 (1,2) and (2,4) now are saturated, then we will switch them out. Reflected loops are (2,1), (4,2) connected to the marginal scheme.

Iteration 3
The increasing of the flow is ½, but the total one P=1½ un P'=16. The expenses are 32(from the 1st iteration+40(from the 2nd iteration)+(1/2)*168=156. Loops (1,2), (2,4) are (3,4) switched out and reflected loops (4,3) and (3,1) are added.

Iteration 4
After the 4th iteration in the first node 16 comes, but from the 4th goes out 1½. The minimum expenses are 12*2+4*20+4*12+2*2=156

4. The structure of the problem realisation algorihtm

Solving the problem of information logistics was solved with the algorithm of Jenes and Bomik on the searching for the shortest way with losses can be applied for the solutions of the real problems. This paper considers the example of electric energy delivery to consumers with the minimum expenses (Figure 2). The essence of the problem is to find the shortest way with the minimum expenses till the point of destination:

$$V_j = \min[V'_{ij}] = \min[(V_i+c_{ij})/A_{ij}] \quad V'_{ij}=(V_i+c_{ij})/A_{ij},$$

where: V'_{ij} - are the delivery expenses in the node j through loop (i,j)

and maximum flow value is: $P_t = \min 1 \pounds k < m [E \min k P_{mr} = k_{ar}]$

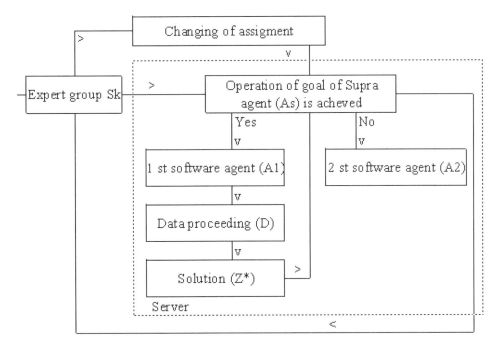

Figure 2. Scheme of interaction of Supra software agent and software agent

The realisation of the task is foreseen for the internet environment, for the development of a bank web-page based on the solutions of e-commerce. Web-page is very easy to be used as inputting the information at any time we can know how much electric energy is taken into account, the expenses for the way, or the problem could be solved from the state budget planning point of view – how much the contribution for obtaining the electric energy should be, that each consumer obtains a corresponding electric energy, knowing the delivery expenses.

The development of the algorithm envisages the searching of the shortest way of electric energy delivery with the minimum post expenses, when the road capacity and distance are given. Programming languages Apache, PHP and MySQL are applied to develop the program in a home page.

5. Experimental checking of the algorithm

The development of the algorithm envisaged searching for the shortest way with minimum post expenses for the delivery of the electric energy to consumers, when the road capacity and distance are given (Figure 3, 4).

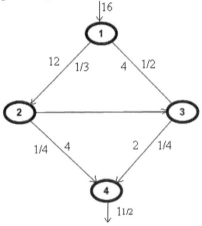

Figure 3. The graph of money flow

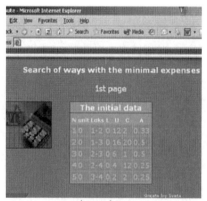

input data

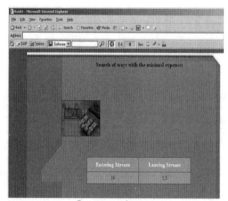

Oooutput data

Figure 4. Input and output flows flow

6. Conclusions

The paper considers the development of the algorithm of minimum expenses loop with losses for a server. The problem of information logistics was solved with the algorithm of Jenes and Bomik on the searching for the shortest way with losses. The algorithm is developed for a real server with the help of MySQL, Apache and PHP programming languages. The algorithm is applied for the solution of a problem of raw materials delivery to electric power producers with the minimum expenses.

References

1. Fishburn P., Fishburn P.C.: *Decision and Value Theory*. New York, Wiley, 1964.
2. Jakson M.A.: *Principles of Program Design*. London: Academic Press, 1975.
3. Kunicina N., Levchenkovs A., Ribickis L.: *Algorithm for software agents to power supply modeling in Baltic region*. EPE - PEMC, 11 th International power electronics and motion control conference, 2004.
4. Kunicina N.: *Development of modelling methods of software agents in electric power supplay and transport logistics systems*. Abstract of Promotional Paper, Riga, 2004.
5. Meystel A.: *Intelligent systems: architecture, design and control*. Alexander M. Mystel, James S. Albus, New York, Wilwy – Interorice, 2002.
6. Patel M., Honavar V., Balakrishnan K., (edit.): *Advances in the evolutionary synthesis of intelligent agents*. Cambridge, Mass. MIT Press., London., 2001.
7. Sauhats A., Bockarjova M., Dolgicers A., Silarajs M.: *New method for complicated automation systems simulation test EPE – PEMC*. 11 th International power electronics and motion control conference, 2004.
8. Witting, T.: *ARCHON: an architecture for multi agent systems*. Ellis Hordwood limited, England, 1992.
9. Williams E., Harp V.: *Improving logistical procedure within a hospital inpatient pharmacy*. The International Workshop on Harbour, Maritime and Multimodal Logistics Modelling & Simulation (HMS 2003) proceedings, 2003.

LOW COST LASER GUIDANCE SYSTEM FOR DE-MINING ROBOTS

Potgieter J., Bright G., Diegel O., Tinnelly M.J.

Abstract: The growing number of abandoned minefields throughout the world, and the continued use of such weapons, necessitates the development of effective, autonomous detection devices. The current methods of detecting mines rely on the sense of dogs and the very careful treading of human mine technicians It is for this reason that military research is showing a general trend towards more autonomous methods of war fighting and the requirements for more effective methods of mine detecting and breaching. Technologies are developing that will ensure that accurate systems are able to be developed that will not leave the use second-guessing the system's findings. Next to its sensory array, the most important aspect of an accurate system is its navigation method. This navigation system should be able to direct the user to the exact point that the detected object is at. It should also have a high resolution.

1. Introduction

Throughout the world previous war zones are littered with landmines. This problem is one that will simply not disappear, and with the relative ease with which they can be purchased and used, this problem continues to grow.

The current method of detecting these mines relies on the sense of dogs and the very careful treading of human mine technicians. The problem however is growing faster than these methods are able to deal with. For this reason, the demand for faster more effective and autonomous methods is increasing.

Military research is showing a general trend towards more autonomous methods of war fighting, and the requirement for more effective methods of mine detection and breaching. Presently, technologies are developing that will ensure accurate systems are able to be developed that will not leave the user second-guessing the systems findings.

The most important aspect of an accurate system, next to its sensory array, is its navigation method. This navigation system must be able to direct the user to the exact point that the detected object is at. It must also have a high resolution. The overall aim of this project was to develop a LASER Guidance System (LGS) to assist in the provision of autonomous function of an omni-directional robotic platform [1].

Within this aim existed the project objectives that were to:
- develop a system that was cheap, yet accurate to 92%, (the internationally accepted value for resolution), with respect to position and location,
- ensure that this system was small, so as to fit onto a small detection vehicle,
- system was reliable, in order to allow for ease of use in multiple environments,

- system was user friendly so as to allow for ease of use by all levels of technical competencies,
- a system that was easily interfaced to other platforms.

The following paper details these objectives.

2. System theory

The LGS developed relies on reasonably old theory. The basic principle of this system is similar to that used by sailing vessels when taking Horizontal Sextant Angle (HAS) fixes from a ship.

2.1. Theory of operation

The LASER emitter was set on top of two servomotors and was able to rotate 360°. The LASER light was then reflected back to vehicle and the angle at which the reflection was detected, recorded for geometric analysis.

The rotating head of the LASER featured three phototransistors that detected the reflection as a voltage drop. The servos both featured feedback control from encoders; therefore the position(s) of reflection were easily noted. The target points were merely retro-reflector units, similar to that found on any bicycle. Instead of relying on trigonometric calculation to give measurement, the system incorporated the use of, simpler, linear algebra [2].

As can be seen in figure 1, within the system existed X and Y-axes: the Y-axis being that upon which the targets sat, and the X-axis being the distance from the wall. This system operated in inches, something that the original system did as well, and although not metric, this provided the best-calculated values. The LGS operated by calculating the gradient of the lines and from this, was able to accurately determine the position of the robot, which is the intersecting point of the two circles.

The overall maximum of both axes was 127 inches, or 3.23 metres, and therefore by this, it could be stated that for RX, it lay in the interval $0 \leq RX \leq 127$ whilst for RY, it existed within the interval $-127 \leq RY \leq 127$.

As stated earlier, and as figure 1 shows, the robot was situated on the intersection of the two circles, with C23 being the centre of the circle intersecting with Targets 2 and 3, whereas C12 wais the centre of the circle that intersects with Targets 1 and 2. Therefore the coordinates for each circle could be defined as having both an X and Y component. This gave C23 the centre coordinates C23X C23Y, with C12 giving the coordinates C12X and C12Y [3].

The scanner measured the angles between the targets and gave the two angles as A12 and A23 respectively. Utilising a lookup table contained internally in the microprocessor, a Basic Stamp 2 Intelligence Chip (BS2-IC), the system, knowing A12 and A23, was able to calculate the values of C12 and C23.

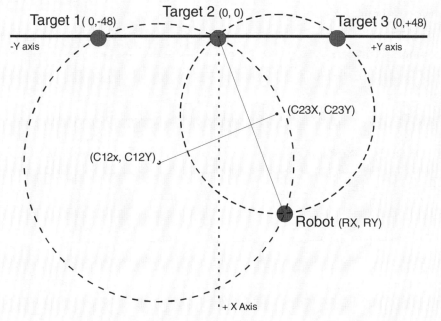

Figure 1. Geometric layout of guidance system [3]

The next process was for the processor to establish the equation, in X intercept form, for the line C12 to C23. This line was also defined as being the perpendicular bisector of line (0, 0) to (RX, RY). The equation for this line was therefore defined as being:

$$x = my + X\text{int} \quad (1)$$

where the slope is then defined as [3]:

$$m = \frac{C23X - C12X}{C23Y - C12Y} = \frac{N}{D} \quad (2)$$

In this equation, N and D represent numerator and denominator respectively. *Xint* is defined as the intercept of the graph with the equation [3]:

$$X\text{int} = (C23X + C12X)/2 \quad (3)$$

Due to the line passing directly through the coordinates (0,0), the X intercept, is equal to zero. This ensures that the equation can be reduced to the following [3]:

$$x = \frac{-1}{m} y \quad (4)$$

Through combination of the equations, we were then able to establish that the coordinate, of where the two lines cross, was at [3]:

$$\frac{-1}{m}y = my + X\text{int} \quad \text{or,} \qquad -2y = \frac{N * X\text{int} 2 * \text{Ta} 12}{(\text{Ta}^2 + N^2)} \qquad (5)$$

where:

N = C23X - C12X, $X\text{int}2 = 2 * X\text{int}$ = C23X+C12X, $Ta^2 = (\text{Ta}12)^2$ [3].

The value of $-2y$ is the Robot's RY coordinate, and thus the RX coordinate can be given as [3]:

$$RX = 2x = -2y\frac{\text{Ta}12}{N} \qquad (6)$$

Thus the sensors position was calculated and able to be given.

This element of the system was noted to have the distinct set back of no obstacle detection, nor any from of navigation for vehicle heading. For this reason a SHARP GP2D120 Infrared ranger and a Devantech CMPS03 Compass module were included in the design.

3. Prototype design

3.1. Hardware

Due to the nature of the task to be performed, the platform used obviously required suspension and general mobility for use in sometimes rough and challenging terrain. For this reason a remote controlled toy car was utilised as the platform. The model chosen was a replica of a Land Rover Freelander, which featured full functional suspension but with a small size and invariably cheap purchase price. Due to availability, modified DC Servo motors were used to drive the platform.

3.2. Onboard controllers

The vehicle contained a reasonable amount of on board intelligence; the BrainStem GP1.0 and the BS2-IC.

a) BrainStem General Purpose 1.0 Module

The Brainstem General Purpose (GP) 1.0 Module was used to control the Devantech CMPS03 Compass Module, the Sharp GP2D120 Module and also the four driving servomotors. This module supports 5 10-bin A/D inputs, 5 flexible Digital Outputs, a GP2D02 port, and 4 high-resolution servo outputs.

The internal architecture control of the chip originates from a Microchip PIC18C252. The chip also contains 32 bytes of on-board RAM with a running speed of 40 MHz. All this yields an overall chip speed of 9000 instructions per second.

The BrainStem can be run in one of three modes; in slave, whilst tethered to a host, in stand alone, or in a reflex mode - a mode that enables the processor to exhibit reflexive actions to inputs.

When run stand-alone mode, the chip was programmed using its own language. This language is known as TEA (Tiny Embedded Application) Language and is similar in syntax to the C or C++ high level languages.

When run in slave mode, the chip is controlled using packet passing of control-bit values. This mode can allow for a more intuitive and advanced method of operation, however does rely on a constant link between the chip and the PC being maintained. For the purpose of this project, the controller was run in slave mode, using a Visual Basic 6® Graphical User Interface (GUI).

b) Basic Stamp 2 Intelligence Chip

The Basic Stamp 2 Intelligence Chip (BS2-IC) controlled the LGS and associated servomotors. This controller interfaced with the RS 232 port of the PC through an in-built RS 232 communication hardware. The chip is based upon a Microchip PIC16C56c with 2 Kilobytes of on board EPROM, a 20MHz processor and 32 byte RAM. This microchip is capable of generating an overall speed of 4000 instructions per second. The chip also has 16 IO ports with 2 dedicated serial IO ports.

This chip was programmed using a language called PBASIC. Its syntax is almost identical to the original forms of the BASIC High-Level Language and thus programming this chip was relatively easy to understand. This chip was run from internally stored programs; however, it was able to interface with RS 232, whilst executing these programs very easily. This chip was run simultaneously as an RS 232 interface and circuit controller.

3.3. Software control

This part of the software development provided suitable challenges, especially when communicating with the BrainStem module. Due to the BrainStem being run in slave mode, communication to it was via data packets, with specifically assigned integers as commands. Specific pattern transmission routines were used to facilitate this. The code followed the same principle as when using the BrainStem console, only with the inclusion of the data packet size. The VB GUI shown in Figure 2 controlled the whole truck in a relatively user-friendly manner [5, 6, 7].

The needle indicated the Compass bearing, whilst motor control is by simple push buttons and the speed slider. The two sliders on the left also gave an indication of the direction and magnitude of speed experienced by the motors. The LASER Navigation module printed a simple text output to the textbox indicating the position in terms of RX, RY, and or whether or not the beacons cannot be located. The IR Ranger output was a simple progress bar that filled the closer an object came.

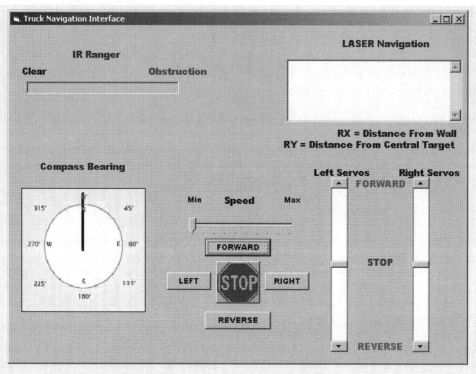

Figure 2. Visual Basic GUI for control of the Truck

4. Performance

The performance of the LGS was very much dependant on the overall ambient light conditions and thus how well the software had been tuned to deal with this. The accuracy of this system was tested in controlled conditions with readings being taken only from a straight-line position.

The reason for this use of a straight line was due to the fact that the algorithm responds its best when in a direct straight line from the centre target. Due to the underlying resolution of the system, it was only tested out to 127 inches, the maximum RX value (refer figure 1).

Investigation showed that optimal accuracy occurred at 65 inches out, with a large dead spot between 0 and 20 inches for values of RX. Figure 3 below shows the average error found for each position. The results presented here are a combined average for both RX and RY over a number of readings taken from experimentation with the system.

Although only measured for accuracy on one axis, this still gave a good appreciation of the overall accuracy for the system. The resolution of this system can be considered as ranging as well from its best value at 65 inches of 99.5% to its worse value of 95.4%.

4.1. Topology

The control topology of the system is displayed accurately in figure 4, and gives and indication of how the system interfaces to a PC. This system worked effectively and provided relatively effective autonomous function to the vehicle although with limited accuracy. The VB GUI shown (refer figure 2) previously greatly assisted with this. The overall performance of the system was therefore well within the required limits placed upon it.

5. Conclusion

The LGS provided the vehicle with accurate navigation. The required system resolution was achieved. The worst-case scenario for the system, in terms of resolution was calculated as being %92 with the best being %99.4.

This system compared well to commercially available systems with similar resolutions. The current price on standard LASER navigation systems is approximately $10 000. The approximate price of development of this system was in the vicinity of $1550. The LGS was small enough to be fitted onto a small mine detection platform. This objective was also successfully met with the entire system, including vehicle being quite small and only having a combined total weight of 1kg without batteries. With batteries the system was heavier, and this was an issue that will need to be investigated in order to reduce the payload. Due to the nature of battery technology, this is an issue that will not be easily solved.

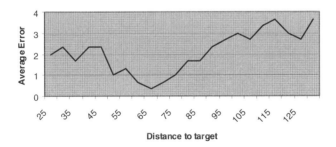

Figure 3. Average Error of System on a Straight Line

The LGS was only tested under laboratory conditions and thus the objective of developing a reliable system was met. The limitations experienced here include the a flimsy sensor head and also the continued requirement for the vehicle to be attached to an umbilical cord. This is another area that will need to be remedied in order to make the system viable for widespread use. The system was not tested under field conditions, however the system did meet the project objectives. Although limitations have been identified in terms of robustness and also inter-operability with other vehicles, the system functioned effectively.

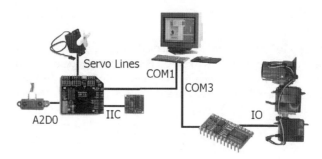

Figure 4. Control Topology

References

1. Fu K.S., Gonzalez R.C., Lee C.S.G. (1987): *Robotics: Control, Sensing, Vision, and Intelligence*. Auckland: McGraw – Hill Book Company.
2. Bolton W. (1999): Mechatronics: *Electrical Control Systems in Mechanical and Electrical Engineering*. Longman Press.
3. Ubersetzig J. (1999): *A Circular Navigation System* [Part 2]. The Robot Builder, Volume 11, number 10. Retrieved 15 June, 2003, from Robotics Society of California Website: http://www.rssc.org
4. Parker L. E., Schultz A. C.(Eds.) (2002): *Multi-Robot Systems: From Swarms to Intelligent Automata*. London, Kluwer Academic Publishers.
5. Bradley J.C., Millspaugh A.C.(2002): *Programming in Visual Basic 6.0*. Sydney: McGraw-Hill.
6. Dietel H.M., Dietel P.J., Nieto T.R.(1999): *Visual Basic 6: How To Program*. New Jersey: Prentice-Hall.

The International Journal of **INGENIUM** 2005 (4)

ENGINEERING ACHIEVEMENTS ACROSS THE GLOBAL VILLAGE

edited by

Janusz SZPYTKO

Mechatronic

Cracow - Glasgow - Radom, 2005

TABLE OF CONTENTS   page

7.   Mechatronic **447**

7.1.   Modular generic controllers for mechatronic systems,
       Bright G., Tlale S., Zyzalo J. 449
7.2.   Mechatronic design of the six legged walking robots,
       Buratowski T., Giergiel M., Uhl T. 457
7.3.   Accurate position control of a pneumatic actuator using DSP,
       Csiszár A., Gyeviki J. 463
7.4.   Micro linear actuator,
       Okabe H., Sakano S. 471
7.5.   An experimental robot controller for force-torque control tasks,
       Oláh I., Tevesz G. 477
7.6.   Multi-level continuous paralleling,
       Riismaa T. 485
7.7.   Control of two legs walking robot with pressure sensoers in the feet,
       Yada S., Okabe H., Sakano S. 493
7.8.   Autonomous hexapod robot for wall climbing tasks: distributed control architecture implementing CAN,
       Tlale N.S., Bright G., Xu W.L. 497
7.9.   Neural network control of robot hand,
       Shaw T., Kakad Y. P., Karwal V., Gandhi J., Hari Y. 505
7.10. The experimental container "GeoFlow" for the fluid science laboratory of the international space station,
       Stein C., DeSilva A.K.M., Harrison D.K. 515
7.11. Embedded controllers for magnetorheological dampers in suspension systems,
       Rosół M., Sapiński B. 523
7.12   Application of FMECA causal-consecutive analysis with an example of IMS digital platform,
       Kocerba A., Tekielak M. 531

All rights reserved. No part of this book may be reproduced, stored in a retrieval system, or transmitted, in any form or by any means, without prior written permission from the Publisher.

The International Journal of INGENIUM
Chief Editor: Professor David K. Harrison, Glasgow Caledonian University, UK

© GCU Glasgow

ISSN 1363-514x

A CIP catalogue record for this publication is available from British Library

Publishing cooperation: Instytut Technologii Eksploatacji – PIB w Radomiu

MODULAR GENERIC CONTROLLERS FOR MECHATRONIC SYSTEMS

Bright G., Tlale S., Zyzalo J.

Abstract: There is a constant need for manufacturers to change their manufacturing processes due to customer demands. There are manufacturing techniques based on the principles of flexible manufacturing and dedicated manufacturing production. This research addressed the control of automated machines by using the Mechatronic concept of modular systems. This includes modular machine controller hardware, software, mechanical design, and Generic "plug'n'play" capability. These conceptual designs allowed rapid reconfiguration of manufacturing machines that increased system simplicity and significantly minimised manufacturing downtime. The main objective of the first stage of research was to develop a modular, Mechatronic, plug and play, prototype controller for controlling two test bed systems: a six-axis robot arm and later a two-axis lathe. Successful control of both systems was accomplished. The lathe system was reconfigurable by having an extra modular axis that could be added and removed. Robot intelligence was achieved through a Mechatronic sensory system. The controller consisted of an industry standard racking system into which controller card modules were inserted. The card was designed to have the ability to control two types of actuation, DC and Stepper motors. These motors were within a limited range of motor voltage ratings. A PIC microprocessor was used on each card to communicate with a PC through the serial port interface. It was also used to communicate with other controller cards in a CAN topology. The PIC microprocessor also gave each card the ability to monitor various Mechatronics sensory equipment. All interfaces used for each system were industry standard so as to accommodate modular design.

1. Introduction

Mechatronic and robotic systems are becoming a common feature in the manufacturing workplace. A number of vendors offer component based manufacturing machines and robots. A problem with the automated systems available today is the lack of interoperability between different vendors' systems and indeed between different generations of automated machinery. Most systems are custom built and are difficult to upgrade [1].

The overall aim of the research is to achieve generic control of automated machinery for Agile Manufacturing by means of plug and play systems. In addition, middleware (Hardware and software communication research and development) for integration and distributed operations for Agile Manufacturing will be developed. Machine vision and smart recognition for Advanced Manufacturing quality control will be researched and implemented. High-level Mechatronic and Robotic programming for hardware integration for Advanced Manufacturing systems undertaken. Figure 1 shows the relationships between the components that will make up the Agile Manufacturing system.

The paper is concerned with the design, prototyping, testing and validation of the Mechatronics hardware layer, (Generic controller cards) and the development of the low-level driver software necessary for control and communication. The research project consisted of two stages. The first one being the hardware design and subsequent component assembly of a modular Mechatronic controller card. Further research resulted in designs of Generic Mechatronic controller cards. These cards were built and tested on primary electro-mechanical hardware (motors, sensors, slides etc.).

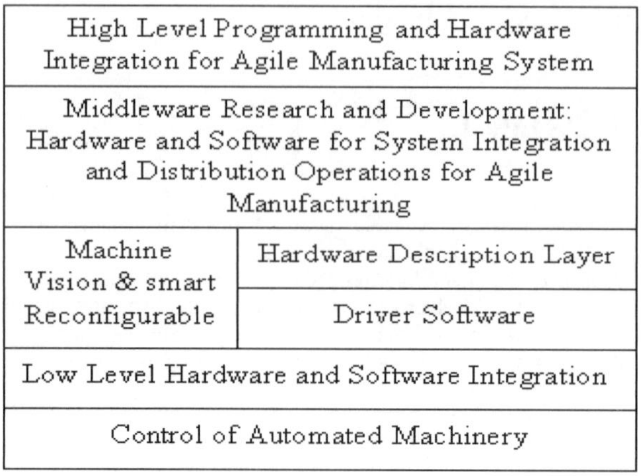

High Level Programming and Hardware Integration for Agile Manufacturing System	
Middleware Research and Development: Hardware and Software for System Integration and Distribution Operations for Agile Manufacturing	
Machine Vision & smart Reconfigurable	Hardware Description Layer
	Driver Software
Low Level Hardware and Software Integration	
Control of Automated Machinery	

Figure 1. Advanced Manufacturing layered system architecture for Agile Manufacturing

The next stage involved software control development. Detailed simulations were performed to simulated machine control. The results were then used to modify the controller of the machine where necessary. Extensive testing then proved the viability of the Generic Mechatronic controller cards for Agile Manufacturing on actual automated machine tools. Low-level software was used to program the automated machinery controlled by the Generic Mechatronic controller cards

2. Industrial robots for agile manufacturing

Agile manufacturing systems are controlled by computer-based technology. Since the advent of the microprocessor, computer-based technologies have made it possible to improve productivity, reduce manufacturing costs, and produce higher quality goods [2]. The development of the microprocessor has seen the use of robots for many applications. Agile manufacturing systems generally consist of a number of automated machine tools and materials-handling systems. These systems are flexible enough to reconfigure their elements or processes in order to produce custom products. Industrial Robotics is an important part of an agile manufacturing process due to the flexibility of robotic arms [3].

There are many brands of industrial robot arms available. A problem that has, however, occurred over the years is a lack of standardisation of the operating systems between the robots. There is little interoperability between different manufacturers' systems and between the different generations of automated machinery. Most robots and associated automated systems are custom built and expensive to upgrade [4]. The main disadvantage of current robots is that a dedicated and expensive controller is usually required for the robot's actuating systems. This proves costly and makes interfacing with the robot complex due to hardware and software variations. It also reduces the flexibility of the machine [5].

The objective of this research was to develop a low-cost modular Mechatronic plug-and-play controller for the control of a 6-axis robot. The system would then be tested with a Puma 560 robot, and with a CNC lathe (2 axis) to demonstrate the system modularity and flexibility. Ultimately, a truly flexible modular Mechatronic controller would not only be low cost, but would also the selection of any particular motorized system through software, and thus have the controller appropriately reconfigure itself. This means that a large-scale agile manufacturing system could be controlled from a single central PC, with fast set-up times and without any mismatches between hardware and software.

A Unimate PUMA 560 series robot arm was donated to the Institute of Technology and Engineering for the project. This robot was a six-axis revolute robot arm. An initial aspect of the project was to become familiar with the PUMA 560's actuation system. This industrial robot came supplied with the entire arm intact but without the control hardware or power supply. Each joint was operated by a 40V brushed permanent magnet DC motor. The motors for the bottom three joints were rated at 160W and the motors in the wrist rated at 80W. Each of the first three joints (waist, shoulder, elbow) was equipped with a 24V electromagnetic brake. All the joint motors were fitted with 250 line sine-cosine incremental encoders giving position feedback to the controller [6].

3. Hardware development

From the specifications of the PUMA 560 robot, it was found that a 40V DC power supply was needed to power the motors. A 24V DC supply was needed to disengage the electromagnetic brakes and a 5V logic supply was necessary to power all the encoder circuits on the motors and the microprocessors used in the controller. A power supply was built for the system, which included a main transformer with a VA rating of 640W. It also included logic power for the encoder circuits and the microprocessors using a computer ATX switch mode power supply unit. This was a convenient logic supply.

The encoders used for each robot motor were quadrature encoders that generated two output signals, with one of them offset 90° with respect to the other. By evaluating which signal was leading, the direction of rotation of the encoder disk could be determined. Based on an investigation into the workings of modern industrial incremental encoders, a square wave output was expected from the encoder receivers [7]. However, due to age of the encoder circuits, only small, 30mV peak-to-peak sine wave outputs were detected. In order for the microprocessor to be able count the encoder increments, the analogue sine wave signal required conversion into a digital pulse train.

Amplification of the signal was accomplished by using a standard LM741 operational amplifier to amplify the differential of each pair of signal lines. An LM339 voltage comparator converted the amplified signal into a pulse train. The circuit as was implemented on each of the encoder lines of the six motors. Modularity was one of the principal project requirements. This objective was to provide a reconfigurable system with the possibility of easily expanding the system to control more motors, or replacing any damaged modules as required. In this project, the system was to be tested on both a Puma 560 6 axis robot, and a 2-axis CNC lathe.

Figure 2. The 19" Racking System with slides

The project was designed around an industry standard 19" rack. Each individual 4U high rack bin could contain up to six separate 2-channel motor control module assemblies (Figure 2). A full rack bin could thus control up to 12 motors. The rack bin had several buses on the rear of the enclosure used for the 5V and 0V logic supply, the 40V and 0V motor power supply, and the I^{2}C communication bus.

The motor control modules could slide into the rack, from the front, as control cards. Each control card module was designed to control two motors. The block diagram for the modular system is shown in Figure 2. All the inputs and outputs of each card were routed to the back of the module so that the module could plug into the rack. The I/O for the microprocessor was wired to the back of the card with ribbon cable and a female DB-25 pin connector was used to plug the I/O into the rack bin. There were six gold plated terminals provided to plug into the back of the 19" rack bin. This provided the interface with all the high power motors wiring and communication bus system.

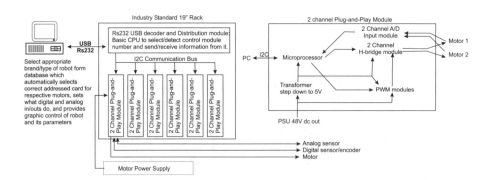

Figure 3. Motor Control Card Block Diagram

All the inputs and outputs of each card were routed to the back of the module so that the module could plug into the rack. The I/O for the microprocessor was wired to the back of the card with ribbon cable and a female DB-25 pin connector was used to plug the I/O into the rack bin. There were six gold plated terminals provided to plug into the back of the 19" rack bin. This provided the interface with all the high power motors wiring and communication bus system.

The H-bridges used for the project were: Devantech MD03s. These H-bridges provided the voltage and amperage requirement to drive the motor system. Rated at 50V 20Amp, the MD03 H-bridge is a medium power motor driver. The MD03 had four control modes. The mode used for the project was the Analogue Mode. This mode could be controlled using digital TTL logic levels, supplied by the microprocessor. The SDL input of the MD03 was used to indicate the direction, logic 0 for reverse direction and logic 1 for forward. The SDA input controlled the amount of voltage sent to the motor. 0V indicated no power and 5V indicated full power. This mode allowed a pulse width modulation (PWM) signal to be used to control the motor speed on the SDA input.

PWM is a technique used by most microprocessors and other controllers to control an output voltage at any value between the power rails. It consists of a pulse train whose duty cycle is varied so that it creates variable "on" and "off" states. The average output value approximates the same percentage as the "on" voltage [8]. In the case of MD03 H-bridge, used for the project, a 0% duty cycle represented a 0V supplied to the motor. A 50% duty cycle represented half of the supply voltage available to the motor. A 100% duty cycle was maximum voltage.

The robot's motors were controlled using PIC18C252 microprocessors. Each PIC was implemented in an off-the-shelf module called the BrainStem Moto 1.0, supplied by Acroname. The BrainStem has the following specifications: 40MHz RISC processor, 2 motion control channels with PWM frequencies from 2.5kHz-5kHz, 5 analogue inputs with 10-bit ADC, 11 digital I/O lines, 1 Mbit I^2C port, I^2C routing, status LED, 368 bytes of user RAM and RS-232 serial port communication. The Motor module was used in Encoder PID mode. The Encoder PID mode made adjustments to the position of the motor, called the set point, based on feedback from the motor's encoder. PID control was used in an algorithm running on the PIC to determine how much PWM was applied over time to the motor in order to maintain or move to a desired position. Proper selection of PID gain constants minimised oscillations of the motor [9]. Figure 4 shows the overall logic flow of the control loop.

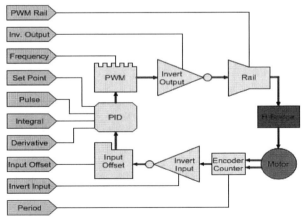

Figure 4. Basic Flow of Encoder PID Mode

The PIC allowed for reduced instruction set computing (RISC). The Moto responded to a limited set of pre-programmed commands. These commands were used to communicate with each microprocessor via a serial cable. The commands were used to retrieve and set data in the microprocessor [10].

Another important feature of the BrainStem Moto was that they could be daisy-chained using an I^2C bus. This allowed control card modules to be placed anywhere on the bus. All that was required was for the I^2C bus address to be set up on each BrainStem before "plugging" in the control card. This allowed communication with any BrainStem on the bus and between all the BrainStems themselves [11].

4. Software development

The Graphical User Interface (GUI) was developed using Visual Basic 6.0. The GUI was developed progressively to test the functionality of the system. It communicated with the microprocessors through RS232 serial communication. The GUI communicated with each of the BrainStems through the first BrainStem Moto module in the chain, which acted as a router [12]. Each packet of data sent from the PC started with an address for the BrainStem the message was intended for. If the packet was not addressed to the first router, it sent the pack on the I^2C bus to the appropriate module.

The main GUI, served the following major functions:
- communication management of packets between BrainStems and the PC,
- inputs for set points for each motor controlled,
- manipulation of the settings of programs running on the PIC. In the case of the BrainStems it included the PID control settings, mode selection, PWM, register monitoring, etc.

The GUI allowed for the control of up to six motors. Each motor's set point could be changed and sent to the respective BrainStem Moto module.

5. System performance

When the entire system was assembled, control of the robot was possible. The system performed well enough to control motion of each joint, (6 joints), of the PUMA robot arm to the specified set points. Though the system did eventually perform correctly and successfully controlled both the Puma 560, and a CNC lathe, there were initially a few problems to overcome. The main problem occurred with the logic power supply. Sporadically, the logic power, which supplied all the encoder circuits and microprocessors, would fail to power the most essential parts of the system. This would result in the unpredictable behaviour of the robot arm and make it unsafe. Further investigation revealed that the ground, 0V, of the logic power supply was floating causing differences in the ground, 0V, of other system components. Using the ground, 0V, of a bench-top power supply, this problem was solved.

The PID control method did not account well for the effects of gravity because the feedback gains of the PID algorithm were fixed. This meant that the system had a very low level of repeatability and accuracy. The encoder PID mode of the BrainStem Moto 1.0, worked well for the wrist joints, but did not perform well for the larger joints. When a new set point was entered for a motor, the PID control loop algorithm outputs maximum voltage to the motor until it neared the new set point based on the feedback from the encoders and the PID gains. It did this without accounting for the effects of inertia and gravity. This meant that, for the shoulder and elbow joints of the robot, a joint would move more rapidly in the down direction than in the up direction. To correct this problem, a velocity control method was implemented.

The BrainStem Moto 1.0 only had a 16-bit set-point number. This was reduced to 15-bits as the most significant bit was used to indicate direction [13]. Due to the resolution of the 250 line incremental encoders, only limited movement of a joint could be completed with each command sent to a BrainStem. The BrainStem program was unable to detect when a movement was complete so that the next movement of the robot could take place, as there were no flags set to indicate a new set point had been achieved. This problem was bypassed by introducing a timer delay between selected movements. The robot joints still required calibration.

Calibration was done each time the robot is turned on. Potentiometers on the motors were connected directly to the analogue inputs of the BrainStem Moto 1.0 that provided a 10-bit analogue-to-digital conversion of the potentiometer value. This value gave an indication of the position of the robot when it was powered up. The modular control system was initially tested and developed around the Puma 560 6 axis robot. After successfully being used to control the Puma, it was then used to control a 2 Axis CNC lathe. All that was required for the changeover were software changes to select the appropriate module addresses, and to retune the PID software algorithms for the Lathe.

Other systems such as dedicated interface cards and DAQ boxes do not have the flexibility of the described system. They are also limited to the number of devices that can be monitored and controlled. The generic Mechatronic controller cards provided the necessary flexibility and control to provide an adaptable and reconfigurable manufacturing system (RMS).

6. Conclusion

The Mechatronic hardware layer was developed for a robot system. It included design, prototyping, testing and validation of electronic hardware and associated driver software. All steps involved in the research project were successfully completed such that the robot was capable of movement through the developed software interface. The generic Mechatronic controller was then successfully used, with only some software reconfiguration, to control a CNC lathe.

BrainStem Moto 1.0 development was required for the system to improve its resolution. Further PIC programming could also improve the repeatability and accuracy of the control system. Research into more sophisticated control techniques is also an area for further development. Future work would include the development of the system so as to integrate it with a computer aided manufacturing (CAM) package for materials handling and assembly. The software will also be developed with a library of available robots and motor driven devices so that the system can easily be configured to drive different devices, thus allowing seamless integration into an Agile manufacturing environment.

Acknowledgements
The financial assistance of the NRF and University of KwaZulu Natal is gratefully acknowledged for the research project

References

1. Agrawal, Jai P., *Power electronic systems: theory and design*, Upper Saddle River, N.J.: Prentice Hall, 2001.
2. Acroname, *BrainStem: Moto.* 1994, Retrieved 14th Nov, 2003, http://www.acroname.com
3. Fu, Gonzalez, & Lee, *Robotics: Control, Sensing, Vision and Intelligence.* Singapore: McGraw-Hill, 1984.
4. Grabowski, R., Navarro-Serment, LE., Paredis, C., KhoslaInstitute, PK., *Heterogeneous Teams of Modular Robots*, In Robot Teams, edited by T Balch and L Parker, Carnegie Mellon University, 2002
5. Krar, S., & Arthur, G., *Exploring Advanced Manufacturing Technologies.* New York: Industrial Press, 2003.
6. Kidd, Paul T.; Agile Manufacturing: Forging New Frontiers, Addison-Wesley, 1994.
7. Goebel, P. (2004), Reconfigurable Manufacturing Systems, *Proceeds of the International Conference on Competitive Manufacturing, COMA'04*, Stellenbosch, South Africa. pp 69-79.2.
8. Kalpakjian, S and Schmid, S.R. (2001), Manufacturing Engineering and Technology, 4 th edition, *Prentice-Hall*, London, UK.
9. Paul G Ranky, A Real-time Manufacturing /Assembly system performance evaluation and control model with integrated sensory feedback processing and visualization, *Assembly Automation Journal*, Emerald publishers, UK, Volume 24, no 2, 2004, pp. 162-167.
10. Potgieter, J and Bright, G. (2002), Modular Mechatronic Control System for Internet Manufacturing. *Proceeds of the 18 th International on CAD/CAM, Robotics and Factories of the future*, Porto, Spain, pp 529 – 536.
11. Preiss, Kenneth; Steven L. Goldman; Roger N. Nagel, *Cooperate to Compete: Building Agile Business Relationships*, Van Nostrand Reinhold, 1996.
12. Valentine, Richard, editor, *Motor Control Electronics Handbook*, McGraw-Hill, Boston, 1998.
13. Wyeth G.F., Kennedy J. and Lillywhite J., *Distributed Control of a Robot Am*, Proceedings of the Australian Conference on Robotics and Automation (ACRA 2000), August 30 - September 1, Melbourne, 2000, pp. 217-222.

MECHATRONIC DESIGN OF THE SIX LEGGED WALKING ROBOTS

Buratowski T., Giergiel M., Uhl T.

Abstract: This document presents the problem related with the design and construction of six legged mobile robots. These mechatronic systems basing on simple algorithms for motion control can be used in many solutions. In this article several kinds of drives have been elaborated on.

1. Introduction

The fast progress in the field of modern technology let for the design and construction of more complex mechatronic devices in an easy way. Following the Robotics development in Poland and abroad we may notice that mobile robots design and group robots control have become a big field of interest. The construction variety of walking mobile robots (Buratowski, 2004; Morecki, 1994) has inspired the authors from the Department of Robotics and Machine Dynamics to initiate several mechatronical projects related with design and manufacture of six legged walking mobile robots. These constructions have different type of drives and walking schemes.

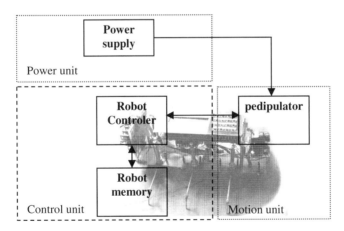

Figure 1. Walking robot scheme

According to literature (Morecki, 1994) we can define walking robots as technical devices oriented to relief locomotion functions of animals and insects equipped with limb. The scheme of a walking robot has been presented in fig.1. In the process of designing walking robots it is very important to assume a proper scheme of motion.

2. Design and manufacture of walking robots

2.1. Scheme of walking

The walking robots movement is a movement sequence of pedipulators (robot legs) in order to relocate from some place to another. Each pediputator movement is divided into step cycles, where the total cycle is when pedipulator is in the same configuration as it was at the beginning of the cycle. It means that robot's walking is a repeated pattern of leg (pedipulator) configuration with regular effect of progress in walking.

Animals and insects have different scheme of walking according to some properties connected with terrain and desired velocity. If we are considering six legged walking robots we can distinguish many schemes of walking, but there are two most popular: changing trisupported walking, wave walking. In the process of the schemes of walking design for six legged walking robots there is a need to keep three legs on the ground all the time, which creates stable tripod. After the scheme of walking assumption we are able to create the basic control algorithm.

2.2. The trisupported walking robot

In our projects we used trisupported walking in the process of robot' design. The forward scheme of walking for robots has been presented in fig. 2.

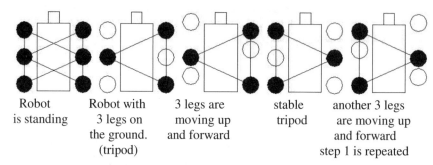

Figure 2. The trisupported scheme of walking for robot

The designed robot with trisupported walking scheme has specially built legs. Each leg can be defined as cybernetic mechanism with two degrees of freedom. As a drive intelligent material, called NiTinol has been used as silicon muscles (Raibert, 1983, Raibert, 1986, Song, 1989). Shape memory effect describes the process of restoring the original shape of a plastically deformed sample by heating it. This is a result of a crystalline phase change known as "thermoelastic martensitic transformation". Below the transformation temperature, Nitinol is martensitic. The soft martensitic microstructure is characterized by "self-accommodating twins", a zigzag like arrangement. Martensite is easily deformed by de-twinning. Heating the material converts the material to its high strength, austenitic condition (Borenstein, 1996; Everett1, 1995; Elahinia, 2002; Flatau, 2002) – figure 3.

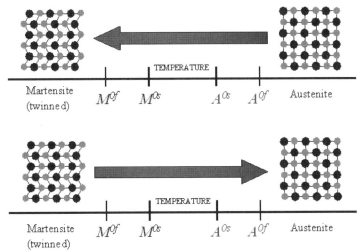

Figure 3. The nitinol phase transformation

A_{Of}: Austenite finish (or final) temperature
A_{Os}: Austenite start temperature
M_{Os}: Martensite start temperature
M_{Of}: Martensite finish (or final) temperature

The transformation from austenite to martensite (cooling) and the reverse cycle from martensite to austenite (heating) does not occur at the same temperature. There is a hysteresis curve for every Nitinol alloy (Gandhi, 1992; Gil, 1999; Gilbertson, 1993; Lewandowski, 2000; Morawiec, 1989; Pelton, 2003; Srinivasan, 2001) that defines the complete transformation cycle. The shape memory effect is repeatable and can typically result in up to 8% strain recovery. Martensite in Nitinol can be stress induced if stress is applied in the temperature range above Af(austenite finish temperature). Less energy is needed to stress-induce and deform martensite than to deform the austenite by conventional mechanisms. Up to 8% strain can be typically accommodated by this process. Since austenite is the stable phase at this temperature under no-load conditions, the material springs back to its original shape when the stress is removed. This extraordinary elasticity is also called "pseudo elasticity" or transformational "super elasticity". The typical curve of a properly processed Nitinol alloy shows the loading and unloading plateaus, recoverable strain available, and the dependence of the loading plateau on the ambient temperature. The loading plateau increases with the ambient temperature. As the material warms above the austenite finish temperature, the distinctive super elastic "flag" curve is evident. Upon cooling, the material displays less elasticity and more deformation until it is cooled to where it is fully martensite; hence, exhibiting the shape memory property and recovering its deformation upon heating. Nitinol alloys are super elastic in a temperature range of approximately 50 degrees above the austenite finish temperature. Alloy composition, material processing, and ambient temperature greatly effect the super elastic properties of the material.

So the robot can move its legs with nitinol by means of the electricity flow. One leg has two nitinol root, one for picking up and one for forward movement (fig. 5). The process of nitinol material shortening is connected with its properties and specifically with the phase austenite and martenzyt transformation. In fig. 4 the nitinol robot has been presented.

Figure 4. The nitinol robot   Figure 5. Designed leg

The control unit is responsible for control signals generation, those signals are connected with electricity flow by nitinol wire. In this robot control unit is based on microcontroller ATMEL AT90S8535. Another very important unit in robot is a power supply. The basic element of this unit is stabilization device LM7805 which supplies voltage. Because of high power consumption related with nitinol (approximately 1 A for each rot) in this robot exterior power supply has been used. The nitinol used makes possible to reduce length by 8 %. The robot velocity is low and it equals approximately 1 m per 60 min. Additionally the robot is equipped with CCD camera for inspection of the area localized at the front of the robot.

2.3. The wave walking robot

If we are taking into consideration the wave walking scheme, we may claim that this movement is very stable, because just one leg is put up. Starting from back leg, this leg is put up and displaced forward. All the remaining ground touching legs are moved backwards. The risen leg is put down and the process is repeated by another leg on the same side. Once the front leg is moved the process is repeated on the opposite side. The wave scheme of walking for the robot has been presented in fig.6 and fig.7.

The analyzed robot has 3 degrees of freedom. It means that the robot has 3 servomechanisms, one is localized in the middle of robot body and is attached with legs 2,5. The legs 1,3 have one drive and are mechanically connected. The same situation is with legs 4,6.

To start the movement sequence leg 2 is put down, in this moment legs 1,3,5 go up.

In this robot interior power supply has been applied. From the bottom of robot's body four batteries have been serially connected (2100 mAh), additionally one 9 V battery is used by sensors. The analyzed robot construction has been presented in fig.8 and its 3D AutoCAD simulation is presented in fig. 9.

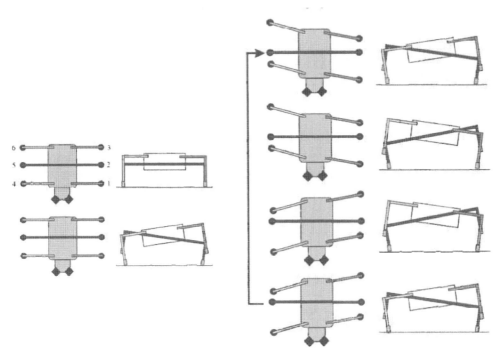

Figure 6. The legs description for robot

Figure 7. The wave scheme of walking for the robot

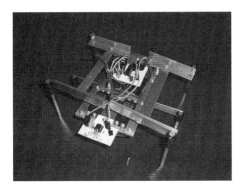

Figure 8. The three degree of freedom robot

Figure 9. 3D AutoCad presentation

The control unit has been placed on the top of the robot's body, the main reason is a easy access to the microcontroller PIC 16F84, with is responsible for robot's movement. This mechatronic device is additionally equipped by infrared sensors responsible for obstacle avoidance. The control algorithm is quite simple. It is connected with the wave scheme of walking and also takes into account the signals from sensors. The analyzed construction of the robot is fully autonomous and uses behavioral control very similar to animals and insects.

3. Summary

Currently experimental research is being carried out which will aid the improvement of some robot construction solutions and the broadening of its sensor possibilities. One of the problems of the presented robot is the slow construction movement. This is the result of the fact that nitinol comes back to the nominal state within about 3 seconds, after the voltage switch off.

However, the nitinol wire diameter diversity and the nitinol appropriate cooling system shall solve this problem and improve the construction as far as locomotion speed is concerned. As referred to the other robot, works are being carried out on the application of supplementary (e.g. sonar) sensors to increase the possibilities connected with the robot's surroundings inspection.

References

1. Buratowski T.: *Podstawy robotyki*, Wyd. Katedry RiDM AGH, Kraków, 2004.
2. Borenstein J, Everett H.R., Feng L.: *Navigating Mobile Robots: Sensors and techniques*, A.K. Peters Ltd, Wellesley, Massachusetts 1996.
3. Everett H.R.: *Sensors for Mobile Robots: Theory and application*, A.K. Peters Ltd, Wellesley, Massachusetts 1995.
4. Elahinia M.H., Ashrafiuon H.: *Nonlinear Control of a Shape Memory Alloy Actuated Manipulator.* Journal of Vibration and Acoustics, Vol.124, 2002.
5. Flatau A.B., Chong K.P.: *Dynamic smart material and structural systems*, Engineering Structures, Vol. 24, 2002.
6. Gandhi M.V, THOMPSON B.S.: *Smart Materials and Structures.* Chapman & Hall, 1992.
7. Gil F.J., PLANELL J.A.: *Thermal efficiencies of NiTiCu shape memory alloy*, Thermochimica Acta, vol. 327, 1999.
8. Gilbertson Roger G.: *Muscle Wires Project Book: A Hands-on Guide to Amazing Robotic Muscles.* Mondo-Tronics, Ill., 1993.
9. Lewandowski D.: *Zastosowania materiałów aktywnych.* Application smart materials, Wrocław, 2000.
10. Morawiec H., Bojarski Z.: *Metale z pamięcią kształtu.* PWN, Warszawa, 1989.
11. Morecki A., Knapczyk J.: *Podstawy robotyki*, Wydawnictwo Naukowo-Techniczne, Warszawa, 1994.
12. Pelton A.R., Duerig T.: *Proceedings of the International Conference on Shape Memory and Superelastic Technologies.* SMST, California, 2003.
13. Raibert Marc H., Sutherland Ivan E.: *Machines that Walk.* Scientific American, January 1983.
14. Raibert Marc H.: *Legged Robots that Balance.* MIT Press, 1986.
15. Song, Shin-Min and Waldron, Kenneth J.: *Machines That Walk: The Adaptive Suspension Vehicle.* MIT Press, January 1989.
16. Srinivasan A.V., McFarland D.M.: *Smart structures.* Cambridge University Press, 2001.
17. http://www.nitinol.com/3tech.htm#Memory NITINOL TECHNOLOGY.
18. www.nitinol.com -"Nitinol Facts".

ACCURATE POSITION CONTROL OF A PNEUMATIC ACTUATOR USING DSP

Csiszár A., Gyeviki J.

Abstract: This paper deals with one of the challenging problems in the field of robot control, namely how to make a pneumatic driven robot manipulator to move as fast as possible with small static positioning error and good tracking accuracy. The control problem in this paper is to keep the accuracy of the position control independent of the payload and piston position. The main contribution of this paper is a design of a robust sliding mode controller implemented on a DSP system.

1. Introduction

As an important driving element, the pneumatic cylinder is widely used in industrial applications for many automation purposes thanks to their variety of advantages, such as: simple, clean, low cost, high speed, high power to weight ratio, easy maintenance and inherent compliance. The most widely used controller is still the PID controller because of its simplicity and ease of implementation, but it isn't good for nonlinear systems with parameters and load variations. Because of control difficulties, caused by the high nonlinearity of pneumatic systems, a robust control method must be applied. There are two main classical directions in the field of robust control. One is the H infinite control for linear systems, and the other is the sliding mode control for nonlinear systems. Another solution is to employ the advanced nonlinear control strategies developed in recent years (soft computing) [6, 7]. Sliding mode control was introduced in the late 1970's [10, 11] as a control design approach for the control of robotic manipulators. In the early 1980's, sliding mode was further introduced for the control of induction motor drives [9]. These initial works were followed by a large number of research papers in robotic manipulator control [2], in motor drive control and power electronics [4]. Some of the experimental work indicated that sliding mode has limitations in practice, due to the need for a high sampling frequency to reduce the high-frequency oscillation phenomenon about the sliding mode manifold - collectively referred to as „chattering". A great deal of energy was invested in empirical techniques to reduce chattering. The design of sliding mode observers provided additional capabilities to sliding mode control.

2. Design of a sliding mode controller

A good introduction into sliding mode control can be found in [5, 8]. The design of a sliding mode controller consists of three main steps. One is the design of the sliding surface, the second step is the design of the control which holds the system trajectory on the sliding surface, and the third and key step is the chattering-free implementation. The purpose of the switching control law is to force the nonlinear plant's state trajectory to this surface and keep on it. The control has discontinuity on this surface that is why some authors call it switching surface. When the plant state trajectory is „above" the surface, a feedback path has one gain and a different gain if the trajectory drops „below" the surface. To introduce the idea of sliding mode control we can consider a single-input, single-output second-order nonlinear dynamic system:

$$\ddot{x} = f(\dot{x}, x) + G(x) \cdot u$$
$$y(t) = x(t)$$
(1)

Where x is the state variable, y is the output signal (position) of the controlled plant, u is the control signal and G is gain of control signal. If x_d denotes the reference state trajectory, then the error between the reference and system states may be defined as $e = x_d - x$. Let $s(\dot{e}, e) = 0$ define the „sliding surface" in the space of the error state. The purpose of sliding mode control law is to force error vector e approach the sliding surface and then move along the sliding surface to the origin (Figure 1) (where ① denotes the approaching phase, ② denotes the sliding phase and ③ denotes the chattering).

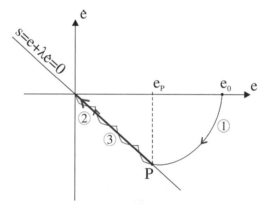

Figure 1. Sliding motion in the state space

The process of sliding mode control can be divided into two phases, that is, the approaching phase with $s(\dot{e}, e) \neq 0$ and the sliding phase with $s(\dot{e}, e) = 0$. In order to guarantee that the trajectory of the error vector e will translate from approaching phase to sliding phase, the control strategy must satisfy the sliding condition:

$$s(\dot{e}, e) \cdot \dot{s}(\dot{e}, e) < 0$$
(2)

This means that e will always go toward the sliding surface. In classical method of sliding mode control the scalar variable is calculated as a linear combination of the error and its derivative.

$$s = e + \lambda \cdot \dot{e}$$
$$\dot{s} = \dot{e} + \lambda \cdot \ddot{e} \quad (3)$$

where λ is a time constant type parameter. The simplest control law that might lead to sliding mode is the relay:

$$u = \delta \cdot sign(s) \quad (4)$$

The relay-type controller does not ensure the existence of sliding mode for the whole state space, and relatively big value of δ is necessary, which might cause a big chattering phenomenon. If the sliding mode exists ($s=0$), then there is a continuous control, know as equivalent control u_{eq} which can hold the system on the sliding surface. In practice, there is no perfect knowledge of the whole system and parameters, so, only $\hat{u}_{eq}$, the estimate of u_{eq}, can be calculated. Since $\hat{u}_{eq}$ does not guarantee convergence to the switching surface, in general, a discontinuous term is usually added to $\hat{u}_{eq}$, thus:

$$u = \hat{u}_{eq} + \delta \cdot sign(s). \quad (5)$$

The role of the discontinuous term in the control law is to hide the effect of the uncertain perturbations and bounded disturbance. The more knowledge is implied in the control law, the smaller discontinuous term is necessary. Usually, all state variables are not measurable, the system parameters are not known and the unmodeled dynamics may cause chattering. The most commonly cited approach to reduce the effects of chattering has been the so called boundary layer control. The discontinuous control law is replaced by a saturation function which approximates the *sign(s)* term in a boundary layer of the sliding manifold *s(t)=0*. For the purpose of obtaining high tracking accuracy, a thin boundary layer is required. However, it risks exciting a high-frequency control input. To reach a better compromise between small chattering and good tracking precision in the presence of parameter uncertainties, various compensation strategies have been proposed. To solve the chattering problem, another solution is the asymptotic state observer. An asymptotic observer can eliminate chattering despite discontinuous control laws. The key idea as proposed by Bondarev et al. (1985) is to generate ideal sliding mode in an auxiliary observer loop rather than in the main control loop (Figure 2). Since the observer structure and all the observer parameters are exactly known, an ideal sliding mode might occur in the observer-controller loop [3].

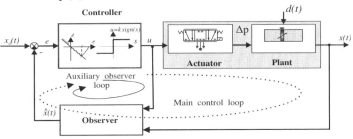

Figure 2. Observer-based solution

3. The servopneumatic positioning system

The system is shown in Figure 3 and Figure 4 (details can be found in [1]). It consists of a double-acting pneumatic rodless cylinder (MECMAN 170 type) with bore of 32 mm, and a stroke of 500 mm, controlled by a five-way servo- distributor (FESTO MPYE-5-1/8 HF-010B type). A linear encoder (LINIMIK MSA 320 type) gives the position.

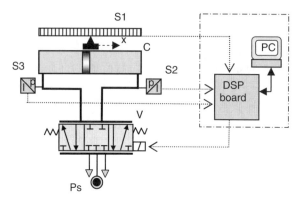

Figure 3. Configuration of pneumatic positioning system

Velocity and acceleration are obtained by numerical derivation. Pressure sensors (Motorola MPX5999D) are set in each chamber. Because of control difficulties caused by the high nonlinearity of pneumatic systems a nonlinear control method must be applied. So we will deal with robust control and a DSP based sliding mode control was designed. We have used the „eZdspTM for TMS320LF2407" DSP target board from Spectrum Digital. The control goal is to move the piston from any initial position to the target position. The orientation of the cylinder is variable as to see the effects of gravity. Using the sliding approach it is possible to minimize the positioning errors.

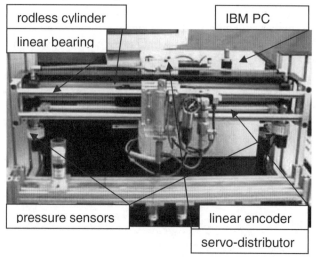

Figure 4. The experimental positioning system

4. Experimental result

The system pressure is set to be 6 bar, the sampling time is 2 ms. In order to analyze the positioning methods a real-time data acquisition program was designed for a PC to capture the system output data through the communication interface between the PC and the DSP controller. The control program is in the DSP program memory. So the DSP controller can operate independently. Since the DSP has a fast operation speed and a large memory, it can be applied in the control loop to increase the sampling frequency and the control accuracy.

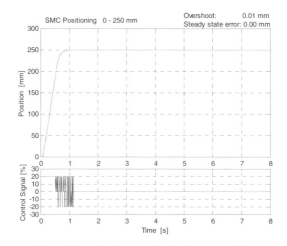

Figure 5. Positioning result with SM controller

In our experiments, we use D/A channel (Analog Devices AD420) for control and incremental encoder channel for position measurements. First, the performance of a well tuned PID controller and SMC are compared in case of a step change in the position reference signal. Because of the well known stick-slip phenomenon, the steady state error is alternating with the value of 3.8 mm. The position error of the DSP based sliding mode control is within ±0.01 mm. The robustness of the proposed SMC is also tested on the vertical position cylinder with mass load disturbances (Figure 6).

Figure 6. Positioning with mass load disturbances

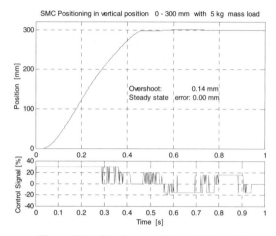

Figure 7. Positioning result in vertical position

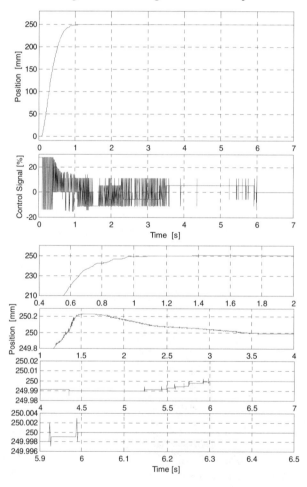

Figure 8. Performance of the controller with boundary layer

The experimental results indicate that the proposed sliding mode controller gives also fast response, good transient performance and it is robust to variations of system parameters and external disturbances, and they do not require accurate modeling. Further works we have done with applying the BTL5-S101 type Micropulse Linear Transducer with 1 µm resolution from Balluff (Figure 8).

As the Figure 8 shows, there is a setting time shorter than 1s, the overshoot is about 0,2 mm and there is no steady state error seen. For the purpose of measuring the tracking error of the piston, a sinusoidal (amplitude of 200 mm) desired input position trajectory is used. The experiment is repeated for 4 different frequencies (1/30, 1/10, 1/5 and 1/2 Hz). The tracking performances shows that the tracking errors are smaller (less then ± 5 mm) at low frequencies, mainly in the points where there is inversion in the direction of movement. Based on the laboratory measurements, we can conclude that the DSP based sliding mode controllers suitable and effective for the position control.

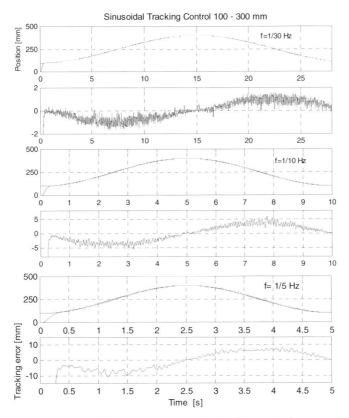

Figure 9. Experimental result for tracking control

References

1. Gyeviki J., Rózsahegyi K.: *Development of a servopneumatic positioning equipment.* MicroCad 2004, pp. 31-36.
2. Harashima F., Ueeshiba T., Hashimoto H.: *Sliding Mode Control for Robotic Manipulators.* 2nd Eur. Conf. on Power Electronics, Grenoble Proc., 1987, pp. 251-256.
3. Korondi P., Hashimoto H., Utkin V.: *Direct Torsion Control of Flexible Shaft based on an Observer Based Discrete-time Sliding Mode.* IEEE Trans. on Industrial Electronics, vol. IE- no. 2, 1998, pp. 291-296.
4. Korondi P., Hashimoto H.: *Park Vector Based Sliding Mode Control.* K.D.Young, Ü. Özgüner (editors), Variable Structure System, Robust and Nonlinear Control, Springer-Verlag, 1999.
5. Korondi P., Hashimoto H.: *Sliding Mode Design for Motion Control.* Studies in Applied Electromagnetics and Mechanics, vol. 16, IOS Press, 2000.
6. Mester G.Y.: *Neuro-Fuzzy-Genetic Controller Design for Robot Manipulators.* Proc.IECON'95, IEEE,Orlando, Florida,USA, Vol. 1, 1995, pp. 87-92.
7. Mester G.Y.: *Neuro-Fuzzy-Genetic Tracking Control of Flexible Joint robots.* Proc. I.Intern.Conf. on Adv. Robotics & Intelligent Aut. Athens, Greece, 1995, pp. 93-98.
8. Noriaki A., Korondi P., Hashimoto H.: *Development of Micromanipulator and Haptic Interface for Networked Micromanipulation.* IEEE/ASME Trans. on Mechatronics. Vol.6, No.4, 2001, pp.417-427.
9. Sabanovic A., Izosimov D.: *Application of sliding modes to induction motor control.* IEEE Trans. Industrial Appl., Vol. IA-17, No. 1, 1981, pp. 41-49.
10. Utkin V.: *Variable Structure Systems with Sliding Mode.* IEEE Trans., Vol. AC-22, No. 2, 1977, pp. 212-222.
11. Young K.D.: *Controller Design for Manipulator Using Theory of Variable Structure Systems.* IEEE Transaction on Systems, Man and Cybernetics, Vol. SMC-8, February 1978, pp. 101-109.

MICRO LINEAR ACTUATOR

Okabe H., Sakano S.

Abstract: The small actuator utilizing the bimorph piezoelectric device had been developed. But, it had been difficult that the actuator produced the large force because the phenomenon of the sliding contact was used as the driving principle. Another driving principle of the actuator has been found on the basis of the above-mentioned research. The hysteresis of the material was utilizing as the driving power of the actuator. It was clarified experimentally that the moving mechanism moved to right and left by the frequency of the power applied to the piezoelectric device, which constitutes the moving mechanism. Afterwards, the theoretical investigation was repeated, and it would be possible to theoretically clarify the moving principle of the moving mechanism. In this paper, while the theoretical investigation is described, it is also shown on the correspondence with the experiments shown in the previous report.

1. Introduction

The researches that are related to the new technologies such as biotechnology and nanotechnology are actively carried out. The microfilming of the actuator is one of the research subjects in these researches. The researches of the actuator using the combination between the various systems of the new drive principles and the new functional materials are advanced. In this study, the imbalance of the force which are occurred by the hysteresis property of the nonlinear vibratory displacement in the supported end of the nonlinear material is used as the power of the actuator. The drive principle is theoretically clarified and the validity of the principle is verified by the experiments.

2. Compositions and principle of the actuator

The operation principle of the actuator is shown in Fig. 1. The piezoelectric device vibrates, when the voltage of the frequency is applied. The vibration displacement has the hysteresis property. The moving body is moved to right and left in proportion to the frequency of the piezoelectric device. The body moves to the left direction in the case of the first resonance and in the second resonance the body moves to the right direction on the figure.

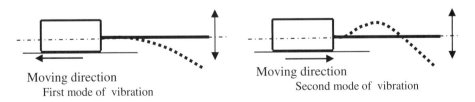

Figure 1. Driving principle of actuator

3. Theoretical analysis

3.1. Assumption for the analysis

The next assumptions are put for the analysis. The model of the nonlinearity of the voltage-displacement in the bimorphic piezoelectric device is shown in Fig. 2. The deformation of the piezoelectric device for the driven voltage is approximated as the deformation of the cantilever beam that receives the uniform load.

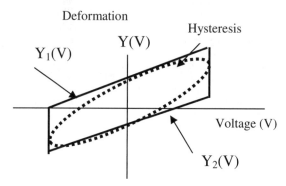

Figure 2. Model of the nonlinear property of the piezoelectric device, *Micro linear actuator*

3.2. Vibration deformation of the uniform load cantilever

The piezoelectric device has the viscoelastic loss in proportion to the vibration speed. When the periodic force works from the supported end of the cantilever beam to (a/l) as shown in Fig. 3, the vibration displacement of the (x/l) position from the supported end is shown as equation (1). ρ is the material density of the piezoelectric device and A is the cross section of the piezoelectric device. And, l is the length of the cantilever. E is Young modulus and ξ is the coefficient of the viscoelasticity:

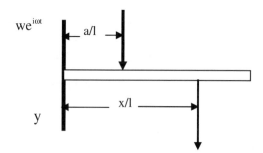

Figure 3. Vibration in cantilever beam

$$y(\frac{x}{l}, t) = \frac{w}{\rho A} \sum_{n=1}^{\infty} \frac{X_n(\frac{a}{l}) X_n(\frac{x}{l})}{\sqrt{(\omega_n^2 - \omega^2)^2 + (2C\omega)^2}} e^{i\omega t} \quad (1)$$

$$X_n\left(\frac{x}{l}\right) = \frac{1}{\lambda_n} \frac{\cos(\lambda_n \frac{x}{l}) + \cosh(\lambda_n \frac{x}{l})}{\sin(\lambda_n \frac{x}{l}) + \sinh(\lambda_n \frac{x}{l})} \tag{2}$$

$$\omega_n^2 = \frac{\lambda_n^4}{l^4} \frac{EI}{\rho A} \tag{3}$$

$$2C = \frac{\lambda_n^4}{l^4} \xi \frac{I}{\rho A} \tag{4}$$

X_n (x/l) is called the standard function and it shows the attitude of the vibration. λ_n is the solutions of the vibration equation. The vibration displacement in the tip of the cantilever beam becomes the following equation, when load w is uniformly distributed in the beam:

$$y(l,t) = \frac{2w}{\rho A} \sum_{n=1}^{\infty} \frac{1}{\lambda_n} \frac{\cos \lambda_n + \cosh \lambda_n}{\sin \lambda_n + \sinh \lambda_n} \times \frac{X_n(1)}{\sqrt{(\omega_n^2 - \omega^2)^2 + (2C\omega)^2}} e^{i\omega t} \tag{5}$$

The next equation is obtained, when the equation is further arranged. y_{st} is the deflection of the tip of the beam in uniformly adding the static load w to the cantilever beam. Micro linear actuator:

$$\frac{y(l,t)}{y_{st}} = 16 \sum_{n=1}^{\infty} \frac{1}{\lambda_n^5} \frac{\cos \lambda_n + \cosh \lambda_n}{\sin \lambda_n + \sinh \lambda_n} \times \frac{X_n(1)}{\sqrt{\left[1-\left(\frac{\omega}{\omega_n}\right)^2\right]^2 + \left(\frac{2C}{\omega_n}\right)^2 \left(\frac{\omega}{\omega_n}\right)^2}} \tag{6}$$

3.3. Comparison of theory and experiment

The comparison between the experiment and theory in the impressing the 60 V voltages to the bimorphic piezoelectric device is shown in Fig. 4. The theory agrees with the experiment almost as shown in the Figure 4.

4. Driving force of the actuator

4.1. Calculation of the driving force

The straight line as following will approximate the nonlinearity of voltage-displacement on the bimorphic piezoelectric device shown in Figure 3, where δ* is the displacement in proportion to the maximum voltage applied to the piezoelectric device:

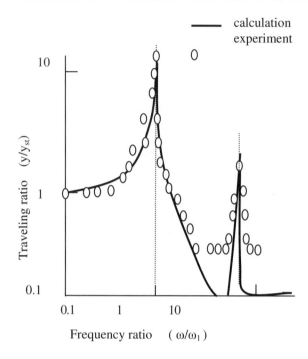

Figure 4. Frequency response of piezoelectric device (theory and experiment)

$$Y_1(V) = \frac{1}{\delta^*}(aV - b) \tag{7}$$

$$Y_2(V) = \frac{1}{\delta^*}(aV + b) \tag{8}$$

As shown in Fig. 5, when the force $wY_i(v)e^{i\omega t}$ in dx works, the force of the fixed end of the cantilever becomes the following equation. All force is calculated by integrating from the tip to the fixed end of the cantilever beam.

$$\begin{aligned}F_i(x,V) &= wY_i(V)e^{i\omega t}\sin\alpha dx \approx wF_i(V)e^{i\omega t}\alpha dx \\ &= wY_i(V)e^{i\omega t}\frac{dy(x)}{dx}dx\end{aligned} \tag{9}$$

$$F_i(V) = 2\frac{[wY_i(V)]^2 l^4}{EI}\sum_{n=1}^{\infty}\frac{1}{\lambda_n^5}\frac{\cos\lambda_n + \cosh\lambda_n}{\sin\lambda_n + \sinh\lambda_n}$$

$$\times \frac{X_n(1)e^{i\omega t}}{\sqrt{\left[1-(\frac{\omega}{\omega_n})^2\right]^2 + (\frac{2C}{\omega_n})^2(\frac{\omega}{\omega_n})^2}} \tag{10}$$

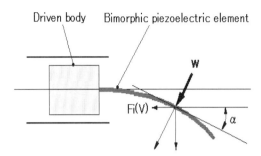

Figure 5. Load and force, *Micro linear actuator*

$F_1(v)-F_2(v)$ becomes the force in which the piezoelectric device pushes the moving body during one period of the vibration by the hysteresis of the piezoelectric device.

$$F_1(V) - F_2(V) = \frac{-4w^2 l^4}{EI\delta^{*2}} ab\, v * \sum_{n=1}^{\infty} \frac{1}{\lambda_n^5}$$

$$\times \frac{\cos\lambda_n + \cosh\lambda_n}{\sin\lambda_n + \sinh\lambda_n} \frac{X_n(1)e^{i\omega t}}{\sqrt{\left[1-\left(\frac{\omega}{\omega_n}\right)^2\right]^2 + \left(\frac{2C}{\omega_n}\right)^2\left(\frac{\omega}{\omega_n}\right)^2}} \tag{11}$$

The relation of w and δ* becomes the next equation:

$$w = \frac{8EI}{l^4} \delta^* \tag{12}$$

$$F_1(V) - F_2(V) = \frac{-256EI}{l^4} ab\, v * \sum_{n=1}^{\infty} \frac{1}{\lambda_n^5} \times$$

$$\frac{\cos\lambda_n + \cosh\lambda_n}{\sin\lambda_n + \sinh\lambda_n} \frac{X_n(1)e^{i\omega t}}{\sqrt{\left[1-\left(\frac{\omega}{\omega_n}\right)^2\right]^2 + \left(\frac{2C}{\omega_n}\right)^2\left(\frac{\omega}{\omega_n}\right)^2}} \tag{13}$$

4.2. Calculation of moving speed of the moving body

Fig.ure6 shows the comparison of the theory and the experiment. As shown in Figure, the theory and the experiment agree well. The moving body moves to the left direction in the first resonance and in the second resonance the moving body moves to the right direction

5. Conclusions

The driving principle of the micro actuator using the material nonlinearity was reported. The driving principle was verified by the experiments, while it was examined analytically. When the microfilming of the mechanism is attempted, the generation of the desired driving force can be expected on the proposed actuator.

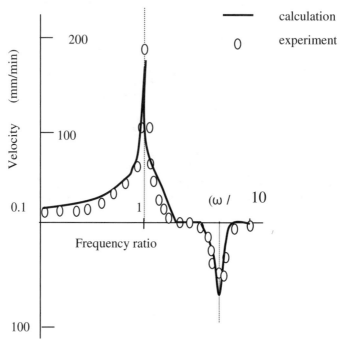

Figure 6. Moving body velocity (calculation and experiment), *Micro linear actuator*

References

1. Onoe H., Matsumoto K., Shimoyama I.: *Three dimensional micro self assembly using hydrophotobic interaction controlled by self assembled monolayers*. Journal of Microelectromechanical system, Vol. 13, No. 4, 2004, pp. 603-611.
2. Takeuchi S., Shimoyama I.: *Dynamic of a microflight mechanism with magnetic rotational wings in an alternating magnetic field*. Journal of Microelectromechanical systems, Vol. 11, No. 5, 2002, pp. 181-187.
3. Miki N.: *Study on micro-flying robots*. Advanced Robotics, Vol. 13, No. 3, 1999, pp. 245-246.
4. Shimoyama I., Yasuda T., Miura H., Kubo Y., Ezura Y.: *Mobile microrobots*. Robotica, Vol. 14, 1996, pp. 469-476.
5. Sudo S., Orikawa, Honda T.: *Locomotive characteristics of swimming mechanism propelled by alternating magnetic field*. International Journal of Applied Electromagnetic and Mechanics, Vol. 19, 2004, pp. 263-267.
6. Sudo S., Segawa S., Honda T.: *Magnetic swimming mechanism in a viscous liquid*. Proc. of the Symposium on Smart Materials for Engineering and Biomedical Applications, 2004, pp. 181-189.
7. Guo S. X.: *A new type of underwater microrobot using ICPF actuator*. A Journal of Chinese Association of Automation and Robot, Vol. 20, 1998, pp. 561-569.
8. Guo S. X.: *A new type of micropump using an ICPF actuator*. Information, Vol. 1, 1998, pp. 109-115.

AN EXPERIMENTAL ROBOT CONTROLLER FOR FORCE-TORQUE CONTROL TASKS

Oláh I., Tevesz G.

Abstract: This paper presents the overview of the hardware architecture of the Experimental Robot Controller built at the Department of Automation and Applied Informatics, BTU. The flexible software system used for operating robot controller is shown. The ongoing development to integrate a six-component force-torque sensor and integrate it into the control system is presented in details.

1. Introduction

At the Department of Automation and Applied Informatics an experimental robot control system has been developed. This plant consists of a NOKIA-PUMA 560 manipulator, the power electronics developed and built at the department, and an IBM-PC based multiprocessor system. The software system of this experimental robot controller is based on QNX Neutrino operating system. The purpose of this research is to study modern robot control algorithms and their realization in a real environment. The project focuses on the problems of multiprocessor systems, including the task distribution and communication. Another part of this research is integrating a six-component force-torque sensor into the robot control system and making use of this information in new robot control algorithms. This paper introduces the original hardware system. Additionally, the six-component force-torque sensor and the control extensions are shown. The features and system services of the new QNX Neutrino operating system is presented briefly. The processing components of this multiprocessor robot control system with its external interfaces are discussed during the overview of the hardware, and some further system level development possibilities is also outlined. The final part of the study gives the summary of the architectural and communication requirements of a hybrid position and force control system in the above environment.

2. The experimental robot controller

2.1. The Advanced Robot Controller Card

The most important part of the experimental robot control system is the ARC card [1]. The block scheme of this module is shown in Figure 1. Each card consists of two microprocessors: the so-called pre-processing unit is an i386EX microcontroller, while the second, the so-called joint processor is a TMS320C31 DSP. On each card four whole featured interfaces are available for electronic drives (joints or axes). Considering the complex task of robot control and depending on the implemented algorithms, the usage of two or three ARC cards was planned. In this way each card provides the connection and control task for three or two joints. The main blocks of the cards are as follows:

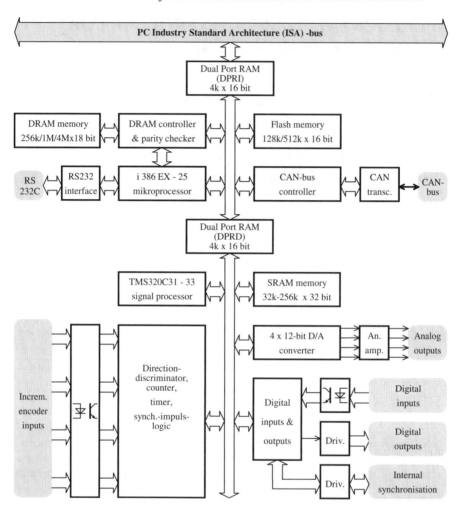

Figure 1. Block scheme of the ARC board

The preprocessing unit is one of the embedded microprocessor of Intel (i386EX). This device provides a 100% compatible environment with PCs integrating the basic peripheral functions of a main board. This 32 bit, i80386 based microprocessor serves not only with the MMU for the possibility of Virtual Protected Mode but contains special enhancements like the missed watch-dog circuit. The memory of this unit consists of a FLASH block (256 Kbytes or 1 Mbyte) and the usual DRAMs (the capacity is depending on the used SIM modules: from 512 Kbytes up to 8 Mbytes). Currently, the non-volatile chip contains a BIOS with reduced capabilities, but the implementation of an embedded operating system is in progress. The present functionality of the software running on this processor is to forward information between the main processor and the joint processor.

The joint processing units are high-speed CMOS 32-bit floating-point single-chip Digital Signal Processors (DSP) - TMS320C31. This DSP has high performance in the system with a capability of 16 MIPS and 32 MFLOPS. The tasks of this joint processor are:
- taking the position signals and calculating the speed and acceleration,
- credibility check using null impulses,
- supervising the position, speed and acceleration limits,
- producing the current set point values for servo amplifiers,
- providing the synchronization in starting and stopping the motor axles.

The high performance of the joint processor is supported by the high-speed static RAM (32-128 kwords). The required access time (18-20 nsecs) is unreachable in case of non-volatile memory so this device uses the built in Boot Loader to load its own program system from the FLASH of the preprocessing unit over the DPRAM. The FLASH memory of the preprocessor contains the programs for both processors on the card. To achieve the highest possible response time the signals of the position, speed and acceleration sensor incremental encoders are connected directly to the DSPs through intelligent interfaces (direction discriminator, counter and null impulse logic). The analogue circuits for current set point and the servo amplifier controlling digital input and output signals are connected to the joint processor directly, too. A three-wire synchronization channel facilitates the simultaneous movement of the axes connected to different ARC cards.

2.2. The six component force-torque sensor interface

The Experimental Robot Controller (ERC) was extended with a six-axis force-torque sensor. This equipment consists of a PC extension card that provides parallel communication with the controlling electronics. The sensor head is built between the last joint of the robot and the end effector.

The sensor consists of a metal spring and several electrical conductors, whose resistances change with deformation (strain gages). The strain gages are normally applied in multiplies of four and wired into a Wheatstone bridge configuration. If the bridges are excited with a fixed voltage source, the output voltage of the bridges is proportional to the applied force/torque. With the appropriate number of bridge circuits (six) the force and torque values can be measured in three dimensions.

The controlling electronics of the sensor excites the bridge circuits and processes the analogue bridge outputs signals. This unit – called 'MiniForce' [3] – performs the analogue-digital conversion of the bridge signals, and it makes a few preprocessing steps like digital filtering and matrix compensation, which eliminates the crosstalk effect between the bridge circuits. Finally, it passes the result via a 16-bit parallel communication channel (RS422 differential lines).

The newest hardware development related to the ERC is the new PCI bus interface card containing a TMS320C32 signal processor. This board communicates with the control electronics (MiniForce), and it can be utilized to complete some preprocessing steps on the measured data. The block scheme of the interface card is shown in Figure 2.

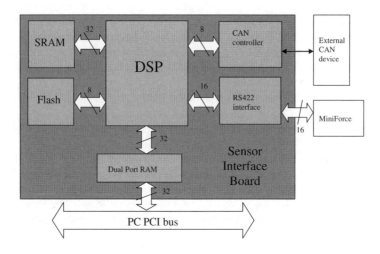

Figure 2. The sensor interface board

2.3. The software system of the controller

The present hardware of the Experimental Robot Controller contains six microprocessors. This complex multiprocessor system provides enough computing power to research modern control algorithms. The continuous improvement both on the hardware side and, simultaneously, the software side brought considerable changes in the resources of the robot control system. The first system had an Intel 80486 DX2 50MHz main processor. The ARC boards had an Intel 80386EX and a Texas Instruments TMS320C31 digital signal processor on them. The system was lent to continuous development: the host processor is now an Intel Celeron processor working at 633MHz, and the force-torque signals can be received through a PCI-based interface card. Table 1 shows the operational frequencies of the processing units and approximates the available bandwidth of the communication channels. In the early versions a highly distributed model was used to perform robot control tasks. As the host became more and more powerful, more tasks have been moved to this module and additionally, the host is now capable to run a newer version of the operating system. Most of the extended resources are available for more sophisticated robot control algorithms; however, some of them are used by the new operating system and its services.

Table 1. Processor frequencies and channel bandwidths

Processing Unit	Operational frequency
Host CPU (Intel Celeron)	633 MHz
Intel 80386EX	25 MHz
TMS 320C31	33 MHz
Communication Channel	Approximate Available Bandwidth
PCI bus on Host	264/132 Mbyte/sec (33MHz)
Host CPU ↔ i80368EX	5 Mbyte/sec
i80368EX ↔ TMS320C31	10 Mbyte/sec

QNX v4 real-time operating system was the initial operating software running on the host CPU. The overall computing power available now allowed upgrading it to the new version called QNX Neutrino / Momentics System (QNX v6) [3]. The new version has many changes based on the experience of the previous version; it supports several platforms, symmetric multiprocessing (SMP) and embedded solutions. Both versions are based on microkernel architecture. In this architecture the principal operating system components run as regular processes in their own separate address space. The microkernel provides the basic services such as *process manager services, thread services, scheduling services, synchronization services, signal services, message passing services* and *timer services*. The high level – process level overview and the flow of the information are shown in the Figure 3. The six processes hide the more than twenty threads that run in a synchronized manner and provide the full functionality of the robot.

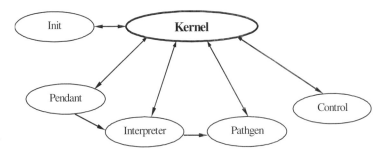

Figure 3. Software processes and communication

On the Intel 80386EX microprocessors an application specific program is running. It provides the necessary communication between the main processors and the joint processors. There is an ongoing development to embed the QNX Neutrino operating system into this environment. Since the operating system supports embedding, a successful development would provide more flexibility and the resources of these units could be exploited more.

The Texas signal processors are operated by an application specific program as well. The software of the sensor interface card is under development. As it shown by the test already executed it will provide the necessary speed and this module can be integrated into the complex robot control system.

3. The hybrid position and force control

The next assignment of the research project is to add the force control capabilities to the present position control system. It means that the necessary system variables should be implemented, and a new command should be integrated into the current environment. The principles of the force and torque control are as follows [2].

The force-torque control limits the freedom of the movement of the end effector. This limitation can be represented by the position specification matrix. It facilitates free motion along the appropriate axis:

$$\Sigma_F = \begin{bmatrix} \sigma_{F,x} & 0 & 0 \\ 0 & \sigma_{F,y} & 0 \\ 0 & 0 & \sigma_{F,z} \end{bmatrix}, \text{ where } \sigma_{F,i} = \begin{cases} 0 & \text{restricted motion} \\ 1 & \text{free motion} \end{cases} \quad (1)$$

The direction of force control is defined by the force specification matrix:

$$\tilde{\Sigma}_F = I - \Sigma_F \quad (2)$$

When not only the position but the orientation is limited, then similar equations describe the orientation and the moment. The first matrix is the rotation specification, while the second one is the moment specification one:

$$\Sigma_\tau = \begin{bmatrix} \sigma_{\tau,x} & 0 & 0 \\ 0 & \sigma_{\tau,y} & 0 \\ 0 & 0 & \sigma_{\tau,z} \end{bmatrix}, \text{ where } \sigma_{\tau,i} = \begin{cases} 0 & \text{restricted rotation} \\ 1 & \text{free rotation} \end{cases} \quad (3)$$

$$\tilde{\Sigma}_F = I - \Sigma_F \quad (4)$$

Using the equations above the generalized task specification matrices can give the description of the tasks related to the motion and contact forces. (A_F and A_τ are the orientation transformation between reference frame and tool frame):

$$S = \begin{bmatrix} A_F^T \Sigma_F A_F & 0 \\ 0 & A_\tau^T \Sigma_\tau A_\tau \end{bmatrix} \quad (5)$$

$$\tilde{S} = \begin{bmatrix} A_F^T \tilde{\Sigma}_F A_F & 0 \\ 0 & A_\tau^T \tilde{\Sigma}_\tau A_\tau \end{bmatrix} = I - S \quad (6)$$

The S and $\tilde{S}$ operators describe operations in the reference frame. The given position and orientation are transformed first into the target frames. In this frame the limitations and the specifications of the forces and torques can easily be considered, and the result is transformed back to the original frame. In the simplest case both the generalized task specification matrices and the orientation transformation matrices are constant. The task becomes more complex if the specification matrices vary. In this case the solution can contain some sections, since the elements of these matrices are discrete. The most complex situation is when the transformation matrices are not constant.

To take the result into the consideration the non-linear dynamic model of the robot should be transformed according to the dynamic decoupling:

$$H(q)\ddot{q} + h_{ccgs}(q,\dot{q}) = H(q)\ddot{q} + h_{ccs}(q,\dot{q}) + h_g(q) = \tau \quad (7)$$

If the robot was not redundant and not singular:

$$\dot{x} = J\dot{q} \quad (8)$$

$$\ddot{x} = J\ddot{q} + \dot{J}\dot{q} = J\ddot{q} + \alpha(q,\dot{q}) \quad (9)$$

$$\tau = J^T F \tag{10}$$

Substituting equations (8)-(10) in (7):

$$HJ^{-1}(\ddot{x} - \alpha) + h_{ccs} + h_g = J^T F \tag{11}$$

Introducing the following notation, the previous equation can be written in the form of (13):

$$H^* = J^{-T} H J^{-1};\ h^*_{ccs} = J^{-T} h_{ccs} - H^* \alpha;\ h^*_g = J^{-T} h_g \tag{12}$$

$$H^*(x)\ddot{x} + h^*_{ccs}(\dot{x}, x) + h^*_g(x) = F \tag{13}$$

This equation is similar to the method of the non-linear decoupling. To use this method the estimation of the H^*, h^*_{ccs}, h^*_g parameters are necessary. With the help of these estimations (stability and control errors) the method can be used for free motion and for continuous trajectory planning.

The previous result can be used in the case of limited motion or force-torque control:

$$F := F_{kin} + F_{active} + F_{ccgs} \tag{14}$$

These components contain the effect of the S and $\tilde{S}$ matrices corresponding to the operation space method:

$$F_{kin} = \hat{H}^*(x) S F^*_{kin} \tag{15}$$

$$F_{active} = \tilde{S} F^*_{active} + \hat{H}^*(x) \tilde{S} F^*_{damp} \tag{16}$$

$$F_{ccgs} = \hat{h}^*_{ccs}(\dot{x}, x) + \hat{h}^*_g(x) \tag{17}$$

Based on the equations above the joint moments evolved by the controller can be written in the following form:

$$\tau = J^T(q)\{\hat{H}^*(q)[SF^*_{mozg} + \tilde{S}F^*_{csill}] + \tilde{S}F^*_{aktic}\} + \hat{h}_{css}(\dot{q}, q) + \hat{h}_g(q) - J^T \hat{H}^*(q)\alpha(\dot{q}, q) \tag{18}$$

$$\tau = H(q)J^{-1}(q)\{SF^*_{mozg} + \tilde{S}F^*_{csill} - a(\dot{q}, q)\} + J^T \tilde{S} F^*_{aktiv} + \hat{h}_{css}(\dot{q}, q) + \hat{h}_g(q) \tag{19}$$

The resulting equations are quite complex, and they demand considerable computing power. The available resources should be used efficiently in order to complete the given task in a timely manner. The continuous evolution of the hardware resources and the capacity growth will help in building better and better controllers to complete the task of force and torque control. The control system that is going to be realized in the Experimental Robot Controller is shown in Figure 4.

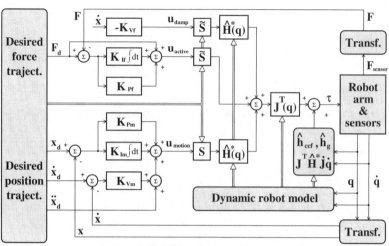

Figure 4. Hybrid position and force control

Acknowledgement

The project of studying modern robot control algorithms and their realization in a real environment is supported by Hungarian Research Found (OTKA, grant No. T029072, grant No. T042634).

References

1. Tevesz G., Bézi I., Oláh I. (1997): *A Low-cost Robot Controller and its Software Problems.* Periodica Polytechnica Ser. El. Eng. Vol. 41, No. 3., pp. 239-249.
2. Lantos B. (1997): *Robotok irányítása (*Robot control). Akadémiai Kiadó, Budapest (in Hungarian).
3. Oláh I., Tevesz G. (2002): *Software Problems of an Experimental Robot Controller Based on QNX Real-Time Operating System.* Periodica Polytechnica Ser. El. Eng. Vol. 46, No. 3-4., pp. 151-161.

MULTI-LEVEL CONTINUOUS PARALLELING

Riismaa T.

Abstract: A method of description and optimization of the continuous structure of hierarchical robot based processing system is presented. Two types of variable parameters are defined: for level size and for relations between adjacent levels. The structure of the system is defined as a finite sequence of density functions of distributions. Each distribution will correspond to the connections or relations between this and previous level and shows how the size of previous level is distributed between the sizes of this level. Corresponding optimization problem is a calculus of variations problem. It is shown that the given choice of variable parameters and the statement of the optimization problem as a double minimum problem enable to construct methods for finding a global optimum of total loss function and select the corresponding structure.

1. Introduction

In robot based multi-level distribution system each element is a supplier for lower level elements and a customer for one higher-level element. The zero-level elements are only customers and the unique top-level element is only supplier. In optimization of multi-level distribution process the choice of optimal number of supplier-customer robots on each level is mathematically complicated problem what must be largely simplified. The robot based assembling process (The academic example in part 4) as well as a broad class of design and implementation problems, such as component selection in production systems, reconfiguration of manufacturing structures, optimization of the hierarchy of decision making systems, multi-level aggregation, etc. can be mathematically stated as a multi-level partitioning problem (Mesarovic et al., 1970; Riismaa, 1993; Laslier, 1997).

For discrete approach (Riismaa, 1993; Riismaa, 2002) the set of feasible structures for such class of systems is defined as a set of hierarchies and the graph theory representation of this set is constructed. Two types of modeling parameters are defined: for level size and for relations of adjacent levels. It is shown that the given choice of variable parameters and the statement of the optimization problem as a double minimum problem enable to construct methods for finding a global optimum of total loss function and select the corresponding structure (Riismaa, 1993; Riismaa, 2002). Arising integer programming problem is extended to the continuous convex programming problem (Riismaa, 2002). For finding the global optimum the method of local search is constructed, where on each step of iteration the value of objective function on ends of unit cube is calculated (Riismaa et al., 2003).

The discrete statement of structural selection procedure (Riismaa, 2002) is based on the full set of hierarchical trees of feasible structure what could be composed from the given set of elements (Riismaa, 1993). In the context of this approach an element is considered as a logical part of the processing system that is carrying out an identifiable mission and obeys the necessary functionality and autonomy (Berio and Vernadat, 1999; Mesarovic and Takahara, 1975).

The main difficulty from point of view of optimization is that the number of subsets of partitioning is a variable parameter. For corresponding optimization problem it means that for goal function the number of summands, the upper limit for summation (integer valued parameter) is a variable parameter. The finding of solution considered this nonlinear integer programming problem is complicated mathematically. This is a reason why continuous statements considered in this paper are more useful. Summations signs in goal function of discrete statement are substituted with integrals in continuous initial statement. Limits of integrations may have non-integer values too. Constructed in such a way models have good mathematical qualities from point of view of optimization.

The continuous approach considered in this paper the structure of the processing system is defined as a finite sequence of density functions of distributions. Each distribution will correspond to the connections between this and previous level and shows how the size of previous level is distributed between the sizes of this level (Riismaa, 2004 a, b, c). Examples of problems of this class are aggregation problems, structuring of decision-making systems, database structuring, and the problems of multiple distribution or centralization, multi-level tournament systems, multi-level distribution systems, optimal clustering problems.

2. Discrete optimal multi-level paralleling

Consider s-level hierarchies, where nodes on level i are selected from the given nonempty and disjoint sets and all selected nodes are connected with selected nodes on adjacent levels. All oriented trees of this kind form the feasible set of hierarchies (Riismaa, 1993; Randvee et al., 2000; Littover et al., 2001).

Suppose $Y_i = (y^i_{jr})$ is an adjacent matrix of levels i and $i-1$ $(i=1,...,s)$: $y^i_{jr} = 1$ if j-th element on level i and r-th element on level i-1 connected, $y^i_{jr} = 0$ otherwise. Suppose m_0 is the number of 0-level elements (level of object). All hierarchies with adjacent matrixes $\{Y_1,...,Y_s\}$ from the described set of hierarchies satisfy the condition (Riismaa, 2002):

$$Y_s \cdot ... \cdot Y_1 = \underbrace{(1,...,1)}_{m_0} \ . \tag{1}$$

The general optimization problem is stated as a problem of selecting the feasible structure which correspond to the minimum of total loss given in the separable-additive form:

$$\min\left\{ \sum_{i=1}^{s}\sum_{j=1}^{m_i} h_i\left(\sum_{r=1}^{m_{i-1}} d^i_{jr} y^i_{jr} \right) \middle| Y_s \cdot ... \cdot Y_1 = \underbrace{(1,...,1)}_{m_0} \right\} \text{ over } Y_1,..., Y_s, \tag{2}$$

where: $h_i(\cdot)$ is an increasing loss function on *i*-th level and d^i_{jr} is the element of $m_i \times m_{i-1}$ matrix D_i for the cost of connection between the *i*-th and (*i*-1)-th level (Littover et al.,2001; Riismaa and Randvee, 2002). The meaning of functions $h_i(k)$ depends on the type of the particular system.

By the optimization of the structure of multi-level tournament system the loss inside the *j*-th tournament on *i*-th level is $h_i(k) = a_i k(k-1)$, where k is the number of participants. Mathematically this problem is an integer programming problem with a non-continuous objective function and with a finite feasible set. For solving this kind of nonlinear integer programming problems only ineffective classical methods are known. For solving the problem (2) in (Riismaa and Randvee, 2002; Riismaa, 2003; Littover et al., 1999) a class of recursive algorithms, where the recursion index is the level number, is constructed.

In earlier papers (Riismaa, 1993; Riismaa, 2002) an important special case is considered where the connection cost between the adjacent levels is the property of supreme level. In this case the total loss depends only on the sums:

$$\sum_{r=1}^{m_{j-1}} y^i_{jr} = k_{ij} \tag{3}$$

Recall that:

$$\sum_{j=1}^{p_i} k_{ij} = p_{i-1}, \quad (i = 1,...,s) \tag{4}$$

where: p_i is the number of nodes on *i*-th level and k_{ij} is the number of edges outgoing from the *j*-th node on *i*-th level.

In this case the optimization procedure can be reorganized into two sequential stages:

$$\min\left\{\sum_{i=1}^{s} g_i(p_{i-1}, p_i) \mid 1 \le p_i \le p_{i-1}; p_i \in N\right\} \text{ over } p_1,...,p_{s-1} \tag{5}$$

where:

$$g_i(p_{i-1}, p_i) = \min\left\{\sum_{j=1}^{p_i} h_i(k_{ij}) \mid \sum_{j=1}^{p_i} k_{ij} = p_{i-1}\right\} \text{ over } k_{i1},...,k_{ip_i} \tag{6}$$

Free variables of the inner minimization are used to describe the connections between the adjacency levels. Free variables of the outer minimization are used for the representation of the number of elements at each level. For solving this reduced problem in earlier papers (Riismaa, 2002; Riismaa et al., 2003) two different types of numerical methods are constructed. To the first type belong recursive algorithms where the recursion index is the number of level. The second type algorithms for solving (5) - (6) are iteration methods of local search (Riismaa, 2002). For finding the global optimum on each step of iteration the value of goal function on ends of some kind of unit cube is calculated.

3. Continuous optimal multi-level paralleling

In the continuous case the structure of the system is defined as a finite sequence density functions of distributions

$$\{f_1(x),...,f_s(x)\}, \qquad f_i : R^+ \to R^+ (i = 1,...,s), \qquad (7)$$

where: R^+ is the set of non-negative real numbers.

The i-th distribution will correspond to the connections between the levels i and $i-1$. The parameter $z_i > 0 (i = 0,...,s)$ is the size of level i. The function $f_i(x)$ describe how the size z_{i-1} of level i-1 is distributed between the size z_i of level i:

$$\int_0^{z_i} f_i(x) dx = z_{i-1} \quad (i = 1,...,s) \qquad (8)$$

Suppose the size of lowest level z_0 and size of highest level z_s are fixed. The statement of the problem is as follows:

$$\min_{z_1,...,z_{s-1}} \left\{ \sum_{i=1}^{s} q_i(z_{i-1}, z_i) \mid z_i \in R^+ (i = 1,...,s) \right\} \qquad (9)$$

where:

$$q_i(z_{i-1}, z_i) = \min_{f_i(z)} \left\{ \int_0^{z_i} h_i(x, f_i(x)) dx \mid \int_0^{z_i} f_i(x) dx = z_{i-1} \right\} (i = 1,...,s-1). \qquad (10)$$

Here $h_i(x, z)$ describe connections between the elements of level i and of level i-1. The meaning of these functions depends on the type of the particular system.

3.1. Continuous horizontally homogeneous hierarchies and the condition of optimality

For continuous statement the hierarchy is called horizontally homogeneous if :

$$h_i(x, z) = h_i(z) \quad (i = 1,...,s) . \qquad (11)$$

This condition substantially simplify the problem (9) - (10) and enables to find the solution for inner optimization problem (10) analytically:

$$f_i^*(x) = \frac{z_{i-1}}{z_i}, \quad x \in [0, z_i] \; (i = 1,...,s). \qquad (12)$$

This relation will mean, that the size z_{i-1} of level i-1 is distributed between the size z_i of level i uniformly. To use (12) the outer problem (9) transform to continuous programming problem:

$$\min_{z_1,\ldots,z_{s-1}} \left\{ \sum_{i=1}^{s} z_i \cdot h_i\left(\frac{z_{i-1}}{z_i}\right) \middle| z_i \in R^+ (i=1,\ldots,s) \right\}. \tag{13}$$

To calculate the derivatives of objective functions and equalizing these to zero the conditions of optimality are as follows:

$$h_i\left(\frac{z_{i-1}}{z_i}\right) - \frac{z_{i-1}}{z_i} \cdot \frac{dh_i\left(\frac{z_{i-1}}{z_i}\right)}{dz_i} + \frac{dh_{i+1}\left(\frac{z_i}{z_{i+1}}\right)}{dz_i} = 0 (i=1,\ldots,s-1) \tag{14}$$

If $h_i(x)$ $(i=1,\ldots,s)$ are convex functions, then (13) is convex programming problem. The convexity of goal function of problem (13) is controllable immediately.

3.2. Monotony of Level Sizes of Optimal Horizontally Homogeneous Hierarchies

3.2.1. Theorem

Suppose $\left(z_1^*(z_0^1),\ldots,z_{s-1}^*(z_0^1)\right)$ is the solution of problem (13), where z_0^1 is the 0-level size (the given size of object). If $z_0^1 \leq z_0^2$, there is the solution $\left(z_1^*(z_0^2),\ldots,z_{s-1}^*(z_0^2)\right)$ of problem (13) for this 0-level size z_0^2 and:

$$z_i^*(z_0^1) \leq z_i^*(z_0^2) \quad (i=1,\ldots,s-1). \tag{15}$$

4. Academic example: optimization the structure of multi-level robot based processing system

Consider the processing of n product units (Riismaa, 2002). In case of one processing unit the overall processing and waiting time for all n units is proportional to n^2 and is a quickly increasing function. For this reason the hierarchical system of processing can be suitable. From zero-level (level of object) the units will be distributed between p_1 first-level robots and processed (aggregated, packed etc.) by these robots. After that the units will be distributed between p_2 second-level robots and processed further and so on. From p_{s-1} (s - 1)-level the units will be sent to the unique s-level robot and processed finally. The cost of processing and waiting on level i is approximately

$$g_i(p_{i-1}, p_i) = (d_i l_{i-1} p_{i-1}/p_i)^2 p_i + a_i p_i \quad (i=1,\ldots,s) \tag{16}$$

Here l_i is the number of aggregates produced by one robot on level i (a number of boxes for packing robot), d_i is a loss unit inside the level i, and a_i is the cost of i-th level robot. The variable parameters are the number of robots on each level p_i ($i = 1, \ldots, s$).

The goal is to minimize the loss (processing time, waiting time, the cost of robots) over all levels:

$$\min \sum_{i=1}^{s}((d_i l_{i-1})^2 \left(\left(p_i \left(\left[\frac{p_{i-1}}{p_i} \right] + 1 \right) - p_{i-1} \right) \left[\frac{p_{i-1}}{p_i} \right]^2 + \left(p_{i-1} - p_i \left[\frac{p_{i-1}}{p_i} \right] \right) \left(\left[\frac{p_{i-1}}{p_i} \right] + 1 \right)^2 \right) + a_i p_i) \quad (17)$$

over natural $p_i (i=1,...,s)$. Here $[p]$ is the integer part of p. The goal function of this discrete programming problem is discrete-convex (Riismaa, 1993).

This structure description and optimization problem may be considered as a discrete modelling problem where the set of feasible structures is a finite set of hierarchies (Riismaa, 1993; Riismaa and Randvee, 2002). For solving this integer programming problem it is possible to use recursive algorithms presented in (Riismaa, 2002; Riismaa and Randvee 1997). It is possible to extend this function to convex function (Riismaa, 2003; Rockafellar, 1970) and use for solving this convex programming problem the method of local search described in (Riismaa, 2003). But more successful is to use the method of continuous modelling described in this paper. For this statement each level is described by size, which may have non-integer values too. Now it is possible to reduce this optimisation problem to a simple one:

$$\min \left\{ \sum_{i=1}^{s} (d_i l_{i-1} p_{i-1})^2 / p_i + a_i p_i \right\} \quad (18)$$

over non-negative $p_i (i=1,...,s)$, which is an easily solvable convex (continuous) programming task.

5. Conclusion

Many discrete or finite hierarchical structuring problems can be formulated mathematically as multi-level partitioning procedure of a finite set of nonempty subsets (Riismaa, 1993; Laslier, 1997). This partitioning procedure may be considered as a hierarchy where to the subsets of partitioning correspond nodes of hierarchy and the relation of containing of subsets defines arcs of hierarchy. The feasible set of structures is a set of hierarchies (oriented trees) corresponding to the full set of multi-level partitioning of given finite set (Riismaa, 1993).

Each tree from this set is represented by the sequence of Boolean matrices, where each of these matrices is an adjacency matrix of neighboring levels. To guarantee the feasibility of the representation, the sequence of Boolean matrices must satisfy some conditions – a set of linear and nonlinear equalities and inequalities. Examples of problems of this class are aggregation problems, structuring of decision-making systems, database structuring, and the problems of multiple distribution or centralization, multi-level tournament systems, multi-level distribution systems.

Described formalism enables to state the problem as a double-stage discrete optimization problem and construct some classes of solution methods (Riismaa, 2002). Variable parameters of the inner minimization problem are used for the description of connections between adjacency levels. Variable parameters of the outer minimization problem are used for the presentation of the number of elements at each level. This double-stage minimization approach guarantees that the involved mathematical properties – possibility to extend the objective function to the convex function - enable us to construct algorithms for finding the global optimum (Riismaa, 2002; Riismaa et al.,2003; Riismaa and Vaarmann, 2003; Riismaa and Randvee, 2003).

For finding the global optimum the method of local search is constructed, where on each step of iteration the value of objective function on ends of some kind of unit cube is calculated (Riismaa et al., 2003). The approach is illustrated by a multi-level robot-based production system example.

The main difficulty from point of view of optimization is that the number of subsets of partitioning is a variable parameter. For corresponding optimization problem it means that for goal function the number of summands (integer valued parameter) is a variable parameter. The finding of solution considered this nonlinear integer programming problem is complicated mathematically. This is a reason why continuous statements considered in this paper are more useful.

In the continuous case the structure of the system is defined as a finite sequence of density functions of distributions. Each distribution will correspond to the connections between this and previous level and shows how the size of previous level is distributed between the sizes of this level. For this continuous statement the size of level may have non-integer values too.

Corresponding optimization problem is a calculus of variations problem. Some reduced variants of this problem have good mathematical properties and is solved analytically.

Acknowlegements

The Estonian Science Foundation and FP5 IST accompanying measures project IST-2001-37592 (eVikings2) have supported this work

References

1. Berio G, Vernadat F.B.: *New Developments in Enterprise Modeling Using CIM-OSA.* Computers in Industry 1999; 40: 99-114.
2. Laslier J.F.: *Tournament Solutions and Majority Voting.* Springer-Verlag, Berlin, Heidelberg, New York, 1997.
3. Littover M., Randvee I., Riismaa T., Vain J.: *Recursive integer programming technique for structural optimization.* Proc. 4-th International Scientific Colloquium on CAx Techniques, H. Ostholt, ed. Bielefeld, Germany: Fachhochschule, Bielefeld, 1999.

4. Littover M., Randvee I., Riismaa T., Vain J.: *Optimization of the structure of multi-parameter multi-level selection*. Proc. 17th Int. CARs&FOF, G. Bright and W. Janssens, eds. Durban, University of Natal, 2001.
5. Mesarovic M.D., Macko D., Takahara Y.: *Theory of Hierarchical Multi-Level Systems*. New York: Academic Press, 1970.
6. Mesarovic M.D., Takahara Y.: *General System Theory: Mathematical Foundations*. New York: Academic Press, 1975.
7. Randvee I., Riismaa T., Vain J.: *Optimization of holonic structures*. Workshop on Production Planning and Control, A. Artiba ed. Mons, Ateliers de la FUCaM, 2000.
8. Riismaa T.: *Description and optimization of hierarchical systems*. Remote Control, 1993; 12: 146-152.
9. Riismaa T.: *Optimization of the structure of multi-level parallel processing*. Proc. IASTED International Conference in Applied Informatics, M. H. Hamza ed. Innsbruck, Anaheim, Calgary, Zurich, ACTA Press, 2002.
10. Riismaa T.: *Optimization of the structure of fuzzy multi-level decision-making system*. Proc. International Conf. on Modelling and Simulation of Business Systems, H. Pranevicius et al. eds. Vilnius, Kaunas University of Technology Press, 2003.
11. Riismaa T.: *Description and optimization of discrete and continuous hierarchical environment*. Preprints. 3th Internacional IFAC/IEEE/ACCA Conference on Management and Control of Production and Logistics, G. Lefranc ed. Santiago, Chile: Universidad de Las Americas, 2004.
12. Riismaa T.: *Maritime activity as a processing in hierarchical environment*. Proc. 7th International Workshop on Harbour, Maritime and Multimodal Logistics Modelling & Simulation & Applied Modelling and Simulation Workshop, A. G. Bruzzone et al. eds. Rio de Janeiro, Universidade Federal do Rio de Janeiro, 2004.
13. Riismaa T.: *Description and optimization of discrete and continuous hierarchical structures*. Proc. 20th Int. Conference on CARs&FOF, M. Marquez ed. San Cristobal, Venezuela: Nadie Nos Edita Editores, 2004.
14. Riismaa T., Randvee I.: *Restructuration of control scenarios*. Proc. 23rd EUROMICRO Conference New Frontiers of Information Technology, P. Milligan and P. Corr eds. Budapest, IEEE Computer Society, Los Alamitos, California, 1997.
15. Riismaa T., Randvee I.: *Recursive algorithms for optimization of multi-level selection*. Proc, 18th International Conference on CAD/CAM, Robotics and Factories of the Future, J. J. P. Ferreira ed. Porto, INESC Porto, 2002.
16. Riismaa T., Randvee I.: Description and optimization of the structure of multi-level parallel selection. Proc. 19th International Conference on CAD/CAM, Robotics and Factories of the Future, Wan Abdul Rahman Wan Harun ed. Kuala Lumpur, SIRIM, Berhad, 2003.
17. Riismaa T., Randvee I., Vain J.: *Optimization of the structure of multi-level parallel assembling*. Proc. 7th IFAC Workshop on Intelligent Manufacturing Systems, L. Monostori et al. eds. Budapest, Oxford, England Elsevier Science Ltd, 2003.
18. Riismaa T., Vaarmann O.: *Optimal decomposition of large-scale systems*. Proc. International Workshop on Harbour, Maritime and Multimodal Logistics Modelling & Simulation, Y. Merkuryev et al. eds. Riga, Riga Technical University, 2003.
19. Rockafellar R.T.: *Convex analysis*. New Jersey, Princeton University Press, 1970.

CONTROL OF TWO LEGS WALKING ROBOT WITH PRESSURE SENSOERS IN THE FEET

Yada S., Okabe H., Sakano S.

Abstract: This paper studies the basic mechanism of the dynamic walking robot from the stable walking point of view. Two legs walking robots are very unstable and it is very difficult to control the walking. We use the pressure sensors in the soles of the robot. The robot walks stably using the output signals of the sensors. We show the mechanism of the robot and the dynamic property of the two legs walking robot.

1. Introduction

The researches of humanoid robot are actively carried out recently, and the many reports about the biped robot are announced. In the background of the development of the humanoid robots, the robots have been developed as the industrial robots but we require the useful robots for human life. Especially, the image of the humanoid robots in TV and animation of the cartoon is stronger in Japan, and the developments of humanoid robots are actively carried out. The biped walking robots such as "ASIMO", "QRIO", "HOAP-2", etc., have been developed and they are marketed. Generally, the walking of the biped robots is unstable, and many problems have remained in realizing the walking like human. In this study, the pressure sensitive sensors are installed in the soles of the foot. The control for the stable walking is carried out using the output signals from the each pressure sensors. The research is advanced with the aim of the stable walking without falling, even if the robot is put in what kinds of the situation. The robot can walk smoothly and stably without the difficult control rules. The structure of the walking robot and the walking control using pressure sensitive sensors are discussed in this paper.

2. Outline of robot

2.1. Structure of robot

The lower half of the robot consists of the actuators and the bearings in order to put the importance for the robot walking. And, the bust of the robot becomes frame construction. The model of the lower half of the robot body is shown in Fig. 1. The servomotors for the radio control are used as the actuators.

Though the servomotors for the radio control are the motors that are used for the small biped walking robots and are cheaply to easy to obtain, but have the defect with the small torque. As shown in Fig.1, the degree of freedom of one foot is 6 and the total freedom of half lower body is 12. The material of the robot frame is the aluminum. The produced two legs walking robot is shown in Fig. 2.

Figure 1. Lower half of the robot body model, control of two legs walking robot

Figure 2. Two legs walking robot

2.2. Control circuit

We use the microcomputer "H8" for the control of the robot. It is possible to obtain the development tolls of the "H8" program freely and easily. The signals of the Pulse Width Modulation (PWM) are used for the control of the servomotors. It is possible to deal with the 5 channels of PWM in the microcomputer "H8". In this study, the software carries out all controls. The jitters may be generated in the case of the interruption operation in this reason. Then, the logic IC is used in order to prevent the jitters. C language is used as the program for the control. The control circuit is shown in Fig. 3.

Figure 3. Control circuit

2.3. Sensor

The super thin pressure sensor is used as the tool of control. The resistance of the pressure sensitive element changes with the inverse proportion to the force that is added to the sensor. The op-amp is used in order to convert the resistance change of the sensor with the change of the voltage. The total 8 sensors control the walking operation of the robot at the one-foot sole using the 4 sensors. Fig. 4 shows the sensor using the control of the robot. Fig.5 shows the sensors in the sole of the foot.

Figure 4. Pressure sensor

Figure 5. Sensors in the sole

3. Experiments

The situations of the walking are shown in Fig.6. The problem for the transfer of center of the gravity in the walking robot during the walking is remained a little. The smooth walking will become possible by the correction of the programming.

Figure 6. Experiment of walking, control of two legs walking robot

4. Conclusions

In this study, the realization of the biped-walking robot is described. The robot with the pressure sensitive sensors in the sole of the foot is manufactured and the walking experiments of the robot are practiced. The robot walking is almost smooth. In the future, the improvement of the control programming will be advanced and the smoother walking will become possible.

References

1. Svinin M., Uchiyama M.: *Contribution to inverse kinematics of flexible robot arm*. JSME Int. Journal, Series C, Vol. 37, No. 4, 1994, pp. 755-764.
2. Hirai K., Hirose M., Haikawa Y., Takenaka T.: *The development of Honda humanoid robot*. Proc. of the 1998 IEEE International Conference on Robotics & Automation, 1998, pp. 1321-1326.
3. Kuroki Y., Ishida T., Yamaguchi J, Fujita M., Doi T.: *A small biped entertainment robot*. Proc. of the IEEE-RAS International Conference on Humanoid Robots, 2001, pp. 181-186.
4. Yamaguchi J., Soga E., Inoue S., Takahashi A.: *Development of a bipedal humanoid robot control method of whole body cooperative dynamic biped walking*. Proc. of the 1999 IEEE International Conference on Robotics & Automation, 1999, pp. 368-374.
5. Vukobratovic M., Stokic C.: *Dynamic control of unstable locomotion robot*. Math. Biosci., vol. 24, 1975, pp. 129-157.
6. Kajita S., Tani K.: *Study of dynamic biped locomotion on rugged terrain*. Derivation and application of the linear inverted pendulum mode. Proc. of IEEE International Conference on Robotics & Automation, 1991, pp. 1405-1411.
7. Kajita S., Tani K.: *Adaptive gait control of a biped robot based on real time sensing of the ground*. Proc. of IEEE International Conference on Robotics & Automation, 1996, pp. 570-577.
8. Raibert M. H.: *Legged Robots that Balance*. MIT Press, 1986
9. Song S. M., Waldon K. J.: *Machines that walk: The adaptive suspension vehicle*. MIT Press, 1989.
10. Everett H. R.: *Sensors for mobile robots: Theory and application*. Wellesley, Massachusetts, 1995.

AUTONOMOUS HEXAPOD ROBOT FOR WALL CLIMBING TASKS: DISTRIBUTED CONTROL ARCHITECTURE IMPLEMENTING CAN

Tlale N.S., Bright G., Xu W.L.

Abstract: A hexapod, wall-climbing robot was designed using a distributed controller. It could be used for applications that require service robots to move along vertical planes e.g. wall painting, window washing, non-destructive testing (NDT), surveillance, etc. A controller of a wall-climbing robot that has many actuators sensors is difficult to design because of safety issues. If a central controller is used, a failure in one part of the system can result in a failure of another system. This can result in the robot's safety mechanisms failing and the robot falling from the wall. This paper describes the design of a wall-climbing robot that implements many actuators and sensors: the hexapod wall-climbing robot. Controller Area Network (CAN) was implemented as a distributed controller of such a robot. Modular Mechatronics were used to design and assemble the robot. This improved the control, flexibility and mobility of the system.

1. Introduction

The International Service Robot Association (ISRA) defines service robots as machines that sense, think, and act to benefit or extend human capabilities and to increase human productivity [1]. A service robot can automate repetitive tasks, such as coffee making and delivery in an office block. Moreover, tasks that are carried out in harsh or dangerous environmental conditions for humans, for example inspection of ship hull, are good candidates for automation implementing service robots.

Design of the drive system of service robots for tasks that require robots to move on horizontal or near horizontally inclined surfaces is straightforward. Wheels or articulated arms/legs can be used. The earth's gravitational pull is only a concern when the horizontal incline is too steep. However, some tasks require service robots that are able to move along the vertical plane, against gravity. Examples of such tasks are window washing, wall painting, large airplane inspection, ship's hull inspection, etc.

The drive system of service robots in this class is complex due the fact that they have to overcome gravity. Safety is another issue that should not be overlooked, that is, service robots should not fall on people while they are moving on the vertical incline against gravity. Safety of the robots is dependent on the mechanical safety mechanisms and the controller design of the robot. Central controllers complicate the development of the control system because:
- many software components have to be developed that interact with each other which sometimes do not behave in the manner expected,
- wiring of the robot to one central controller is difficult,
- error troubleshooting is difficult.

Distributed controllers do not have the disadvantages of the central controllers. They only become expensive when a lot of wiring is used to connect the different parts of the distributed control system. In this paper, a Controller Area Network was used which implemented a distributed controller for a hexapod wall-climbing robot. It used a two-wire bus to connect the distributed controllers.

2. Some present day wall climbing service robots

One of the recent robots to be designed is the micro-robot "The Crawler", designed in the Robotics and Automation Laboratory of Michigan University (ref. Figure 1) [2]. It uses two suction cups and has a light payload. Another robot is the MRWALLSPECT-II, which uses six suction cups, three operating at one time [3]. It was designed in the Mechatronics Laboratory of SungKyunKwan University. The robot is slow and cannot be used for applications requiring high or moderate speeds.

Figure 1. "The Crawler", designed in the Robotics and Automation Laboratory of Michigan University [2]

The most recent successful wall-climbing robot is The Climber, which is commercial manufactured by Clarifying Technologies Ltd (refer Figure 2) [4]. The Climber is mostly suited to robotic applications such as law enforcement, security and inspection due to its low payload of 0.45kg (excluding the weight of the robot itself). The Climber can easily scale many smooth and uneven vertical surfaces – wallboard, plaster, brick, cinder block, and siding are negotiable for this versatile robot. Although it can transition from a horizontal (ground) surface to a vertical one, and back, it cannot go from a vertical to an inverted surface or directly around corners unassisted.

The Climber does not use suction cups to adhere and climb; instead, it uses a patented 'vortex attractor' technology to pull itself to surfaces (refer Fig 2). By combining the VRAM with a versatile six-wheeled position-traction drive train, the robot can travel and maneuver on horizontal, vertical, and even inverted surfaces with ease. The nature of the vortex effect also makes it forgiving of changing surface types. Exhaustive list of current day wall climbing robots and their applications can be found in [5, 6, 7].

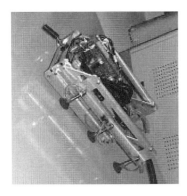 

Figure 2. MRWALLSPECT-II, designed in the Mechatronics Laboratory of SungKyunKwan University [4] and the Climber with a Surveillance Camera Module on the Wall [5]

One unit of the Climber costs USD12,500. It has high power requirements because the motor creating the suction must always be turned on to create the suction force. Longer mission times and more suction force can be obtained by simply adding additional battery packs. However, this will increase the mass of the robot.

3. Modular mechanical design of the wall climbing robot

For motion stability purposes, the design of the service robot described in this paper was inspired by insects' anatomy. Hexapod mechanism, which had been extensively publicized, was implemented. The robot's components were constructed out of aluminum and plastics in order to minimize the weight of the robot. Each of the six articulated legs consisted of three joints which were each actuated by a DC servo-motor (refer Figure 3). The first motor (motor 1) was attached to the base, which in turn was attached to the body-frame of the hexapod robot. Its purpose was to move the rest of the leg forwards and/or backwards when the robot was walking. The second motor's (motor 2) housing was driven by the first motor.

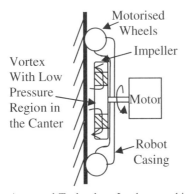

Figure 3. 'Vortex Attractor' Technology Implemented in The Climber [8]

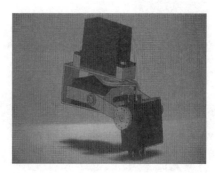

Figure 4. Articulated-leg of Hexapod Robot for Wall Climbing

The purpose of the second motor was to move the third motor and the suction cup up and/or down when the robot was walking. The suction cup holder was attached to the third motor. The third motor's (motor 3) purpose was to also to move the suction cup up and/or down during the locomotion of the robot. The second and the third motors acted together to achieve maximum contact between the suction cup and the surface that was being walked upon when the robot was walking.

Each articulated-leg had a suction cup, which helped the robot to grasp onto smooth surfaces. This gave the robot the capability to walk on vertical surfaces if enough vacuum pressure was marinated in the suction cups. Two miniature suction pumps were used to create and maintain vacuum in the suction cups when they made contact with a surface. The pressure in each suction cup was regulated by miniaturized three-port micro-valves. The valves could set the pressure in the vacuum cups to be either vacuum or atmospheric pressure.

Six articulated-legs are mounted on a body frame. The body frame also housed the electronics, sensors and all the other associated equipment. The complete robot can be seen in Figure 5.

Figure 5. Wall Climbing Hexapod Robot

4. Sensors and actuator architecture

The developed robot consisted of a total of eighteen DC motors with their driver (amplifier) circuits i.e. the tip of each leg was fitted with a touch sensor to sense when the suction cup had made contact with i.e. three DC motors per leg.

Limit switches were used in each leg to stop motor 3 from over rotating and colliding with motor 2. Each leg was also fitted with a pressure sensor to monitor the pressure in the pneumatic system of each leg. The pressure in each suction cup of each valve was controlled by a micro-switch.

The front and rear of the wall-climbing robot were fitted with two ultrasonic sensors each, while the sides were fitted with one ultrasonic sensor each. Ultrasonic sensors were used for object detection. A CCD camera was fitted at the front of the wall-climbing robot. It was used for data collection.

The overall structure of the actuators and sensors was implemented. The developed robot consisted of twenty motors, six micro-valves, six pressure sensors, six touch sensors, six limit switches, six ultrasonic sensors and one CCD camera. The number of sensor and actuators that was used required that a distributed controller be used because:
- a central controller would have difficult wiring, and thus difficult system development,
- if the central controller failed, all the robots functions would fail. For safety reasons, this was not desirable.

Distributed controller architecture was implemented. PIC18F442 were used as local controllers were used to control local processes such as motor control, pressure control, etc. The pressure in each suction cup of each valve is controlled by a micro-switch.

5. Controller area network (can) for distributed control architecture

The robot control architecture was divided in time-critical and event-based control strategies. Control functions such as navigation and motion planning were time critical, while control functions such as path planning and object avoidance were event-based. In order to achieve reliable and adequate control of the robot, Controller Area Network (CAN) was implemented CAN comprise of only layers 2 (Data Link Layer) and 1 (Physical Layer) of the International Standardization Organization/Open Systems Interconnect (ISO/OSI) hierarchical layered structure [9], [10]. CAN allowed time critical sensor information and control signals to be transmitted with determinism [11]. PIC18F442 microcontrollers were used as CAN nodes, with MCP2515 CAN controller from Microchip [12].

5.1. The structure of CAN nodes

Nodes 1-6 were used to control the individual movement of each of the six articulated legs of the hexapod wall-climbing robot. The synchronous movement of the articulated legs in order to produce the stable and the desired motion of the hexapod wall climbing was the function of Node 9. Path planning, user interface and task co-ordination were other functions of Node 9. Node 8 was used to collect data for a particular mission. In this case, a

CCD camera was used on Node 8 to collect data for visual images. Serial bus EEPROM was used to save the large amount of data collected from the CCD camera. Data from the serial EEPROM could be transferred to Node 9 when the bus activity was low. Node 7 was used for obstacle avoidance.

6. Discussion

A hexapod mechanism was used to develop a wall-climbing robot. A variety of sensors and actuators were used in the construction of the robot. Adequate transmission of sensory and control signals was achieved using a Controller Area Network (CAN). A maximum number of bits that could be transmitted in one CAN message implementing the standard data frame was 130 bits including the stuff bits. The CRC code implemented had a hamming distance of 6 i.e. for every five consecutive equal bits a bit of opposite value would be added to the message. With this it was possible to detect up to six single bit errors that were scattered about the message, or burst errors up to length of 15 bits. This error checking capability was handled internally by MCP2515.

The maximum delay time for a CAN message could be calculated under normal circumstances only for the message with highest priority [15]. The delay times for the other messages could not be calculated because of the bus access mechanism of the CAN i.e. Carrier Sense Multiple Access with Collision Detection and Arbitration on Message Priority (CSMA/CD +AMP). The delay time for the bus access to a message with highest priority was 130 bit times. CAN bit time was set to 16 T_Q (Section 5) with T_Q = 500 ns. Therefore the maximum delay for a message with highest priority was:

$$T_{max\_delay} = 130 * 16 * T_Q = 1.04 \text{ μs}$$

This delay was satisfactory because the message with the highest priority was the pressure leakage message. Events that resulted in pressure leakage conditions occurred within far longer time periods than 1.04 μs.

The other cause of message delay was error conditions. The MCP2515 CAN controller network carried out error handling. The following were sources of errors, which could be detected by the MCP2515 CAN controller: bit errors, bit stuffing errors, CRC errors, form errors and acknowledgement errors. Every node on the CAN bus which detected an error would send an error frame, increasing the delay for time critical messages. An error frame had the same message format and number of bits as a normal CAN message frame, except that the error flag replaces the CRC field. Relevant error counters (receive or transmit error counters) of the nodes would be incremented every time an appropriate error was received. Depending on the values of the counters, the nodes would switch to different states in order to limit errors or time delays of messages on the bus. The three states of the nodes were: Error Active (error counter <127), Error Passive (127< error counter<255) and Bus Off (error counter >127). Thus, the maximum time delay that could be experienced by a message on the CAN bus with nine nodes, due to consecutive errors according on the CAN bus, was:

T_{error_delay} ~ no. of CAN nodes * Error Frame Time *127
~ 8 * (1.04 *10^{-6})*127
~ 1.06 ms

This time, T_{error_delay}, represented the worst-case scenario that was hardly experienced in real life. It was satisfactory when it was compared with the safety critical processes of the robot. The developed hexapod wall climbing robot was setup so as to result into a total halt of the system when the Error Passive state was very close to be reached, only if the was no pressure leakage detected or the robot was upright on a horizontal surface. Error flags register was used to determine the status of the different nodes. Robot activity would resume when the bus error conditions were clear. If one of the node was close to the Error Passive mode, all the nodes were programmed so as to not to send any messages on the bus for 50 ms. This ensured that the no node experienced the Bus Off mode.

7. Conclusion

A hexapod mechanism was used to develop a wall-climbing robot. A variety of sensors and actuators were used in the construction of the robot. Adequate transmission of sensory and control signals was achieved using a Controller Area Network (CAN). Performance of CAN was satisfactory. It resulted in message delay times that were far less than the robots process time. The overall performance of the Autonomous Hexapod robot was acceptable especially for a system that was researched, designed and assembled at very low cost. Modular Mechatronics improved the control, flexibility and mobility of the system.

Acknowledgements

The financial assistance of the NRF and University of KwaZulu Natal is gratefully acknowledged for the research project.

References

1. Amos A., Strasser R., Trächtler A.: *Migration from CAN to TTCAN for a distributed control system*. 9th International CAN Conference Proceedings, Munich, 2003.
2. International Standard ISO 11898: Road Vehicles – Interchange of Digital Information – Controller Area Network (CAN) for High Speed Communication. ISO Reference Number ISO 11898: 1993(E), 1993 – 11 -15.
3. International Standard ISO 11519 -2: Road vehicles – Low Speed Serial Data Communication – Part 2: Low-speed Controller Area Network (CAN). ISO Reference Number ISO 11519-2:19994 (E), 1993 – 11 -15.
4. Pransky J.: *Service robots – how we should define them*? Service Robot: An Intern. Journal Volume 2 · Number 1 · 1996 · pp. 4–5.
5. Xiao J., Minor M., Dulimarta H., Xi N., Mukherjee R., Tummala R. L.: *Modelling and Control of an Under-actuated Miniature Crawler Robot*. Proceedings of the 2001 IEEE/RSJ International Conference on Intelligent Robots and Systems, Maui, Hawaii, Oct. 29-Nov. 03, 2001.

6. Microchip Technology Inc., "Stand-Alone CAN Controller With SPI^{TM} Interface", document 21801b, Chandler, USA, 2003.
7. http://www.clarifyingtech.com/public/robots/robots_public.html
8. http://www.microchip.com/
9. http://www.can-cia.de/
10. "Special Issue: Climbing and walking robots", Industrial Robot: An International Journal, Volume 31 Number 2 2004
11. Kang T., Kim H., Son T., Choi H.: *Design of Quadruped Walking and Climbing Robot*. IEEE/RSJ International Conference on Intelligent Robots and Systems, 2003.
12. Ultimate Real Robots Magazine "Robots in Action: Climbing Robots", Issue 36, Eaglemoss Publication, 2003, or www.realrobots.com.
13. Ultimate Real Robots Magazine "Robots in Action: Climbing Robots", Issue 37, Eaglemoss Publication, 2003, or www.realrobots.com.
14. Lawrenz W.: *CAN System Engineering: From Theory to Applications*. Springer-Verlag, New York, 1997.

NEURAL NETWORK CONTROL OF ROBOT HAND

Shaw T., Kakad Y. P., Karwal V., Gandhi J., Hari Y.

Abstract: This paper investigates if shape memory alloy wire (SMAs) can accurately operate a robotic hand. The robotic hand is developed with space exploration in mind. Therefore, minimizing the weight of the robotic hand is critical. Current, more popular methods of operating robotic hands are mechanical actuators such as stepper motors. Stepper motors are heavy in comparison to shape memory alloys. Shape memory alloys can lift masses much greater than itself. In a space exploration application, the robotic hand will be picking up various samples from the environment. Some of these samples will be delicate, requiring fine finger control to avoid damaging the sample. Shape memory alloys have complex non linear characteristics that make developing an autonomous control system difficult. To overcome the uncertainties of the shape memory alloy's characteristics, a neural network is developed. Neural networks can identify an unknown system and learn to control through observation. The neural network will minimize the time needed to update the control system for maneuvering various sample objects.

1. Introduction

In all space explorations weight is a critical consideration so there is a continuous research to find ways to reduce the weight of equipment and supplies on space flights. An example is the food storage technology. By the method of dehydration, the removal of almost all water from the food, weight is reduced. So even the small amounts of water removed from food is a significant reduction in space exploration.

Most robotic applications use heavy electric motors for actuation of joints. Finding an alternative actuating mechanism that is light weight yet powerful would provide immense benefit to space operations. By reducing the weight, the lesser is the effort needed to propel the space shuttle into orbit. The focus of this paper is to provide an alternative actuator to replace the bulky and heavy step motor actuators.

This paper will show that the shape memory alloys (SMAs) can be an alternative to stepper motors in robotics. The alloy used is of nickel and titanium. Nickel Titanium (Ni-Ti) alloys have a large pulling force to weight ratio. Twenty centimeters of 50 micron nickel-titanium wire can lift as much as 100 grams of mass a distance of one centimeter. As the diameter of the wire increases, so does its pulling force. Its ratio of holding torque to weight ratio is approximately 43.84, a significant increase over the stepper motors [1].

Ni-Ti alloys have complex, non-linear response characteristics that make it difficult to model. Due to difficulty in modeling it becomes very difficult to develop a controller. Therefore, to bypass the difficulty of modeling the nickel-titanium alloy and the associated robotic hand, neural networks are employed to control the system. The neural network are then trained. After the training is completed, the neural network is then ready to control the robotic hand system autonomously [2].

2. Shape memory alloys

Shape memory alloys (SMAs) have a unique characteristic to memorize shapes. They can be trained to have one-way memory or two-way memory. SMAs have two phases: martensite and austenite. The martensite phase occurs when the SMA is cooled below a predetermined temperature. When the SMA is heated beyond a set temperature, it enters the austenite phase. If the SMA is trained for one-way memory then the martensite phase is the alloy's flexible state. The SMA can be manipulated easily and when bent it remains bent. As heat is applied and the SMA approaches its austenite phase, it returns to the memorized shape. With two-way memory, the SMA has a memorized shape in the austenite phase and martensite phase. The shape can be different in each phase [3, 4].

Nickel-titanium alloy was the first significant discovery of the shape memory effect (SME). The crystal structure of Ni-Ti alloy is rhombohedral in shape. There are precipitate particle of Ni-Ti that create strain fields in the matrix of the material. The SME can be achived because of these strain fields. When Ni-Ti cools, a shearing effect in the crystal matrix occurs leading to low symmetry. This phase is martensitic and alloys the alloy to be flexible. During the transformation to martensite, large strain is produced. To relieve this strain, the alloy performs twinning, a form of lattice invariant shear. Twinning is closely related to SME. The heating of the Ni-Ti to the austenite phase brings the alloy strucure to a high symmetry, in effect recalling a memorized shape [3, 4].

There have been many applications of SMAs across many fields of study. SMAs have been used in fittings and seals, clothing, plumbing and heating, medical, and robotics. Some of the first applications of SMAs were as fittings and seals for piping and electrical connectors. The SMA would be designed to have an activation temperature equal that of the ambient temperature. Therefore, the SMA remain in tension around the fitting or seal. The advantage is for small spaces where wrenches or other tools cannot reach [4].

There are two forms in which the SMA can become an actuator for robtics. The first is by stretching, thus deforming, straight SMA wire. When heated, the SMA wire recovers its original length. The force at which it can recover is called the recovery force of the SMA wire. Recovery force was earlier referred to as pulling force. The second form is to create a coil with the SMA wire. With the compressed coil being the original shape, the SMA can be useful over a larger range than in the straight wire form.

3. Modeling the shape memory alloy wire

To design a controller for the robotic hand, the behavior characteristics of shape memory alloy wire is needed. As discussed, SMAs respond to changes in temperature, mainly heat. The source of heat can either be external or internal. External sources of heat can be hot air or hot water. The internal heat source is generated by electrical resistance. By passing current through the wire, heat is generated. This is preferred over external heat sources because of the ease of controlling the temperature of the SMA wire.When the stretched SMA wire changes from the martensite phase to the austenite phase, it shortens in length. The response to increased temperature is non-linear. This non-linear effect makes the development of a control system difficult, hence the use of neural networks. Along with a non-linear response to temperature changes, there is a strong hysteresis effect.

Another factor that makes the SMA wire difficult to control is its reponse to different stresses. A study was conducted to show how the SMA wire responded when different masses were used to stretch the wire. The masses were free weights, therefore the force applied was constant via the equation

$$f = ma \qquad (1)$$

The factors like differeing force and temperature offsets create non-linear responses and non-repeatable results. In a system where the force on the SMA wire is not changing, the system response would be predictable and repeatable. For the robotic hand, the force on each finger will differ depending on the position of the finger and the mass within the hand's grasp. Therefore, the response will not be repeatable and not very predictable. This prevents classical linear control systems from working. Non-linear control systems may adequately perform the task, but all the system parameters would need to known; which is a difficult task

The mathematical model of the SMA wire obtained experimentally by applying various step changes to the current in the SMA wire. The response of the wire was found to be linear, with a time-exponential step response [6]. The experimentally obtained mathematical model is described by equation (2).

$$L(k) = L_T \left[L(k-1) - L_T \right] e^{-\frac{t}{\tau}} \qquad (2)$$

The NiTi Wire Length Model for equation (2) is represented in Figure 4.

4. Development of a robotic hand

To use the SMA wire effectively as actuators of a robotic hand, the forces constraints of the wire have to be considered. The limiting factor of the SMA wire is the diameter. Although larger diamter SMA wire provides greater recovery force, it also requires greater electrical current to heat it and greater deformation force. A larger amperage requirement also means a larger power supply is needed, both a cost and weight issue. If the deformation force becomes too large, putting the SMA wire into position becomes more difficult [1].

The first step is to determine the material from which to make the hand. For space exploration, a light weight yet strong material is needed, a material that can withstand the extreme temperature swings and other extreme environmental conditions. Although many metals meets these requirements, the prototype material has to be easy to machine. Delrin is an inexpensive material that is easy to machines and can withstand some extreme conditions.

The robotic hand will be actuated by the SMA wire's recovery force. Therefore, the hand is designed to use cables. In the manufacturing process a lamination style design was used.

Each finger is made of four parts that are sandwiched together. The final design is shown in Figure 1.

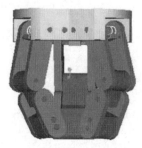

Figure 1. Final Design of the Robotic Hand

Since the hand is operated using cables, an understanding of how the motions are created is essential. A simple model using hinged joints was therefore created. Wires were attached twice across each joint. One attachment was above the joint and the other below the joint for a total of four wires per finger. Each section of the finger was operated simply by pulling one wire per joint at any given time. This provided a reference design from which improvements could be made.

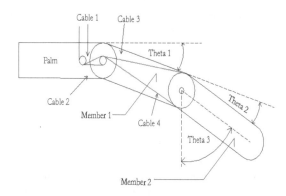

Figure 2. Schematic of one robotic finger

A rough force diagram was then drawn up. The tensions across each joint were obtained from the force diagram. To close the angle between two joints, initially a cable was connected across the joint forming a triangle. To maximize the usable space close to the joints, the cable must be connected as close to the joint as possible. The price of doing this is that the tension required to create movement about the joint becomes larger. Examining the force diagram provided insight into where to connect the cable so as to maximize space and still perform its task.

Thus, during the process of developing the robot hand, six different configurations were considered in order to minimize weight as well as friction in the joints. The final design configuration chosen includes hollowed members and bearings for the joints, and is shown in figure 3.

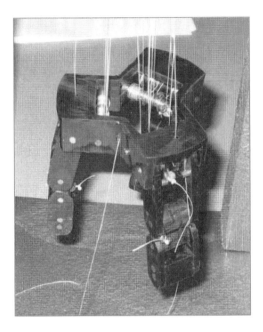

Figure 3. Suspended Mounting Configuration of Robotic Hand

5. Modeling the robotic hand

The model of the robotic hand can be generalized to just movement of the fingers. There are three points of the finger that need to be tracked, the origin or first joint, the second joint, and the finger tip or end effector. With these three points and using straight lines between them, the finger position can be mapped.

The fingers of the robotic hand have two degrees of freedom (DoF). A degree of freedom is an objects motion is a single direction, either rotational or lateral. Each joint in the finger has one rotational DoF, hence the two DoF. To determine the position of the end effector, two angles are needed. The first angle is at the first joint relative to the x-axis. The second angle is measured from the axis formed by the first segment. These angles, in radians, are Q_1 and Q_2. The position of the second joint relative to the first joint or origin is found using trigonometry. Finding the end effector position requires translating the second angle to the origin. To do this, the Devanit-Hartenberg model is used.

The Devanit-Hartenberg model translates the xyz axes from one origin reference to another, hence the position of the end effector relative to the origin. To re-emphasize, this model is used because only the two angles described earlier are known; the angle of the end effector position to the origin or first joint is not known. The transformation of Eq. (4) is the form of

Denavit-Hartenberg model. Q_1 is the angle at the ith joint and ai is the length of the ith link. Eq. (4) is used for both angles and their product forms the transformation of Eq. 5

$$A_i = \begin{bmatrix} \cos(\theta_i) & -\sin(\theta_i) & 0 & a_i \cos(\theta_i) \\ \sin(\theta_i) & \cos(\theta_i) & 0 & a_i \sin(\theta_i) \\ 0 & 0 & 1 & 0 \\ 0 & 0 & 0 & 1 \end{bmatrix} \quad (3)$$

$$T = A_1 * A_2 = \begin{bmatrix} \cos(\theta_1 + \theta_2) & -\sin(\theta_1 + \theta_2) & 0 & a_1 \cos(\theta_1) + a_2 \cos(\theta_1 + \theta_2) \\ \sin(\theta_1 + \theta_2) & \cos(\theta_1 + \theta_2) & 0 & a_1 \sin(\theta_1) + a_2 \sin(\theta_1 + \theta_2) \\ 0 & 0 & 1 & 0 \\ 0 & 0 & 0 & 1 \end{bmatrix} \quad (4)$$

The last step is to translate the length of the NiTi wire to the angle between the finger links. More important than the actual length of the NiTi wire is the change in length of the NiTi wire needed to pass the joints through their full range of motion. For the model, the input is a percentage of the full range of motion which is tranlated to the delta length of the NiTi wire. This delta length is entered as L_T in Eq. 2. The resulting length from Eq. 2 is tranlated to the angle of the joint by:

$$\theta = \frac{L}{R} \quad (5)$$

where:
    L   is the length of cable around the joint and R is the radius if the joint. All angles in radians.

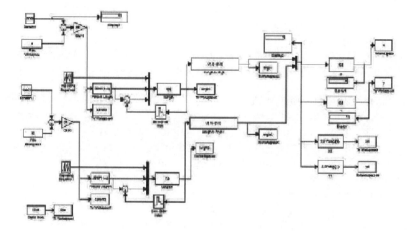

Figure 4. Simulink Model of One Finger

6. Neural network

This architecture is the most popular structure in practice due to its non-parametric, non-linear mapping between input and output [5, 7]. Networks with this architecture are known as universal approximators, including multilayer feed-forward neural networks employing sigmoidal hidden unit activations. These networks can approximate not only an unknown function but also its derivative [6].

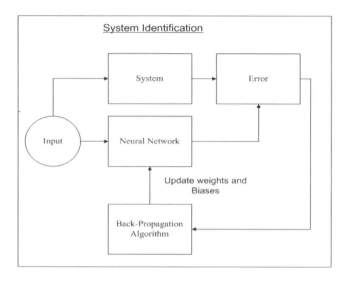

Figure 5. System Identification with Neural Network

Feed-forward neural networks include one or more layers of hidden units between the input and output layers. As its name suggests, the output of each node propagates from the input to the output side. All connections point from input towards output. See figure 6.
Multiple layers of neurons with nonlinear activation functions allow this type of neural network to learn nonlinear or linear relationships between input and output vectors. Each input has an appropriate weighting W. The sum of the weighted inputs and the bias B forms the input to the transfer function. Any differentiable activation function f may be used to generate the outputs. Three of the most commonly used activation functions are purelin $f(x) = x$, log-sigmoid $f(x) = \left(1+e^{-x}\right)^{-1}$,

and tan-sigmoid: $f(x) = \tanh\frac{x}{2} = \dfrac{1-e^{-x}}{1+e^{-x}}$. (6)

The hyperbolic tangent (Tan-sigmoid) and logistic (Log-sigmoid) functions approximate the signum and step functions, respectively, but provide smooth, nonzero derivatives with respect to the input signals. These two activation functions are called squashing functions since their outputs are squashed to the range [0, 1] or [-1, 1]. They are also called sigmoidal functions because of their S-shaped curves that exhibit smoothness while retaining the correct asymptotic properties.

The activation functions f_h of the hidden units must be differentiable nonlinear functions (typically, f_h is the logistic or hyperbolic tangent function $f_h(x) = \tanh(x)$. If f_h is linear, then one can always collapse the net to a single layer and thus lose the universal approximation / mapping capabilities. Each unit of the output layer is assumed to have the same activation function.

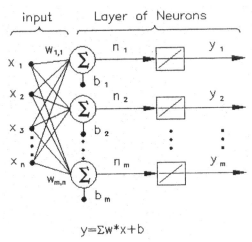

y=Σw*x+b

Figure 6. Feed Forward Neural Network

The node function for the output layer is a weighted sum with no squashing functions. This is equivalent to a situation in which the activation function is a purelin function, and output nodes. In this section, basic concepts necessary in the development of the learning algorithm are presented. This algorithm is the gradient-based Levenberg-Marquardt Method. Learning in a neural network is normally accomplished through an adaptive procedure, known as a learning rule, whereby the weights of the network are incrementally adjusted so as to improve a predefined performance measure over time. In other words, the process of learning is best viewed as an optimization process. More precisely, the learning process can be viewed as a "search" in a multidimensional parameter (weight) space for a solution, which gradually optimizes an objective (cost) function.

6.1. Forward-Propagation and Back-propagation

During training, a forward pass takes place. The network computes an output based on its current inputs. Each node i computes a weighted sum a_i of its inputs and passes this through a nonlinearity to obtain the node output y_i (figure 7). The error between actual and desired network outputs is given by:

$$E = \frac{1}{2}\sum_p \sum_i (d_{pi} - y_{pi})^2 \qquad (7)$$

where:
    p indexes the patterns in the training set, i indexes the output nodes, and d_{pi} and y_{pi} are, respectively, the desired target and actual network output for the i th output node on the p th pattern.

Figure 7 shows the learning process.

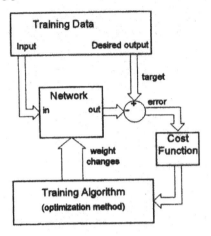

Figure 7. Learning Process

7. Conclusion

The research reported in this paper provides an innovative method of actuating space robotic hand fingers. The neural network provides an effective technique of controlling the operation of this experimental hand. The current research work is to implement the operation of this type of hand for all joints of this class of robotic hand using a hierarchical scheme with individual neural network based control algorithm for each joint operation and with an overall supervisory neural network to coordinate all joint movements.

Acknowledgements

This work was supported in part by NASA sponsored N. Carolina Space Grant Consortium.

References

1. Shaw T., Timothy S.: *Study of Wire Shape Memory Alloys and Their Operating Ranges*. Submitted to University of North Carolina at Charlotte, Department of Electrical and Computer Engineering, 2002.
2. Srinivasan A. V., McFarland D.M.:. *Smart Structures: Analysis and Design*. Cambridge: Cambridge UP, 2001.
3. Otsuka K., Wayman C.M. (eds.): *Shape Memory Materials*. Cambridge: Cambridge UP, 1998.
4. Shaw T., Kakad Y.P., Barry G.: *Neural network control of a shape memory alloy wire actuated robot hand*. Sherlock, Cars & Fof, 2004.
5. Luenberger D.G.: *Linear and Nonlinear Programming*. Addison Wesley, 1984.
6. Demuth H., Beale M.: *Neural Networks Toolbox Users Guide*. Mathworks, 1997.
7. Fu L.: *Neural Networks in Computer Intelligence*, Prentice Hall, 1994.

THE EXPERIMENTAL CONTAINER GEOFLOW FOR THE FLUID SCIENCE LABORATORY OF THE INTERNATIONAL SPACE STATION

Stein C., DeSilva A.K.M., Harrison D.K.

Abstract: The Experimental Container "GeoFlow" is a facility for the operation in the so-called Fluid Science Laboratory on board the International Space Station. The facility contains an experiment that allows the investigation of fluid flow patterns within a spherical gap, driven by variable stimuli under microgravity conditions. The provision of the experiment infrastructure within the Experimental Container incorporates hardware features like flow visualisation via an optical system, generation and transmission of high voltage to the rotating fluid cell, thermal control of the entire assembly and customisation of the fluid cell for allowing parameter variation. In the following the design concept of the Experimental Container Engineering Model as a result of a number of performance and life tests is presented.

1. Introduction

Currently the Experimental Container "GeoFlow" is developed and built by EADS Space Transportation GmbH under ESA contract in the frame of the Fluid Science Laboratory (FSL) Experimental Container program. The experiment is developed under the scientific coordination of Prof. Egbers, Department of Aerodynamics and Fluid Mechanics, Brandenburg University Cottbus, Germany. This Experimental Container GeoFlow is designed to operate an experiment that allows the investigation of fluid flows within a spherical gap, driven by an imposed electro-hydrodynamic and thermal field. In order to operate the experiment in the Fluid Science Laboratory it must be housed in a standardised envelope, which provides the necessary interfaces. As such the Experimental Container is an almost autonomous system which is commanded and supported by the Fluid Science Laboratory.

The Experimental Container was designed to generate various flow patterns under the influence of a variable central force field, induced across the fluid, with a variation of the temperature gradient over the fluid gap so that these can be used for future analyses. After Egbers (1999) the fluid flow is mainly influenced by Coriolis forces, caused by rotation, and by buoyancy forces, caused by a central force field. The stability of flow patterns depend on a number of parameters like the radius ratio inner and outer sphere, the temperature difference, the fluid viscosity, the strength of the force field or the rotation rate. To allow the investigation of parameter influences the most important parameters are variable during the mission. An electrohydrodynamics field is used to produce a symmetric central directed force field that acts similar to the gravity field on planets (Sitte, 2001). A thermal field is applied by producing a thermal gradient across the experiment cell. The facility is provided to be able to visualise the flow by means of an interferometer of the Fluid Science Laboratory. This is used to document the development and the final state of the flow. The provision of the experiment infrastructure within a given Experimental Container has also to incorporate hardware features like interferometry via an adaptation optical system, generation and transmission of high voltage to the rotating fluid cell assembly, thermal control of the entire assembly, precise alignment of the cell to the optical axis and customisation of the fluid cell assembly for parameter variation. The critical parts of the Experimental Container were tested and proofed to be successfully in the realisation of the scientific and mechanical specifications.

The FSL is one of several laboratory racks that will be accommodated in the so-called Columbus module, the major European contribution to the International Space Station (ISS), which will be launched in 2007. The FSL is a multi-user facility for scientific experiments, build up of different modular sub-units for the control and monitoring of the individual exchangeable experimental modules called Experimental Container, which is a common envelope for all experiments with standard interface options to the facility itself in order to be accommodated within FSL. This entails, that each experiment has a specific sets of requirements, such as mass, shape, optical field of view, observation direction, electrical, thermal and functional interfaces.

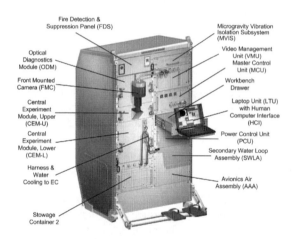

Figure 1. The Fluid Science Laboratory

2. The GeoFlow design concept

The Experimental Container GeoFlow is composed of a number of sub-systems which constitute the infrastructure of the experiment cell. The major infrastructure sub-systems are:
- experimental container outer housing,
- experimental container controller unit,
- high voltage power supply,
- heat exchanger turret,
- adaptation optics.

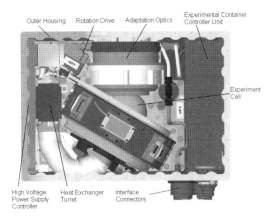

Figure 2. A CAD picture of the current Experimental Container seen from the top, shown without outer housing cover

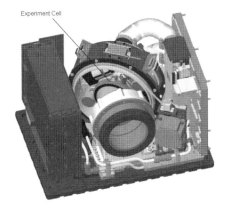

Figure 3. A CAD picture of the current Experimental Container seen from the rear side, shown without outer housing

2.1. The experiment cell

The experiment cell is composed of a polished sphere made of Tungsten Carbide surrounded by two concentric spherical shells of optical glass, each having a thickness of 4mm. Each glass shell is composed of two hemispheres, which are mounted mechanically. The glass shells form two separated gaps, filled with silicone oil. The inner gap constitutes the observed area while the outer gap is used for temperature gradient control. The thermal gradient is induced by heating the inner metal sphere and cooling with a fluid flow across the outer gap of the cell. The outer diameter of the outer glass shell is 80 mm, the outer diameter of the inner glass shell is 67 mm and the diameter of the inner sphere is 27 mm. The outer glass shell has a diameter of 80 mm to be able to observe the pole and the equator at the same time with the maximum usable field of view of a diameter of 80 mm predetermined by the FSL facility. During subsequent missions the radius ratio will be changed by changing the radius of the inner sphere while keeping the radius of the inner glass shell constant.

The induced dielectrophoretic effect is used to generate the central directed force field between the inner sphere and the inner glass shell. To allow reliable conductivity for the high voltage, the inner side of the inner glass shell is coated with transparent Indium Tin Oxide (ITO). It constitutes the outer electrode of a spherical capacitor with the experiment fluid as dielectric fluid. The coat is connected to ground potential. The inner sphere constitutes the inner electrode and is connected to the central high voltage contact. Besides the use for heating and as electrode, the inner sphere is also used as spherical mirror for the optical observation.

The Tungsten-Carbide, used as material for the inner sphere, has the advantage of high and homogenous heat conduction and low heat expansion which is imperative for the experiment realisation. The disadvantage is that the material cannot be polished to optical quality as it is a sintered material, leading to a reduction of the accuracy in the interferometrical results. Due to the demanded numerous exchanges of the experiment cell after each return, that are required for variation of the experiment fluid viscosity and/or the radius ratio, the possibility must be given to carry out this task in a quick on-ground operation, to be able to launch with the next mission increment. To allow the exchangeability without having to open any fluid circuits, a closed unit containing the experiment cell and its thermalisation loops is necessary (see Figure 4).

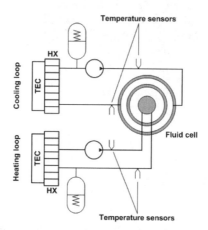

Figure 4. A view of the experiment cell fluid loops

To be able to use the interferometrical observation, the plane wave coming from the FSL facility needs to be transformed to a spherical and vice versa. This results in the use of an adaptation optics. Since the experiment cell shall be rotated during observation with a stationary interferometer the consequences on the precision are tremendous. The observation technique is also responsible for the experiment cell and adaptation optics position inside the Experimental Container. The optical axis which is used is in the middle of the containers back side. This only leaves marginal space on the sides of the experiment cell.

Due to the requirement of rotating the experiment cell, the use of a slip ring for power and data transmission to and from the experiment cell is obvious. This also results in a slip ring for the high voltage in the central axis of the cell. The rotation also has impact on the microgravity disturbance, which is very restricted. This results in a setup that provides balancing possibilities.

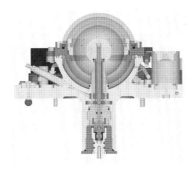

Figure 5. A sectional view of the experiment cell

2.2. The experimental container outer housing

In order to accommodate the experiment in the FSL, it has to be installed in a FSL standardised hermetically sealed envelope which couples the experiment via standard interfaces to the facility. The dimensions of the Outer Housing are predetermined by the CEM-L of the FSL facility which is the accommodation of the Experimental Container. The volume is approximately 30 l with outer dimensions of height = 270 mm, width = 400 mm and depth = 280 mm. The container wall is machined from a solid aluminium block to avoid the problems of sealing in the corners of the housing; the cover and the bottom are removable. The container is designed for a maximal design pressure of up to 2 bars. To ensure the structural strength at a minimum mass this requires a rib-structure on the inner side of the housing. FEM-analysis has shown that a wall thickness of 3 mm in areas without and 10 mm in areas with ribs is sufficient to meet the structural requirements. For the observation of the experiment via FSL, a glass window with a diameter of 115 mm is mounted the rear side of the housing. This results from a maximal quadratically field of view of 80 x 80 mm² which is predetermined by the FSL observation possibilities.

2.3. The experimental container controller unit

The Experimental Container Controller Unit (ECCU) is an electronic sub-assembly of the Experimental Container, which provides and handles the electrical interface between the FSL Master Control Unit (MCU) and the experiment. It constitutes thus the core of the electrical interface between the Experimental Container and the FSL. The communication with the FSL is set up as Master-Slave combination, the FSL acting as the requesting part (Master) and the ECCU as responding part (Slave). When the FSL sends commands to the Experimental Container the ECCU will respond by collecting and sending defined bundled housekeeping or scientific data from the experiment over the same communication line to the FSL. Internally the ECCU acts as Master for processing of the internal signals. The

ECCU also provides electrical power distribution from the FSL Power Control Unit (PCU) to the Experimental Container, actuators handling inside the Experimental Container and health checks of the Experimental Container internal hardware. The ECCU enables the Experimental Container to be an almost autonomous facility that is supported and commanded by the FSL.

2.4. The high voltage power supply

As mentioned before the high voltage for the GeoFlow experiment is needed to generate a central symmetrical force field inside the experiment cell. The high voltage has to be alternating because otherwise extrinsic particles in the experiment fluid would chain in a constant field and evoke an arc discharge, which might damage the experiment cell. Since there is no high voltage from the FSL available it needs to be generated Experimental Container internally. The challenging problems of the development are provoked by the boundary conditions of the experiment.

The highest voltage in the Experimental Container is 28V DC. From this supply the high AC voltage with an adjustable value from 1kVrms up to 10kVrms needs to be generated. The requirement is given that the inner sphere diameter of the experiment cell might be changed after a mission which changes the load condition. The high voltage accuracy has to be in the order of 2%. The possible area of the sub-system inside the Experimental Container is very limited, meaning that the sub-system must be highly optimised. The maximum heat dissipation towards the cooling loop has to be below 30W. The high voltage is generated by the High Voltage Power Supply (HVPS) which consists of the high voltage control electronics, which is commanded by the ECCU, and the high voltage transformer.

The principle of the high voltage generation inside the Experimental Container is based on a resonant tank circuit whereat the experiment cell forms a spherical capacitor with the experiment fluid as dielectric and the high voltage transformer is the inductor. The change of the inner sphere diameter of the Fluid Cell evokes a capacitance change of this capacitor. This leads to the problem that the load and with that the resonant condition changes when a different Fluid Cell inner sphere is implemented. For being able to supply the Fluid Cell with an accurate adjustable AC voltage under different conditions the HVPS needs to be able to adjust itself to different conditions. In order to react to possible discharges, which might result in safety-critical conditions, the sub-system is equipped with different indicators which are able to switch off the HVPS in case of a malfunction.

2.5. The heat exchanger turret

The Heat Exchanger Turret is the counterpart of the experiment cell heat exchanger. Together they form a cooling loop that is able to transport the dissipated heat from the experiment cell to the Experimental Container water cooling loop. While the experiment cell heat exchanger is a gas-to-gas heat exchanger, the Heat Exchanger Turret forms a gas-to-water heat exchanger.

This heat exchanger consists of a cooling tower to which fans are coupled to generate the gas flow. The gas flow through the cooling tower is cooled down by TECs that can be controlled to allow the accurate adjustment of the experiment cell structure temperature.

The TECs are coupled to water cooled cold plates that are connected to the Experimental Container water cooling loop and with that to the FSL water cooling loop. The following schematic shows the principle setup of the gas loop.

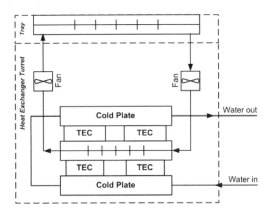

Figure 6. The principle setup of the heat exchanger turret gas loop

3. Conclusion

During earlier development phases the sensitivity of the optical system, the performance of the Heat Exchanger Turret and the reliability of the high voltage slip ring have been tested with breadboards and gained results that lead to confidence for the Engineering Model design. To achieve a state that can be used for designing and manufacturing of a reliable Flight Model more investigations have to be made. All parts have to be proven to survive the launching loads which involve stress from vibration and thermal loads. This is demanded to be done by calculation simulation and testing under prescribed conditions. The currently manufactured engineering model is in the first place used as a prototype and later on, in an updated version, as a reference model to the Flight Model for possible error search.

Acknowledgements

The current Phase C/D of the Experimental Container GeoFlow is performed under ESA contract 118300/04/NL/JS. We would like to thank especially Mr. DeWandre and Mr Mundorf as responsible technical officers from ESA/ESTEC for their unique support during the development phases.

References

1. Egbers C., Brasch W., Sitte B. Immohr J., Schmidt J-R.: *Estimates on diagnostic methods for investigations of thermal convection between spherical shells in space.* Meas. Sci. Technol. 10, 866-877, 1999.
2. Sitte B., Immohr J., Hinrichs O., Maier R., Egbers C., Rath H.: *Rayleigh-Bénard convection in dielectrophoretic force fields.* 12th International Couette-Taylor Workshop, September 2001.

EMBEDDED CONTROLLERS FOR MR DAMPERS IN SUSPENSION SYSTEMS

Rosół M., Sapiński B.

Abstract: The study is concerned with real-time control of magnetorheological (MR) dampers employed for vibration suppression in the suspension systems to be tested. MR dampers control utilizes the following embedded controllers: 8-bit ST52E420 (STMicroelectronics), 16-bit M16C6N (Mitsubishi Corporation) and 32-bit MPC555 (Motorola Corporation). These controllers were employed in MR suspension systems of: a driver's seat (1 DOF system) and pitch-plane vehicle models (2 DOF and 3 DOF systems) respectively. For each suspension we consider: a model of the system and experimental setup, control schemes implemented in the embedded controller and the experimental data.

1. Introduction

The MR damper incorporated in a suspension system acts as an interface between the electronic control unit and the mechanical structure of the suspension. The force produced by the damper is adjusted by a controller that may be programmed with any number of control schemes. The controller determines the required damping force utilizing the assumed control scheme. Such control system uses external power only to adjust the damping and to operate an embedded controller and set of sensors. Usually the aim of damping control is to minimize the mean or the root-mean square value of acceleration of the plant to be protected, taking into account the existing constraints. Accordingly, the energy transmitted from the vibration source to the protected plant is reduced.

In the study we considered the suspension systems of a driver's seat (1 DOF system) and pitch plane vehicle models (2 DOF and 3 DOF systems) equipped with the conventional springs and MR dampers. For MR dampers control we employed embedded controllers: 8-bit ST52E420, 16-bit M16C6N and 32-bit MPC555 in which various control schemes were implemented. In a driver's seat suspension we used the ST52E420 that is a programmable controller capable of executing, both Boolean and fuzzy algorithms (STMicroelectronics, 2000). The ST52E420 is equipped with a fuzzy core responsible for fuzzyfication, inference (following the Mamdani procedure) and defuzzyfication. In the 2 DOF pitch-plane suspension model we employed the M16C6N controller. The M16C6N accommodates several processing units (CPU) to execute arithmetic/logic operations. It also includes a peripheral unit of CAN (Control Area Network) module which is a serial communications protocol that efficiently supports distributed real-time control with a very high level of security (Renesas, 2003). In the 3 DOF pitch-plane suspension model we used the MPC555 controller. The MPC555 incorporated in phyCORE-MPC555 development board is a high efficiency embedded controller comprising 32-bit PowerPC core with 64-bit floating-point unit and dual CAN 2.0B controller modules (Motorola, 2000).

In the experiments we employed MR dampers of RD-1005-3 series (Lord, 2003), LVDTs of PSz20 series (Peltron, 2004) and the current driver engineered by the authors.

2. Models of suspension systems and experimental setups

The diagram of a driver's seat suspension model is shown in Fig. 1. The model has 1 DOF due to removed cushion of the seat. The designations in Fig. 1 are as follows: k_1 – spring stiffness; c_r – damping of the MR damper; m – mass of a driver and seat; x_0 – displacement excitation acting upon a seat (driver); x_1 – seat (driver) displacement.

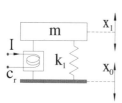

Figure 1. Model of a driver's seat suspension

Figure 2. Driver's seat suspension in the experimental setup

The suspension of the seat to be tested, equipped with the MR damper and specially designed spring, is shown in Fig. 2. We assumed the following values of system parameters: k_1=36861 N/m, m=112 kg. The value of c_r results from the current applied in the MR damper coil.

The diagram of the 2 DOF and 3 DOF pitch-plane suspension model is depicted in Fig. 3. In the 2 DOF pitch-plane model the suspended body is simulated by a rigid rectangle-intersection beam of mass m, moment of inertia I, total length L, width a, height b and centre of gravity (c.o.g.) in P_g. The beam is supported at points P_f and P_r by two identical spring–MR damper sets (suspension-sets), which are subject to bottom displacement excitations similar to these acting upon conventional vehicle suspensions. Distances from P_g to P_f and from P_g to P_r are denoted by l_f and l_r. Elasticity factors of the front and rear springs are denoted by k_f and k_r. Similarly, the i_f and i_r denote currents in the front (d_f) and rear (d_r) MR damper coils. Excitations applied to the bottom of the front and rear suspension-sets are denoted by w_f and w_r, displacements of points P_f and P_r by x_f and x_r. The model possesses 2 DOFs: vertical (bounce) displacement x and pitch displacement φ of the beam's c.o.g (Sapiński, Martynowicz, 2005). We assumed the following values of system parameters: l_f=0.7 m, l_r=0.7 m, L=1.5 m, a=0.124 m, b=0.173 m, m=253.3 kg, I=49.2 kgm^2, $k_{sf}(k_{sr})$=42016 N/m.

It is apparent that the 3 DOF pitch-plane model comprises the 2 DOF model described above and, additionally, a mass m_s (modelling the driver's seat), suspended in P_g on a spring–MR damper set. The elasticity factor of the spring is k_s, current in the MR damper d_s coil is i_s. The displacement of m_s in relation to beam's c.o.g. is x_s.

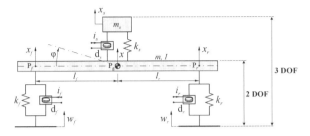

Figure 3. 2 DOF/3 DOF pitch-plane suspension model

The experimental setups of 2 DOF and 3 DOF pitch-plane suspension models to be tested are shown in Fig. 4.

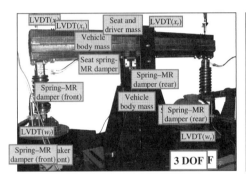

Figure 4. 2 DOF and 3 DOF pitch-plane suspension model in the experimental setup

3. Control schemes

Control schemes for suspension systems described in section 2 are outlined below. Each control scheme was implemented in an individual software-hardware environment. As regards the driver's seat , the chief aim of the control was to minimize the acceleration vibrations transmitted to the seat. Control algorithm was developed in Fuzzy Studio environment, supported by STMicroelectronics. The structure of the fuzzy controller (CON1) is shown in Fig. 5. The CON1 has five rules. The input membership functions are triangular-shaped while the output membership functions are of the singleton type. Input-output graph of the CON1 is depicted in Fig. 6.

The aim of control in the 2 DOF pitch-plane suspension model was to minimize vibration acceleration of the beam's front part. In this case only the front side of the pitch-plane model was controlled (the current applied to the rear MR damper coil was set as 0.00 A). We tested three controllers, one *on-off* (CON2) and two continuous controllers (CON3, CON4). The control schemes are governed by the following formulas:

$$\text{CON2: } i_f = \begin{cases} c_2, & \dot{x}_f \left(\dot{x}_f - \dot{w}_f \right) \geq 0 \\ 0, & \dot{x}_f \left(\dot{x}_f - \dot{w}_f \right) < 0 \end{cases} \quad \text{CON3: } i_f = \begin{cases} c_3 \left| \dot{x}_f \right|, & \dot{x}_f \left(\dot{x}_f - \dot{w}_f \right) \geq 0 \\ 0, & \dot{x}_f \left(\dot{x}_f - \dot{w}_f \right) < 0 \end{cases} \quad (1)$$

CON4: $i_f = \begin{cases} c_4|\dot{x}_f - \dot{w}_f|, & \dot{x}_f(\dot{x}_f - \dot{w}_f) \geq 0 \\ 0, & \dot{x}_f(\dot{x}_f - \dot{w}_f) < 0 \end{cases}$

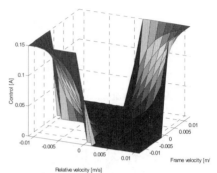

Figure 5. Structure of CON1: $\dot{x}_1$ - velocity of seat, $(\dot{x}_1 - \dot{x}_0)$ - relative velocity (the difference between the seat velocity and the shaker base velocity, PWM - control signal

Figure 6. Graph of CON1

The output signal from CON1, CON2 and CON3 is considered as current (i_f) in the front MR damper coil. The values of the c_2, c_3 and c_6 depend on the maximal value of current applied to the MR damper coil. These values were taken to be: c_2=0.2, c_3=14, c_4=14. Control schemes were implemented in High-performance Embedded Workshop (HEW) environment, supported by Renesas Technology, using C++ language (Renesas, 2003). The aim of control in the 3 DOF pitch-plane suspension model was to minimize vibration acceleration of the driver's seat. We controlled only the current in the coil (i_s) of the MR damper incorporated in the driver's seat suspension (the currents applied to the front and rear MR damper coils were fixed). The control schemes were implemented in the MATLAB/Simulink environment, supported by Embedded Target for Motorola MPC555 with a Real Time Workshop Embedded Coder. Application for MPC555 was generated in Real Time mode execution for rapid prototyping. This mode of operation for Embedded Target for Motorola MPC555 enables the use of controlled developed in real-time in Simulink to perform embedded control.

We tested two controllers, *on-off* (CON5) and sky-hook controller (CON6) with control schemes governed by the following formulas:

CON5: $i_s = \begin{cases} c_5, & \dot{x}_s(\dot{x}_s - \dot{x}) \geq 0 \\ 0, & \dot{x}_s(\dot{x}_s - \dot{x}) < 0 \end{cases}$ CON6: $i_s = \begin{cases} c_6|\dot{x}_s|, & \dot{x}_s(\dot{x}_s - \dot{x}) \geq 0 \\ 0, & \dot{x}_s(\dot{x}_s - \dot{x}) < 0 \end{cases}$ (2)

The output signal from CON5 and CON6 is considered as i_s (current in the MR damper coil of a driver's seat). The values of the c_5 and c_6 depend on the maximal value of current applied in the MR damper coil. In the study these values were taken to be: c_5=0.2, c_6=7.

4. Experiments

The CON1 designed for the MR damper in the driver's seat was tested in the experimental setup shown in Fig. 7. Experiments were run for the applied sine excitations of the amplitude $\pm 3.0 \times 10^{-3}$ m, over the frequency range (1, 12) Hz and current level (0.00, 0.15) A.

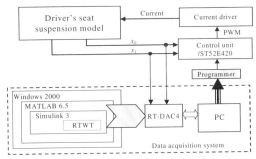

Figure 7. Diagram of the driver's seat experimental setup

The selected results of the tests in the near-resonance frequency range, which was about 5 Hz, are shown in Fig. 8. The time patterns enable us to compare the seat displacement in open loop and feedback system configurations for the sine base excitation with the frequency 5 Hz and 6 Hz. It appears that the acceleration transmissibility decreases at 5 Hz and increases at 6 Hz.

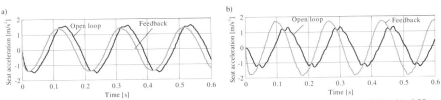

Figure 8. Seat acceleration in the open loop and feedback system: a) 5 Hz, b) 6 Hz

The 2DOF pitch-plane suspension model was tested in open loop and feedback system configurations using networked control system (NCS) with structure depicted in Fig. 9 (Sapiński, Rosół, 2005). It is apparent that the NCS has 3 CAN nodes. Two nodes are used as smart displacement sensors and smart actuators for MR dampers. Each of them has two inputs and one output, where: x_f, x_r – front and rear beam displacements, w_f, w_r – front and rear base excitation displacement, i_f, i_r – front and rear MR dampers current (control signals). The third node acts as the driving unit, executing the controller algorithm. The measurement frames containing the beam and base displacement data are transferred from the sensor nodes to the control node. The control frames are transferred in opposite direction. The measurement data are sent by the timer to handle interrupt procedures. In the NCS system we used CAN in standard format (2.0B) with 11 bit ID. The speed of CAN was set as 1 Mbit/s.

The standard PC with the multi I/O board of RT-DAC4 series and MATLAB/Simulink with Real-Time Workshop (RTW) toolbox and Real-Time Windows Target (RTWT) extension running on Windows 2000 was used only as the data acquisition system. Experiments were run for the applied sine excitations of the amplitude $\pm 3.0 \times 10^{-3}$ m, over the frequency range (1, 9) Hz. The obtained frequency characteristics (acceleration transmissibility) of the front side of the beam in open loop and feedback system configurations are compared in Fig. 10. It is apparent that the best results are achieved for feedback system with the CON3. The performance of the system in the near resonance frequency range (3.5 Hz) was the best though the amplitudes of vibrations with the frequencies in excess of 5 Hz would slightly increase.

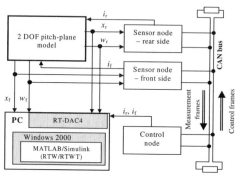

Figure 9. Diagram of the NCS

Figure 10. Acceleration transmissibility: shaker base-beam

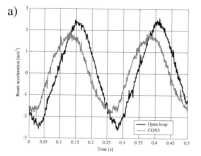
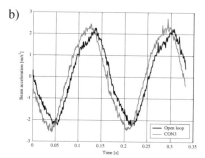

Figure 11. Acceleration of the beam in the open loop and feedback system:
a) f = 4 Hz, b) f = 6 Hz

These observations are confirmed by the registered plots of beam acceleration, obtained for the feedback systems with the controllers CON2, CON3, CON4 for sine excitations 4 and 6 Hz. These particular frequencies were selected to show the system behaviour at the near-resonance frequency and beyond that range. The comparison of plots obtained for the open loop system (when no current is applied) and the feedback system with the CON3 is shown in Fig. 11.

The 3DOF pitch-plane suspension model was tested in the open loop and feedback system configurations using an autonomous control system (ACS) with structure presented in Fig. 12 (Sapiński, Rosół, 2005).

In the open loop configuration the control generated by the ACS (current applied to each MR damper) was kept constant. Therefore experiments were carried out for various current levels in the MR dampers. Results obtained at this stage served as the reference for evaluating the performance of the tested control schemes.

Embedded controllers for MR dampers ... 529

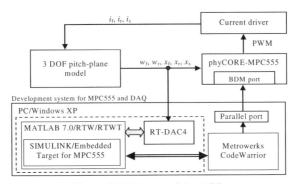

Figure 12. Diagram of the ACS

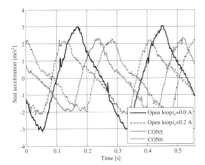

Figure 13. Acceleration of the seat at 3.5 Hz (i_f= 1.0 A, i_r= 1.0 A)

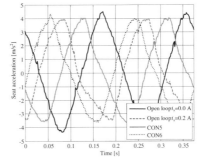

Figure 14. Acceleration of the seat at 5.5 Hz (i_f= 1.0 A, i_r= 1.0 A)

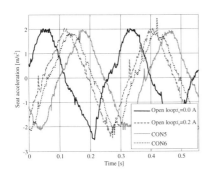

Figure 15. Acceleration of the seat at 3.5 Hz (i_f= 0.0 A, i_r= 0.0 A)

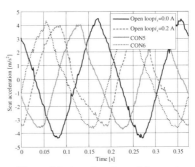

Figure 16. Acceleration of the seat at 5.5 Hz (i_f= 0.0 A, i_r= 0.0 A)

In the feedback system only the MR damper in the driver's seat suspension of the pitch-plane model was controlled. The currents applied to the front and the rear MRA dampers were set as constant. Experiments were run for the applied sine excitations of the amplitude $\pm 3.0 \times 10^{-3}$ m, over the frequency range (3.5, 10) Hz. The lower frequency value of sine excitations depends on parameters of the rear shaker and cannot be changed in our setup.

Selected results of the test are presented in Figs 13-16. The plots show seat acceleration vs. time data in the open loop and feedback system configurations under sine excitations at 3.5 Hz and 5.5 Hz. The time patterns in Fig. 13 and 14 (for $i_f=1.0$ A and $i_r=1.0$ A) prove the good performance of those controllers in the frequency range (3.5, 5.5) Hz. In excess of 5.5 Hz the CON5 and CON6 bring about an increase of the seat acceleration amplitude in relation to an open loop system. Fig. 15 and 16 evidence the good performance of the CON5 and CON6 only for the frequency about 5.5 Hz. At higher frequencies the controllers' action does not bring anticipated effects. This is a result of MR dampers operation in front and rear section of the 3 DOF suspension model.

5. Summary

Three suspension systems 1 DOF, 2 DOF and 3 DOF were engineered and investigated accordingly. Depending on the level of complexity of control schemes, various control embedded systems were employed. For the control of the 1 DOF system the use of 8-bit microcontroller seems to a good solution. In order to control 2 DOF and 3 DOF systems, it is required that more elaborate control schemes be used. Hence, high performance microcontrollers ought to be employed.

Embedded controllers used in the test program enable the hardware-in the loop or processor-in the loop control of the tested object supported by the MATLAB/Simulink. These controllers afford us a mean to easily transfer the well-proven algorithms and apply to the control of a target object. The experimental data reveled good performance of the embedded controllers selected for MR dampers control in the tested applications.

Acknowledgement

This paper was supported by the research grant of the AGH – University of Science and Technology No. 10.10.120.38

References

1. Lord Corporation, *http://www.lord.com*, USA, 2003.
2. Motorola Corporation, *MPC555/MPC556 User's Manual*, USA, 2000.
3. Peltron Ltd., LVDT PSz20 series, *http://www.peltron.home.pl*, 2004.
4. Renesans Technology Corporation, Mitsubishi M16C/6N Group, *http://www.renesas.com*, 2003.
5. Sapiński B., Martynowicz P.: *Vibration control in a pitch-plane suspension model with MR shock absorbers*. Journal of Theoretical and Applied Mechanics, 43, 2005.
6. Sapiński B., Rosół M.: *Autonomous control system for a 3 DOF pitch-plane suspension model with MR shock absorbers*. II ECCOMAS Conference, Portugal, (to be published), 2005.
7. Sapiński B., Rosół M.: *Networked control system for a pitch-plane model of a magnetorheological suspension*. Archive of Control Sciences, (to be published), 2005.
8. STMicroelectronics, Data Sheet of ST52x420 8-bit DuaLogic™ Microcontroller, 2000.

APPLICATION OF FMECA CAUSAL-CONSECUTIVE ANALYSIS WITH AN EXAMPLE OF IMS DIGITAL PLATFORM

Kocerba A., Tekielak M.

Abstract: The paper is describing applications of the FMECA method with use of the CASIP e-platform in practice. The two examples have been discussed: hydraulic valve system, overhead traveling crane.

1. Introduction

The FMECA (*Failure Mode Effect and Critical Analysis*) method makes possible to prosecute quality and quantitative analyses. It has been implemented for the IMS (Intelligent Manufacturing System) platform, with the example of hydraulic valve. The research station is located in CRAN (Centra de Recherché en Automatique de Nancy) on Université Henri Poincaré Nancy I. The FMECA method has been elaborated using CASIP (*Computer Aided Safety and Industrial Productivity*) digital platform. The CASIP environment allows managing preventive-type supervision of device exploitation process [3, 4]. This is a multi-module engineer program, which provides process management and supervision.

There was CASIP Suite module used here, which belongs to program's first level, and makes possible to prosecute the FMECA analysis describing the analysed technical system. The analysis prosecuted with FMECA method was made with the help and courtesy of employees of above-mentioned university: Prof. Benoit Iung, Dr Hervé Panetto and Eng. Michel Braucourt.

Another example of application of FMECA causal-consecutive method is the model of overhead traveling crane in the Department of Machines for Technological and Environmental Protection on AGH University of Science and Technology in Cracow, in the Technological Transportation Group leaded by Prof. Janusz Szpytko.

Examples of FMECA use are easy to find in many publications for instance: [1, 2, 5, 7].

2. Presenatation of IMS platform

The research station consists of two cylindric containers (top and bottom), a pomp, and automated control valve [6] - fig. 1. The system is connected by PCV pipes. The control valve is equipped with servomotor and mechanical transmission. This system enables the flow control. The ball-tap valve applied here controls the fluid flow in the range of 90^0 slew (0^0 – closed, 90^0 – full open).

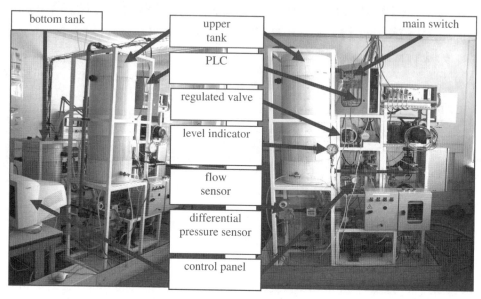

Figure 1. View of the research station

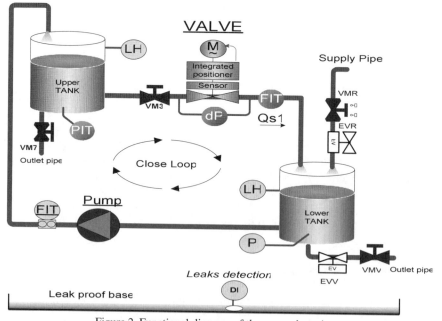

Figure 2. Functional diagram of the research station

The fluid circulation (in the closed loop) starts with the initially fulled containers (fig. 2). The water from the bottom container is pumped up to the top container with the centrifugal pomp, and flows down by gravity. The valve controls the flow power; received values are shown on the control panel. The problem of system algorythm is generating time-differential parameters.

The FMECA analysis digital platform can be used for the analysis of any technical device. The example of such application is the research station in Department of Machines for Technological and Environmental Protection on AGH University of Science and Technology in Cracow, in the Technological Transportation Group. There is a model of overhead traveling crane used for the research (fig. 3), with lifting capacity up to 150 kg, with 3200 x 2300 x 2200 mm of its work space.

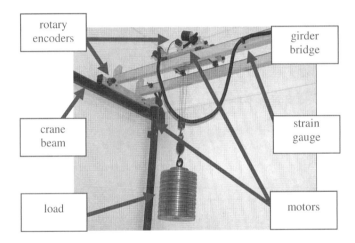

Figure 3. View of overhead crane model

Figure 4 shows the schema of hardware configuration of the model, which allows to supervise the crane at work by CASIP program. The impulse-rotary sensors on non-driven wheels are installed directly on the bridge, on the trolley and on the drum. The sensors are connected with Mitsubishi FX -24MR DS controller, by the high frequencies counter inputs. The strain gauge is attached in the middle of crane bridge, it is connected to work voltage by the stabilized power supply. The voltage signal from the strain gauge is passing directly on the controller analog inputs. The gantry operator gives the commands to the controlled object, using the control panel.

3. Structural analysis of the system

With the help of CASIP program there is a possibility of reduction of FMECA quality and quantitative analysis realization time, and the characteristics of featured installation can be additionally used in CASIP SAM to online-supervising.

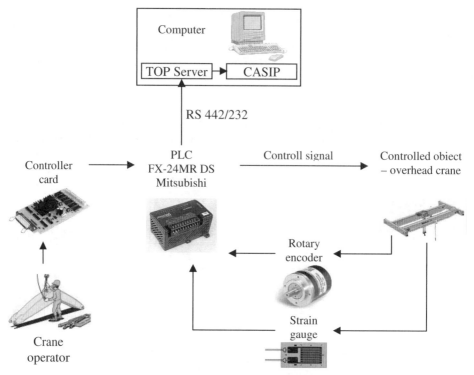

Figure 4. Schema of hardware configuration of the model

Before running the Suite module it is necessary to open the one of available databases (SQL, Oracle, Access) - the appropriate results will be recorded there. After initial preparation of the sheet, the decomposition of analysed installation follows. It rests on describing the dependence between system particular elements, by the „tree" form - from the main element, to the following secondary elements (fig. 5). The user can fold or unfold the whole branch of described installation – describing of the system becomes much easier.

Figure 5. Example of device's list in Suite module

Application of FMECA causal-consecutive analysis ...

The next step of analysis for the particular system elements is arrogating the right functions, then degradation – possible disfunctions of defined function, and then arrogating the possible damages. Additionally, the variables which will be responsible for supervising the appropriate function can be arrogate to the table. The data, prepared in this way give the full FMECA causal-consecutive analysis. An example of that window made in CASIP Suit program is shown on the figure 6.

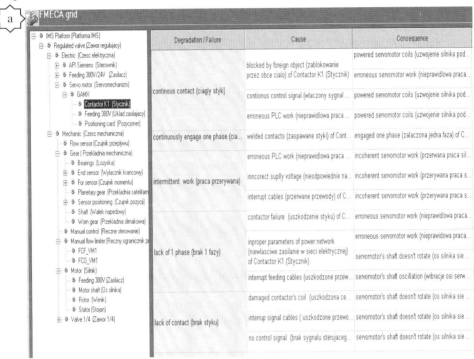

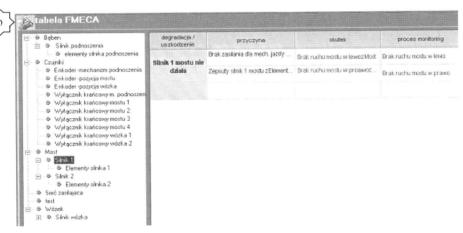

Figure 6. FMECA grid in CASIP Suite for:
a - hydraulic valve, b - overhead crane model

Next, there were monitoring variables added to the particular degradations; these variables will appear as the symptoms in the CASIP SAM module.

4. Summary

Application of FMECA causal-consecutive analysis with an example of Predict IMS digital platform, makes analysis process much faster in practise, both in system decomposition phase, and further causal-consecutive analysis. The result of prosecuted analysis is a causal-consecutive matrix for the particular system elements.

Additionally, the generated analysis is a platform to online-supervising the technical state of the installation, using the other modules included in CASIP software pack, such as CASIP-SAM.

This paper has been financially supported by central budget on science for the years 2005-07 as the research project

References

1. Notash L., Huang L.: *On the design of fault tolerant parallel manipulators*. Mechanism and Machine Theory, Volume 38, p. 85-101, Issue 1, January 2003.
2. Rausand M., Oien K.: *The basic concepts of failure analysis.* Reliability Engineering & System Safety, Volume 53, p. 73-83, Issue 1, July 1996.
3. Szpytko J.: *Kształtowanie procesu eksploatacji środków transportu bliskiego*. Biblioteka Problemów Eksploatacji, ITE, Kraków - Radom, 2004.
4. Szpytko J., Iung B., Leger J.B., Harrison D.K.: *Computer aided safety and productivity system for cranes*. Proceedings of 21st Conference on European Plant Engineering Committee (CETIS, VDEh), 2002.
5. Tixier J., Dusserre G., Salvi O., Gaston D.: *Review of 62 risk analysis methodologies of industrial plants*. Journal of Loss Prevention in the Process Industries, Volume 15, p. 291-303, Issue 4, July 2002.
6. Yu R., Iung B., Panetto H.: *A multi-agents based E-maintenance system with case-based reasoning decision support*. Engineering Applications of Artificial Intelligence, Vol. 16, p. 321-333, Issue 4, June 2003.
7. Szpytko J.: *Integrated decision making supporting the exploitation and control of transport devices*. Monografie, UWND AGH, Kraków, 2004.